MW00760626

PHYSICS RESEARCH AND TECHNOLOGY

PHOTONIC CRYSTALS: FABRICATION, BAND STRUCTURE AND APPLICATIONS

Physics Research and Technology

Additional books in this series can be found on Nova's website
under the Series tab.

Additional E-books in this series can be found on Nova's website
under the E-book tab.

PHYSICS RESEARCH AND TECHNOLOGY

PHOTONIC CRYSTALS: FABRICATION, BAND STRUCTURE AND APPLICATIONS

VENLA E. LAINE
EDITOR

Nova Science Publishers, Inc.
New York

For permission to use material from this book please contact us:
Telephone 631-231-7269; Fax 631-231-8175
Web Site: http://www.novapublishers.com

Additional color graphics may be available in the e-book version of this book.

LIBRARY OF CONGRESS CATALOGING-IN-PUBLICATION DATA

Photonic crystals : fabrication, band structure, and applications / editors, Venla E. Laine.
p. cm.
Includes index.
ISBN 978-1-61668-953-7 (hardcover)
1. Photonic crystals. I. Laine, Venla E.
QD924.P465 2010
548'.83--dc22

2010014111

Published by Nova Science Publishers, Inc. † New York

CONTENTS

PREFACE

Photonic crystals are periodic optical nanostructures that are designed to affect the motion of photons in a similar way that periodicity of a semiconductor crystal affects the motion of electrons. Photonic crystals occur in nature and in various forms have been studied scientifically for the last 100 years. This book gathers and presents topical data in the field of photonic crystals including the phenomena in photonic crystals stipulating the presence of omnidirectional band gaps, i.e., overlapping of stop bands in all directions; the physics of photonic crystal couplers and their applications; photonic crystals and their capability to control the propagation and emission of electromagnetic waves and others.

According to their periodicity in space, PCs can be categorized mainly into one-dimensional (1D), 2D and 3D PCs. All above-mentioned PCs have been fabricated artificially from a variety of materials. To date, the main 3D PCs include opal, inverse opal, woodpile, diamond structure, etc. Among those structures, opal structure comprised of polymer colloidal crystals demonstrates promising advantages over other structures, for example, polymer colloidal is easily processed, and its properties could be easily modulated based on the design of molecular structure. Moreover, polymer colloidal crystal can also act as template for the fabrication of PCs of inorganic, metal and complicate materials, since nanometer sized polymer particles are readily available with a high mono-dispersity or can be synthesized in a controlled way. Plenty of review papers[3-9] have been published about the fabrication of polymer PCs and its special applications. Such as, fabrication of polymer PCs form latex system[3-5] inverse opal structure[6], and micro-phase separation[10]. Polymer PC material is a novel material, and can provide material basis for the development of relative subjects, which covers photo-computing, photo-communication and optic-electric integration. In Chapter 1, the authors review the recent research progress of polymer PCs, mainly concern on the fabrication and application of polymer PCs.

Photonic crystals (PCs) were introduced by E. Yablonovitch [1] and S. John [2] in 1987. In 1990, Ho *et al.* demonstrated theoretically that a diamond structure possesses complete band gaps (CBGs) [3]. Since then, great interest was focused on fabricating three-dimensional (3D) photonic crystals (PCs) in order to obtain CBGs. Several methods were reported, and CBGs were achieved in the range of microwave or submicrowave. Theoretical analysis showed that although CBGs can be obtained by diamond structure, a strict condition should be satisfied, i.e., the modulation of the refractive index of the material used should be larger than 2.0 [4,5]. So, attention was paid to finding the materials with high refractive index. Some scientists tried to fill the templates with high refractive index materials to increase the modulation of the refractive index [6-8], and CBGs were obtained. However, the CBGs achieved in 3D PCs were mostly in microwave or infrared regions [7,9-11]. Holography is a cheap, rapid, convenient, and effective technique for fabricating 3D structures. In 1997, holographic technique was introduced for fabricating the face centered cubic (fcc) structure [12]. Campbell *et al.* actually fabricated the fcc structure with holographic lithography [13]. Several authors reported their works on this topic [4,5,9,14]. Toader *et al.* also showed theoretically a five-beam "umbrella" configuration in the synthesis of a diamond photonic crystal [15]. Because both the value and modulation of the refractive index of the holographic recording materials are commonly low, there would be no CBGs in PCs made directly by holography. For example, the epoxy photoresist generally used for holographic lithography [13,16-18] has n=1.6, which is a little bit too low. They may, however, be used as templates for the production of inverse replica structures by, for example, filling the void with high refractive index and burning out or dissolving the photoresist [13], and a good work was done by Meisel *et al* [19,20]. However, to find materials with large refractive index is not easy, and the special techniques needed are very complicated and expensive. It limits the applications of PCs, especially for industrial productions.

Although some efforts had been made [20], CBGs in the visible range had not yet been achieved by using the materials with low refractive index. Therefore, it is a big challenge to fabricate 3D PCs possessing CBGs in the visible range by using materials with low refractive index, though it is greatly beneficial for the future PC industry. As a first step for achieving this, it is important to obtain very wide band gaps. In the previous investigations [21,22], it was evidently shown that the anisotropy of a photonic band gap in a two-dimensional photonic crystal is dependent on the symmetry of the structure, and as the order of the symmetry increases, it becomes easier to obtain a complete band gap. One would naturally ask whether or not such a method is applicable to 3D PCs. In view of this question, in Chapter 2 the authors proposed a series holographic method for fabricating some special structures by using materials with low refractive index, and the features of the band gap in such structures were then studied experimentally and theoretically.

Chapter 3 introduces photonic crystals for microwave, millimeter wave and submillimeter wave applications. First, the authors present a new type of photonic crystal waveguide for millimeter or submillimeter waves which consists of arrayed hollow dielectric rods. The attenuation constant of the waveguide is smaller than that of a Nonradiative Dielectric (NRD) Waveguide.

Second, metallic photonic crystals consisting of metamaterials are designed. Metamaterials are artificially-constructed materials whose permittivity and/or permeability can be negative. A hollow metal rod with a slit in its sidewall is a metamaterial which plays the role of an LC resonator. In addition, an array of such metal rods can behave as a two-dimensional photonic crystal. Conventional metallic photonic crystals have a photonic bandgap for electromagnetic waves with E-polarization (where the electric field is parallel to the axis of the rods). On the other hand, metamaterials consisting of hollow metal rods with sidewall slits have a bandgap for electromagnetic waves with H-polarization (where the electric field is perpendicular to the axis of the rods) because of their negative permeability. Eigenvalues can be calculated using the two-dimensional finite-difference time-domain (FDTD) method, and metallic metamaterial-based photonic crystals can be designed with independently customized bandgaps for E- and H-polarizations.

Third, the authors analyze propagation losses in metallic photonic crystal waveguides. A conventional metallic waveguide structure whose E-planes are replaced by metal wires can easily confine electromagnetic waves because their electric field direction is parallel to the wires. In this section, the attenuation constants of metallic photonic crystal waveguides, whose E-planes are replaced by arrayed metal rods, are calculated and compared with those of conventional metallic waveguides. Losses are calculated using another FDTD method. The propagation losses are found to decrease when the radii of the rods (relative to the lattice constant) is increased. The bandwidth can be extended by widening the waveguide relative to the lattice constant.

Fourth, the authors investigate a metallic waveguide in which in-line dielectric rods are inserted. In such a situation, the electromagnetic field is equivalent to that produced by a two-dimensional photonic crystal with a lattice constant in one direction equal to the inter-rod spacing, and in the other direction equal to twice the distance between the rods and the sidewalls. The calculations of the fields are performed using the FDTD method. The authors' results showed that the frequency range of a single propagation mode in the waveguide can be extended.

Finally, arrayed dielectric rods in a metallic waveguide are known to change the modes and of course the group velocities of electromagnetic waves. In particular, the authors focus on TE_{10} to TE_{n0} (where n equals 2 or 3) mode converters and the dependence of their behavior on the frequency range. Conventional metallic waveguides only allow propagation at frequencies above the cutoff frequency f_c. The TE_{20} mode, however, is also possible for frequencies higher than $2f_c$. In this investigation, a mode converter is proposed which passes the TE_{10} mode for frequencies less than twice the cutoff frequency, and converts the TE_{10} to the TE_{20} mode for frequencies higher than twice the cutoff frequency; this is achieved by small variations of the group velocity.

In Chapter 4, properties of the directional coupler made by photonic crystals (PCs) are studied by the tight-binding theory (TBT), which considers the coupling between defects of PCs. Based on this theory, the amplitude of the electric field in a photonic crystal waveguide (PCW) can be expressed as an analytic evolution equation. As an identical PCW is inserted with one or several partition rods away, the PC coupler is created. The nearest-neighbor coupling coefficient α between defects of the coupler causes the splitting of dispersion curves, whereas the next-nearest-neighbor coefficient β causes a sinusoidal modulation to dispersion curves. The sign of α determines the parity of fundamental guided modes, which can be either even or odd, and the inequality $|\alpha| < 2|\beta|$ is the criterion for occurring crossed dispersion curves. There is no energy transferred between the PCWs at the frequency of the crossing point, named as the decoupling point, of dispersion curves.

By translating the defect rods along the propagation axis of the coupler, blue shift (red shift) in the frequency of the decoupling point occurs to the square (triangular) lattice. Therefore, the frequency of the decoupling point and coupling length of the coupler can be adjusted by moving defect rods. By applying propagating fields having a frequency at the decoupling point and another frequency where the coupler has an ultra short work region, the authors designed a dual-wavelength demultiplexer with a coupling length of only two wavelengths and power distinguish ratio as high as 15 dB.

An analytic formula can also be derived by the TBT for asymmetric couplers made of two different coupled PCWs. The asymmetric coupler possesses the following properties: (1)

Its dispersion curves will not cross at the "decoupling point" and the electric field would only localize in one PCW of the coupler; (2) The eigenfield at the high (low) dispersion curve always mainly localizes in the PCW that possesses high (low) eigenfrequency, even though the symmetry of eigenmodes has changed. As the field with a given frequency is incident into one of the PCWs of the coupler, both the energy transfer between two waveguides and the corresponding coupling length can be given. If an optical Kerr medium is introduced into a symmetric coupler, the coupler becomes an optical switch through the modification of the refractive index of the coupler by sending a high-intensity field.

The most interesting phenomena in photonic crystals stipulate the presence of omnidirectional band gap, i.e. overlapping of stop bands in all directions. A higher rotational symmetry and an isotropy of quasicrystals in comparison with ordinary crystals give a hope to achieve a gap opening at lower dielectric contrasts. But nonperiodic nature of quasicrystals makes the size of stop bands lower than in the case of an ordinary periodic crystal. Another possibility to get a higher lattice isotropy (symmetry) is to use the so called quasicrystal approximants and periodic structures with a large unit cell. Within such a unit cell quasicrystal approximants are nearly identical to the actual quasicrystal. Outside of that unit cell, they qualitatively resemble the quasicrystal, but in a periodic manner. Chapter 5 is devoted to investigation of a photonic band structure of such lattices.

The transition from periodic structure to nonperiodic one is studied for 2D case by considering quasicrystal approximants of growing period. It is done in order to weigh advantages and disadvantages of quasicrystals. It is shown, that higher isotropy in the 2D case allows one to reduce the refractive index threshold necessary for the gap opening

For the 3D case the authors consider different approximants of icosahedral quasicrystals with 8 and 32 "atoms" per unit cell and Si-34 structure with 34 "atoms" per unit cell. All considered structures demonstrate a photonic gap opening and a high isotropy of the band structure. At the same time there are no structures with refractive index threshold for gap opening lower than the best cases of periodic lattices, in particular, with diamond symmetry.

In Chapter 6, the effect of different controlling parameters such as number of periods of the structure, refractive index contrast, filling fraction and angle of incidence of light on the structure on the output spectrum of one-dimensional (1D) photonic crystal has been shown and discussed. By knowing the effect of these parameters on the output spectrum, one can engineer the output spectrum of the 1D photonic crystal to utilize them in various optical applications such as filters, reflectors etc., in a desired and optimum way.

The transmission properties of guided waves in photonic crystal waveguides containing barriers formed from Kerr nonlinear media are studied theoretically. The photonic crystal waveguides are formed in a two-dimensional (square lattice) photonic crystal of linear dielectric cylinders by cylinder replacement, with replacement cylinders made of linear dielectric media. The channel of the waveguide is along the x-axis of the photonic crystal. Barriers formed of Kerr nonlinear media are introduced into the waveguide by cylinder replacement of waveguide cylinders in the barrier region by cylinders containing Kerr nonlinear media. The transmission of guided modes incident on the Kerr media barriers, from the linear media waveguides, is computed and studied as a function of the parameters characterizing the Kerr nonlinear barrier media. A focus of Chapter 7 is on the excitation of intrinsic localized modes within the barrier media and on the effects on these modes from additional off-channel and in-channel features that are coupled to the barrier media. Systems treated are a simple Kerr nonlinear media barrier formed by replacing waveguide sites by cylinders containing Kerr nonlinear media, Kerr barriers that couple to off-channel impurity features formed by cylinder replacement of off-channel sites of the photonic crystal which are adjacent to the barrier, and Kerr barriers containing an in-channel impurity site. The off-channel and in-channel impurity features may be formed of linear media and/or Kerr media replacement cylinders and are found to give rise to multiple bands of intrinsic localized modes. Suggestions are made for possible technological applications of these types of systems.

Colloidal dispersions with self-assembling properties have been prepared by soap-free emulsion copolymerization of styrene (St) with hydroxyethylmethacrylate (HEMA), using various St/HEMA molar ratios. Investigations by SEM, AFM and reflexion spectra have shown that copolymer particles give colloidal crystals while monodisperse particles of polystyrene, with no HEMA units, do not crystallize. XPS analysis of the copolymer particles has shown that the HEMA hydrophilic units tend to concentrate on the surface of the copolymer particles, while the core is richer in St. The copolymer particles have been separated and redispersed in other liquids than water (ethylene glycol, formamide, ethanol, hexane, and toluene) to investigate the conditions that allow self-assembling after the diluent evaporation. For each case, contact angles have been measured and the adhesion work has been computed. The results in Chapter 8 have shown that the contact angle does not influence the self-assembling process, while the adhesion work is an important parameter. Only dispersions characterized by a value of the adhesion work superior to 87 mJ/m^2 did crystallize after the diluent evaporation.

Taking into account that the current development of the world requires Increasingly devices operating at higher frequencies and bandwidths, and that these requirements are no longer possible to be obtained by electronic, in Chapter 9 the authors propose and analyze an optical logic gate based on nonlinear photonic crystal (PhC). This device uses an optical directional coupler. In the authors' simulations the authors used the following methods: PWE (Plane Wave Expansion), FDTD (Finite-Difference Time-Domain), and COMSOL, which is integrated to their own BiPM (Binary Propagation Method).

During the past two decades, there has been considerable attention paid to materials with periodic modulation of the refractive index. Such structures, called photonic crystals, can have periodicity in one, two or three dimensions and typically utilize a period that is approximately one half of the operating wavelength. One dimensional photonic crystals have been used in practical devices inlcuding distributed feedback lasers [28, 50] and vertical cavity surface emitting lasers [64] which are utilized in applications ranging from optical communication systems to laser printing. Devices based on two and three dimensional photonic crystals remain an ongoing research topic. Chapter 10 discusses the use of two dimensional photonic crystals for integrated photonics applications. In particular the authors focus on the band structure and modal properties of waveguides and cavities that posess non-symmorphic space group symmetry in the form of a glide plane.

In Chapter 11, structural, optical and nonlinear optical properties of strontium tetraborate (SBO) single crystals are summarized. SBO presently is not considered to be a ferroelectric, however, domain structures consisting of alternating oppositely poled domains, strongly resembling those present in such ferroelectrics as potassium titanyl phosphate and lithium niobate, were recently discovered in this crystal. Such geometrical properties as orientation, size, and degree of randomization, of these domain structures are described. The problem of origin of domain structures in SBO is considered, as well as the possibility of their characteristics control. From the point of view of optical properties, domain-structured samples of strontium tetraborate are classified as randomized nonlinear photonic crystals. Comparative study of nonlinear diffraction and random quasi-phase-matching of nanosecond and femtosecond laser pulses in these nonlinear photonic crystals is presented. Prospects of creation of nonlinear optical converters of laser radiation into VUV spectral region based on domain structured strontium tetraborate are discussed.

Photonic crystals are composite materials possessing a periodical modulation of their dielectric constant on the scale length of the visible wavelength. This dielectric periodicity affects light propagation creating allowed and forbidden photon energy regions. Among different types of photonic crystals, the authors addressed their attention to artificial opals i.e. three dimensional self-assembled arrays of spheres packed in a face centred cube crystal lattice. In order to tune the optical properties of such photonic crystals, the authors inserted gold nanoparticles in the interstices between the spheres. The authors call this process doping. In Chapter 12 the authors report on the optical properties of artificial opals doped at various levels with gold nanoparticles (NpAu) stabilized with citrate molecules. By increasing the NpAu doping level, the opal photonic stop band was bathochromically shifted as well as its full width half maximum and intensity were reduced. No effects of nanoparticle coalescence was observed as previously detected for NpAu prepared by laser ablation. In order to improve the mechanical stability of NpAu doped opals, the authors started a preliminary thermal annealing study with in situ monitoring of optical properties. The authors found that thermal degradation of doped samples was reduced with respect to bare opals. In addition, the annealing temperature of NpAu doped opals was much higher than that observed for bare ones.

In the Short Communication, the authors analyze the mathematical structure of the refractive group index (briefly, group index) of a photonic crystal starting from the fact that, in a photonic crystal, the index of refraction as a function of wavelength is a periodic function whose period is the atomic lattice spacing of the crystal so that the expansion in a Fourier series of the refractive-index function is considered. In particular, the case in which the wavelength is sufficiently small (geometrical optics) is examined.

In: Photonic Crystals
Editor: Venla E. Laine, pp. 1-29
ISBN: 978-1-61668-953-7

Chapter 1

FABRICATION AND APPLICATIONS OF POLYMER PHOTONIC CRYSTALS

Jingxia Wang and Yanlin Song*

The Laboratory of New Materials, Key Laboratory of Organic Solids, Institute of Chemistry, Chinese Academy of Sciences

1. Introduction

Photonic crystals (PCs) are defined as a periodic dielectric structure with a refraction index varying on the order of the wavelength of light, which was firstly described by Yablonovitch[1] and John[2] in 1987. Due to the periodic variation of refractive index, the crystals possess unusual optical properties------photonic bandgap (PBG), a range of frequencies, in which light (with certain wavelength) is forbidden to exist within the bulk of the PCs, and light with other wavelength may propagate through PCs uninterruptedly. PCs are also called light-semiconductors due to its optical analogue of the electronic bandgap in semiconductors. The function of special light manipulation endows PCs many potential applications.

According to their periodicity in space, PCs can be categorized mainly into one-dimensional (1D), 2D and 3D PCs. All above-mentioned PCs have been fabricated artificially from a variety of materials. To date, the main 3D PCs include opal, inverse opal, woodpile, diamond structure, etc. Among those structures, opal structure comprised of polymer colloidal crystals demonstrates promising advantages over other structures, for example, polymer colloidal is easily processed, and its properties could be easily modulated based on the design of molecular structure. Moreover, polymer colloidal crystal can also act as template for the fabrication of PCs of inorganic, metal and complicate materials, since nanometer sized polymer particles are readily available with a high mono-dispersity or can be synthesized in a controlled way. Plenty of review papers[3-9] have been published about the fabrication of polymer PCs and its special applications. Such as, fabrication of polymer PCs form latex system[3-5] inverse opal structure[6], and micro-phase separation[10]. Polymer PC material

* E-mail address: ylsong@iccas.ac.cn. (Corresponding author)

is a novel material, and can provide material basis for the development of relative subjects, which covers photo-computing, photo-communication and optic-electric integration. In this paper, we review the recent research progress of polymer PCs, mainly concern on the fabrication and application of polymer PCs.

2. Fabrication of Polymer PCs

There are generally two approaches for the fabrication of polymer PCs: top-down lithography process[11-12] and bottom-up self assembly approach[13-16]. The former includes mechanical -manufacture, micro-, photo-, electric- lithography and holography etc. This approach is easy for the manufacture of PCs with stopband in the ranges of microwave, but it is expensive and time-consuming for the PCs with stopband at visible or near infrared ranges. Thus, polymer PCs are preferably to be fabricated from self-assembly of latex spheres and its inverse opal or microphase separation of block copolymers (BCPs) for the process is simple, low-cost, and easy duplication. Self-assembly approaches that have been applied to the fabrication of 3D PBG crystals include the following: 1) crystallization of spherical or non-spherical building blocks such as monodisperse colloids; 2) template-directed synthesis against a close packed lattice of spherical colloids, the structure of the resultant PCs depend on the PCs template; 3) Phase separation of BCPs. In this section, several typical fabrications of polymer PCs were presented based on various fabrication principles.

2.1. Self-Assembly of Latex Spheres from Gravity

Polymer PCs, in the earliest report, were self-assembled by natural deposition under gravity field[14]. This process takes longer time (several weeks or months), and is cannot control the assembly process, so that the resultant sample consists of mixing crystal structure. Later, the PCs were obtained via centrifugation or filtration under high vacuum, which could greatly accelerate the latex nucleation and transfer processes.[15] To further control the fabrication process, Nagayama *et al.* firstly put forward the fabrication of polymer PCs from convective assembly, and investigated in detail various inter-particle forces governing the packing of two-dimensional latex aggregates,[17-20] and found that the water evaporation and transfer convective flow played a crucial role on the latex arrangement. This method could effectively control crystal procedure by evaporation rate. Subsequently, Jiang[21] further simplified the fabrication process: vertically putting a glass substrate into the latex suspension, the latex sphere will be well-ordered transferred upon the substrate during evaporation procedure, the so-called vertical deposition method greatly boosted the wide application of this self-assembly approach. In this process, the crystal thickness could be well controlled by the concentration of the latex suspension or latex diameter. To meet application requirement of high quality PCs, many researchers took much efforts to further optimize the crystal procedure. Ozin *et al.* introduced the mechanical stirrer[22] or temperature gradient[23] into the constant temperature deposition system.[24] These methods are suitable for the assembly of latex spheres with diameter of larger than 800 nm.[22-25] Gu[26] achieved a fine control of the crystal thickness by using lifting apparatus, the apparatus could effectively control the thickness by modifying the lifting rate or latex concentration.

Fudozui[27] removed the uneven crystal structure by covering the hydrophobic liquid upon the contact line of the substrate. Meng[28] put forward capillary assembly of latex spheres and ultra-fine particles (Figure 1). Recently, they also demonstrated a fabrication of PCs with high quality by pressure controlled isothermal heating vertical deposition method[29], and the films showed deep photonic band gap and steep photonic band edge.

2.2. Self-Assembly of Latex Sphere by Exterior Field

Various exterior fields, such as electric [30-35] or magnetic field, are also used for self-assembly of polymer PCs with special properties. There are mainly three kinds of electric action for the fabrication of colloidal PCs, such as, electrostatic[30], electrophoresis[31-33] or electrochemical[34-35]. Highly charged monodisperse latex particles could form well-ordered 3D arrangement by electrostatic interaction [30] and lateral capillary force. The obtained crystalline structure could be further modified by ion or electrolytes concentration. This slow assembly process required stringent assembly condition. Latex assembly via electryphoresis[31-33] is a rapid and controllable assembly way for the charged latex spheres. The assembly process could be well controlled by electronic strength and viscosity of liquid media. While electrochemical assembly [34-35] method has special advantages, such as rapid assembly and controllable process. This method is especially suitable for complex substrate, and prefers to the semiconductor materials with high refractive index (CdS, CdSe, II-IV, III-V, add IV). The resultant films show high mechanical strength and good heat-resistance. The earliest fabrication of PCs based on electric is reported by Yeh *et al.*[31] They assembled ordered colloidal aggregates by electric-field-induced fluid flow, in which, the highly charged monodisperse latex particles were applied among two electrodes. The polarization of the bead double layer in the external field produce the repulsive interaction between beads, and electroosmetric flow yields attractive interaction among beads. Velev *et al.*[33] developed colloidal crystal from alternating electric fields by two coplanar electrodes (Figure 2). This assembly is driven by forces resulted from the electric field gradient. The assembly process can be controlled via the field strength, frequency, and viscosity of the liquid media.

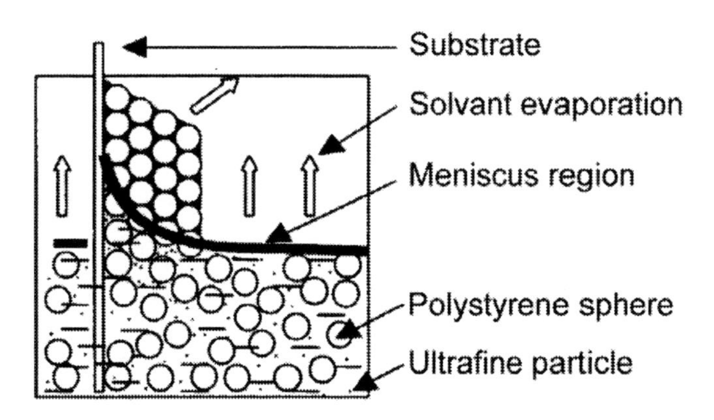

Figure 1. Schematic illustration of capillary assembly of latex spheres and ultra-fine particle. [28] Copyright 2002 American Chemical Society

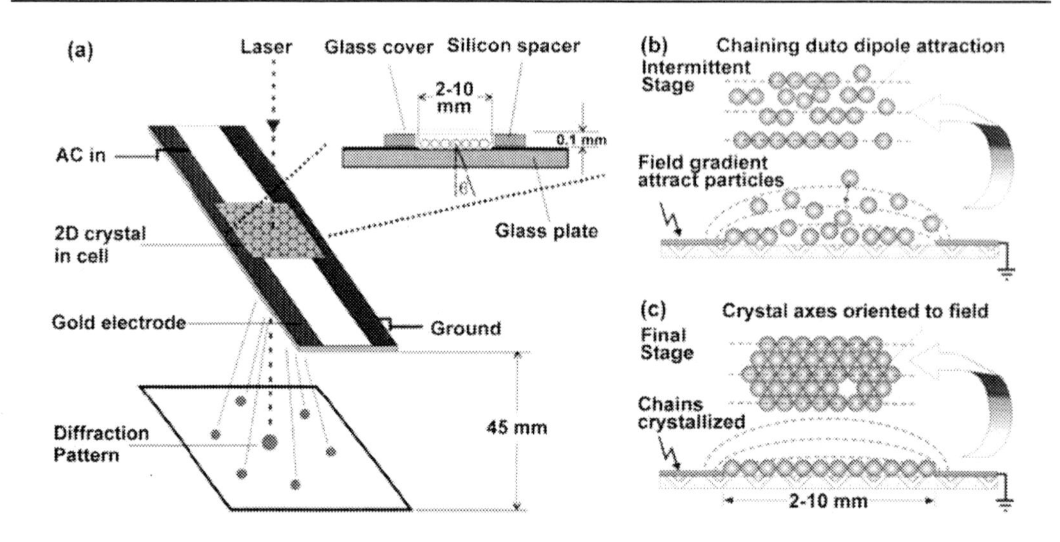

Figure 2. (a) Schematics of the experimental arrangement and dimensions of the experimental cell. (b, c) Schematics of the two-stage mechanism of crystallization deduced from the set of diffraction and microscopy data. (b) Immediately after the field is applied, the particles align in chains due to the induced dipole-dipole interactions. Simultaneously, the gradient-driven dipole-field force pulls the chains and particles toward the surface of the substrate between the electrodes. (c) The particle chains confined on the surface crystallize to form 2D hexagonal crystals aligned by the field. [33] Copyright 2004 American Chemical Society

In addition, self-assembly of latex spheres by magnetic field is another effective approach for the fabrication of PCs with special properties. Typically, Bibette *et al.* reported that emulsion droplets containing γ-Fe_2O_3 nano-particles can form one dimensional chain structure and diffract light in the visible range in the presence of an external magnetic field.[36-37] Asher *et al.* have developed more stable building blocks by embedding superparamagnetic iron oxide nanoparticles into monodisperse polymer colloids through an emulsion polymerization process.[38-39] While Yin *et al.*[40-43] applied pure magnetic materials as building blocks of colloidal crystal, which is favorable to a faster response to the external field. They synthesized a polyacrylate-capped superparamagentic colloidal nanocrystal clusters of magnetite (Fe_3O_4) with tunable size from 30 to 180 nm. The Fe_3O_4 colloidal nano-crystal clusters readily self-assembled in deionized water upon application of external magnetic field[40] , and diffract brilliant visible light when the field strength is changed. Further modification of colloidal nanocrystal cluster particles with a layer of silica allows their dispersion in various polar organic solvents[41] such as methanol, ethanol and ethylene glycol. The modified particles can still self-assemble into ordered structure in these nonaqueous solvents[42] and diffract light upon application of an external magnetic filed.

2.3. Self-Assembly of Latex Spheres Under Physical Confinement [44-52]

The crystal structure fabricated from above-mentioned method is face centered cubic structure. Its stopband is narrow, and the displacement of the crystals structure may close the stopband. Self-assembly of the latex spheres based on template effect [44-47] make it feasible for controllable crystal orientation. Blaaderen[44] fabricated PCs film with uniform thickness by combining microcavity template and evaporation effect. Amos[45] obtained large-area single crystal by applying controlled shear upon the substrate. Cheng[46] applied the template gradients to control the nucleation and growth of hard–sphere colloidal crystals (CCs). Velikov[48] obtained binary colloidal crystals (CCs) with control over the crystal orientation through a simple layer-by-layer process. Well-ordered single binary colloidal crystals with a stoichiometry of large (L) and small (S) particles of LS2 and LS were generated. The formed structure is mainly based on the templating effect of the first layer and the forces exerted by the surface tension of the drying liquid.

Different from above method, Xia[49] designed a special apparatus (Figure 3) for the fabrication of crystalline lattices of mesoscale particles over ~1 cm^2. The apparatus could effectively control the crystal structure. The apparatus includes the sample cell fabricated by sandwiching a photoresist frame between two glass substrates. A tube injecting latex suspension was connected with the upper glass, while water leakage tube was connected to the bottom glass. Photolithography films provide the suitable channel for the latex assembly, while the exterior pressure through injection entrance of upper glass could accelerate latex assembly and water leakage, the additional ultrasonic could induce latex arrangement. This apparatus combines the complicate effects of the filtration, ultrasonic, shake and template, and will benefit for the well-ordered assembly of the latex spheres. Based on this work, Xia[50-51] developed series of complex aggregates of monodisperse colloids with well-defined sizes, shapes and structure by taking advantage of various physical templates, such as, cylindrical hole, square, rectangular, trenches, triangular and noncircular pattern. Ozin[52] also obtained various colloidal crystal by using the predefined silicon template and centrifugation action.

2.4. PCs Fabricated from Micro-Phase Separation

Different from above-mentioned self-assembly approach, fabrication of PCs from microphase separation needs BCPs rather than monodisperse latex particles. BCPs are interesting platform materials for PC applications due to their ability to self-assemble into 1D, 2D and 3D periodic structures and the ability to selectively incorporate additives. The simplest example for this kind of PCs is the multilayer stack, whose wavelength mainly depends on the optical thickness of each layer. BCPs with compositions of 35-65% can self-assemble into a periodic lamellar structure. The period scales are of two-thirds power of the molecular weight. Alternatively, when PCs are made from phase separation of homopolymer-BCP blends, its length scales can be increased via swelling of homopolymer, its stopband can be conveniently tuned in an approximately linear manner by simply adjusting the amount of homopolymer.[53-54] Varying the refractive index of the layers can modify the band structure, and increasing the dielectric contrast will lead to larger bandgaps. Self-assembly from a pair of immiscible homopolymers and relatively high molecular weight BCPs has produced 2D[55] cylindrical domains and 3D-body-centered cubic packed spherical domains. Thomas *et al.* have prepared three known types of micro-domain structures with cubic symmetry: spheres, double gyroid and double diamond[56-57]. The double gyroid micro-domain structure is comprised of two interpenetrating, 3-coordinated labyrinthine networks of the minority lamellar and cylindrical phases for BCPs. Figure 4 exhibits the periodic structure obtained from microphase separation of BCPs of Polystyrene (PS)-b-polyisoprene (PI))[57-

58], the periodic structure of PS and air could be obtained when removing PI by UV irradiation. The complex double gyroid structure could be fabricated by applying triblock polymer (PS-PI-polymethyl methacrylate(PMMA))[59]. The polymer materials not only displays anisotropic dielectric properties but also is able to vary the anisotropy by applied fields, which points the way towards the use of self-assembled block polymeric materials in optical switches, couplers and isolators.[60]

There are four main challenges that need to be overcome in order to achieve necessary photonic properties with BCPs. The primary challenge is to obtain the large domain sizes needed for the optical frequencies of interest. Additional significant concerns for creating useful BCP PCs include long-range domain order and domain orientation, as well as sufficient dielectric contrast between domains.

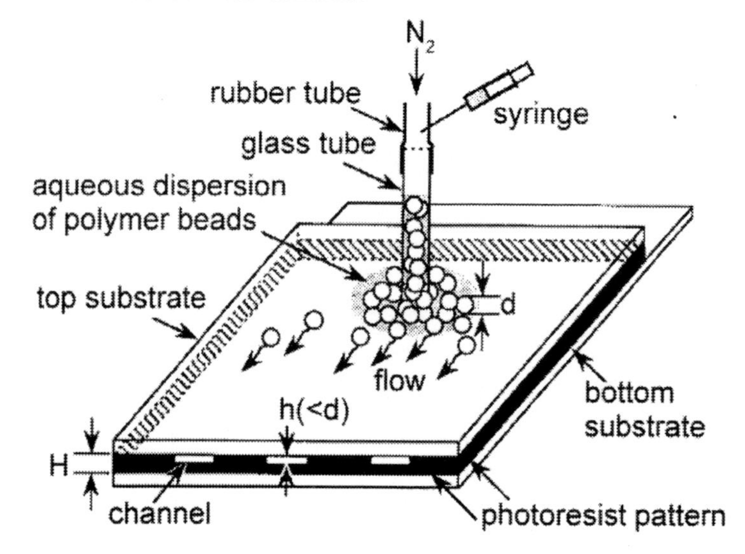

Figure 3. Schematic illustration of experimental set-up for physical confinement approach. [49] Copyright wiley-VCH Verlag GmbH & Co. KGaA. Reproduced with permission.

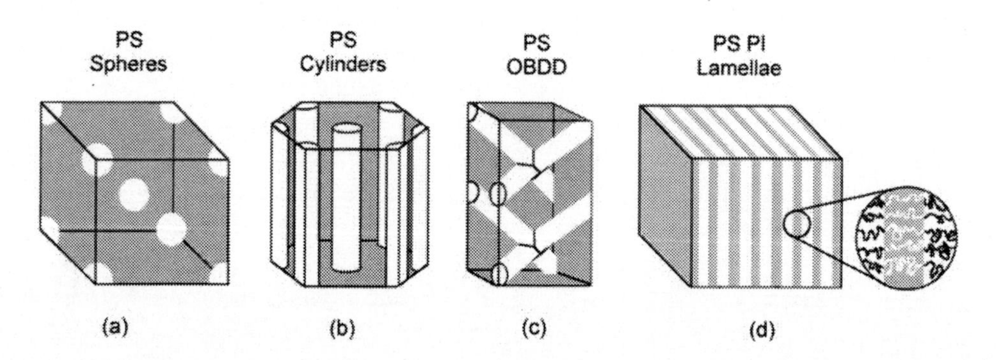

Figure 4. [5] Schematic illustrations of some typical mesoscale periodic structures that could be generated through the phase separation of organic BCPs (in this case, PS±PI). By varying the ratio between the chain lengths (or molecular weights) of these two units, one could obtain long-range ordered phases such as the 3D body-centered cubic lattice of PS spheres; the 2D hexagonal array of PS cylinders; the three-dimensionally ordered bicontinuous double diamond structure; and the 1D lamellae. [5] Copyright Wiley-VCH Verlag GmbH & Co. KGaA. Reproduced with permission.

2.5. Large-Scale Fabrication of Polymer PCs

Although above methods put forward some effective fabrication of polymer PCs with high quality, it is still difficult for the large-scale fabrication of PCs relating to real practical application. For example, it will take long time for large-scale samples, while the long-time soaking in the latex suspension would rust metal substrates, e.g., iron substrate. Regarding of these problems, Jiang *et al.* firstly put forward the fabrication of polymer PCs by spin coating, which is easy for the fabrication of PCs with centimeter-size within minutes.[61] The large-scale sample could be used as template for the inverse opal, and this method is also be used for the fabrication of PCs with special morphology[62]. However, this method needs highly viscous latex suspension, where, the re-dispersion of latex spheres is tedious,[62] and the application of special dispersant would impair optic properties of the resulted films due to the lower refractive-index contrast.[61] Moreover, the size of the resultant films is circumscribed to the area of the top surface of the spin coating instrument. Song *et al.*[63] recently developed an ultrafast fabrication of large-scale polymer PCs by spray coating (Figure 5). They realized a rapid assembly of the latex sphere during short spray process by taking advantage of latex spheres with hydrophobic PS core and hydrophilic PMMA/polyacrylic acid(PAA) shell, and the latex shell with abundant COOH groups resulted in strong hydrogen bonding interaction among latex spheres, which boosted latex arrangement during the ultra-fast spray procedure. As a result, the large-area colloidal PCs could be rapidly fabricated on different substrates by spray coating. The resultant colloidal PCs showed well-ordered latex arrangement and iridescent color. This ultra-fast fabrication procedure will be of great importance for practical application of PCs in the fields of optic devices and functional coatings.

2.6. Fabrication of Patterned Polymer PCs by Printing

Patterned PCs[41, 43, 52, 64-65] have attracted great attention because of their broad applications in optical devices,[66-72] displays,[73-76] and microfluidic devices.[77] Generally, the patterned PCs can be achieved by lithographic approaches[78-79] and self-assembly methods.[80-81] But the traditional lithographic approach was complex, time-consuming and expensive, while the patterns of the PCs fabricated from the self-assembly method, were limited by the pattern of the template.[82] Ink-jet printing, a particularly attractive patterning technique for the manufacture of high performance devices,[83-87] has been utilized for the fabrication of pattern PCs by Moon and Song *et al.* Moon fabricated PC micro-arrays composed of two different-sized latex spheres by using single-orifice ink-jet printing.[88-92] This facile fabrication of patterned PCs by combining the self-assembly and direct write was regarded to be promising for the realization of nano/microperiodic structures for next generation photonic and display devices. To further improve the optic properties of the patterned PCs and accelerate the application of this method. Song *et al.*[93] fabricated large-scale patterned PCs from common ink-jet printer(Figure 6). In this work, the latex suspension with special latex structure was used as ink, and the printing substrates with suitable wettability were applied. The pattern designed from computer software could be directly printed. These work demonstrated the first fabrication of multi-stopbands macroscale patterned PCs from common ink-jet printing. This rapid fabrication of multi-stopbands PCs will be of great significance for extensive applications of PCs. Just recently, Kim and Yin[94-95] fabricated a high-resolution patterning of multiple structural colors within seconds, based on successive tuning and fixing of colour using a single material along with a maskless lithography system. The colour of films is tunable by magnetically changing the periodicity of the nanostructure, and the pattern structure of the films is photochemically immobilized in a polymer network. This simple, controllable and scalable structural colour printing may have a significant impact on colour production for general consumer goods.

3. Polymer PCs with Special Properties

The periodic dielectric structure of polymer PCs endows the films special light manipulation properties, which have potential applications in various optic devices. While application properties of films can be effectively modulated by applying functional polymers. In this section, we mainly refer to some research work about the films' mechanical strength, wettability, and its optic properties.

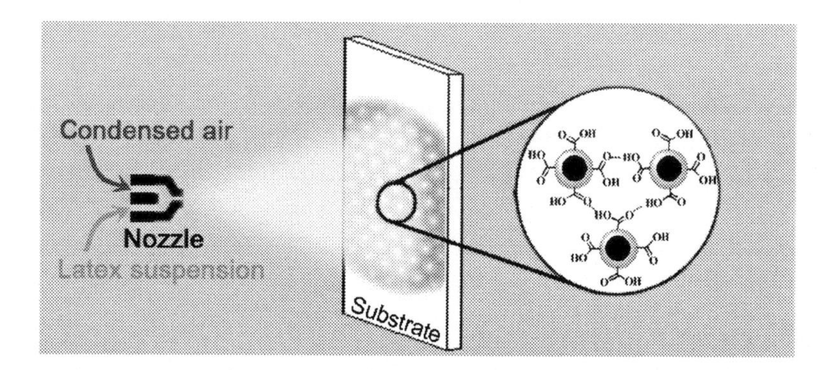

Figure 5. The schematics of fabrication procedure for colloidal PCs by spray coating method (the insert indicated the typical hydrogen bonding interaction resulting from latex spheres with abundant COOH groups). [63] Copyright Wiley-VCH Verlag GmbH & Co. KGaA. produced with permission.

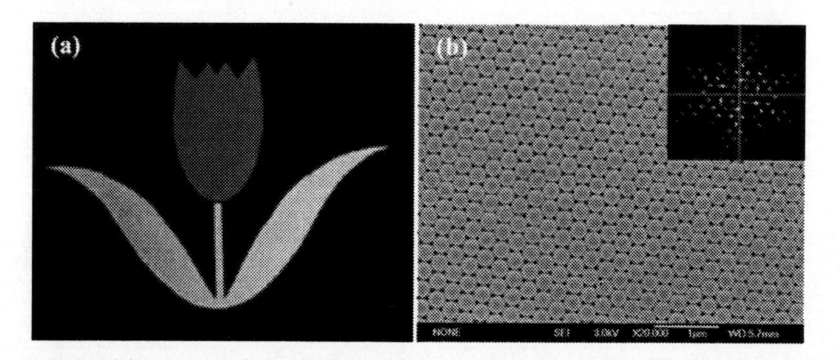

Figure 6. (a) Photo of the PCs with flower-leaf pattern, the scale bar: 1 cm, (b) SEM images of green-leaf (220 nm) pattern region (scale bar: 1 um). [93] Reproduced by permission of The Royal Society of Chemistry.

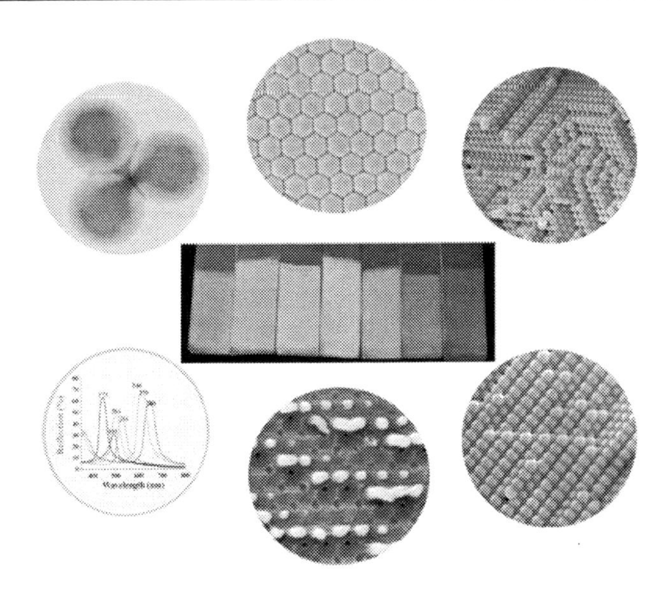

Figure 7. Tough PCs made from latex spheres with hard PS core-flexible PMMA/PAA shell. [102] Copyright Wiley-VCH Verlag GmbH & Co. KGaA. Reproduced with permission.

3.1. Polymer Pcs with High Strength

Generally, polymer PC films suffer poor mechanical strength[24, 96] due to the interstice structure among latex spheres. In order to improve the applicability of PCs under various circumstances, many measures have been taken to enhance the linkage among latex spheres[3, 97-105]. For instance, Xia[97] and Blanco[98] sintered the PC films at high temperature to decrease pore size. Ozin[99] allowed the growth of a continuous silica layer covering the film by chemical vapor deposition. Ruhl[100] and Wang[101] fabricated a tough colloidal PCs with hard PS core structure and soft PMMA/PAA shell (Figure 7), in which, the hard PS core keep well-ordered arrangement, and the adhesion effect of soft and flexible PMMA/PAA shell will enhance the linkage of latex spheres. While the tough closed-cell polyimide inverse opal structure[102] could be fabricated when applying this core-shell PS-PMMA-PAA latex spheres aggregates as PC template. The obtained polyimide PCs with a closed-cell structure had much better mechanical properties than those with open-cell structures.

Some literatures[104-105] used flexible (poly(N-isopropylacrylamide)(PNIPAM) as latex shell to improve the mechanical properties of the CCs. Wu *et al*[106] fabricated flexible and tough PC films from building blocks of a soft polymer core and hard SiO_2 nanoparticle shell. All these methods produced great enhancement for the mechanical properties of the PC films. However, the resultant films show poor solvent-resistance due to the intrinsic properties of polymer materials. This greatly restricts their practical application, especially in some solvent systems where PC films were required. Inspired from weather-resistance properties of natural PCs, such as feather of peacock, wings of butterfly, whose tough properties mainly be derived from their crosslinked network structure, Song *et al.*[107] developed a common method to improve both the mechanical strength and solvent-resistance properties of PCs by photo-crosslinked approach. The photopolymerized monomers were infiltrated into the interstice of latex spheres and subsequent photo-polymerization after UV irradiation. The obtained elastic and flexible crosslinked polymer network could enhance the linkage among latex spheres, which greatly improve both the mechanical strength and solvent resistance of the resultant PCs. Furthermore, this simple photo-crosslinkage enhancement approach can be effectively extended to most PCs, such as, PS and SiO_2 PCs. This approach opens a facile way for improvement of both mechanical strength and solvent resistance of PCs, and shows promising implication for the practical application of PCs.

3.2. Polymer PCs with Special Wettability[108]

There are many PC structures in nature, such as natural opal, hair of seamouse, feather of peacock and wings of butterfly. These PCs have not only structure color from their periodic structure, but show special wettability. Taking a butterfly wing as a example, its wings show self-cleaning properties as lotus, and also the wings exhibit iridescent color upon sun's irradiation. Inspired from these natural phenomena, Gu *et al.*[109-111] fabricated PCs with both structural color and self-cleaning properties based on uniform inverse opal structure over a large area. In this case, the structure color is derived from their periodic micro-porous structure, while the special wettability results from the combination of both hydrophobicity/hydrophilicity surface and roughness of periodic structure.

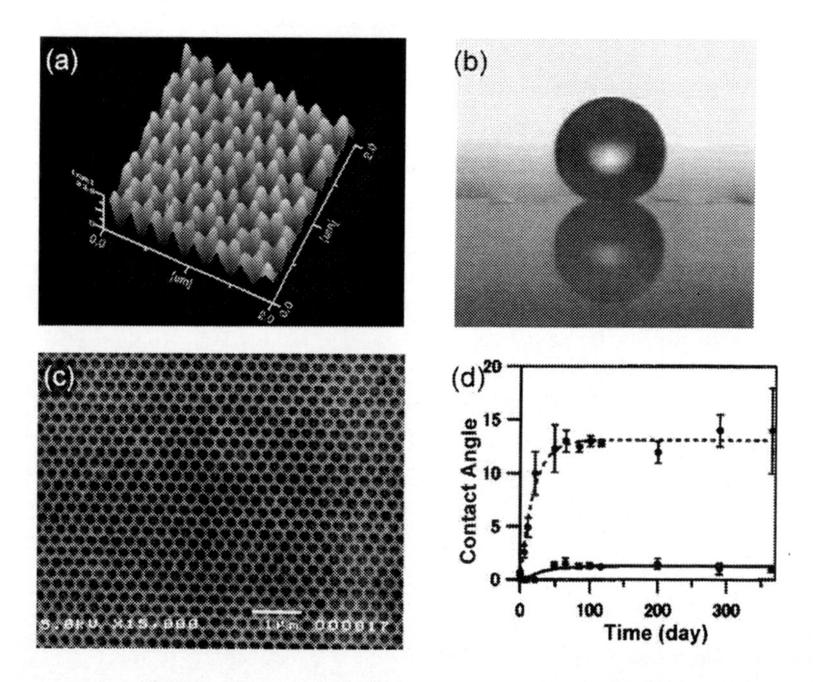

Figure 8. (a) AFM image of as-prepared superhydrophobic inverse opal film, (b) water droplet profile on as-prepared film (a) [109] Copyright Wiley-VCH Verlag GmbH & Co. KGaA. Reproduced with permission. (c) SEM image of as-prepared superhydrophilic film, (d) the change of water CA with storing time, dashed lie is films without pore, while solid line is the film of CCs. Reprinted with permission from [111]. Copyright [2004], American Institute of Physics.

The superhydrophobic film[109] was obtained when coating the inverse opal structure with fluoroalkylsilane by thermal chemical vapor deposition or directly applying of hydrophobic PS PCs[112]. The as-prepared inverse structure demonstrated a hexagonal arrangement of the air spheres encircled by the silica networks (Figure 8a). Such 3-D ordered structure contributes to the distinct stopband, and the roughness surface from periodic structure amplifies the hydrophobicity of fluoroalkylsilane surface, all of these result in a colorful superhydrophobic film shown in Figure 8b. While superhydrophilic surface was obtained based on hydrophilic TiO_2 inverse opal[111] in Figure 8c and 8d. Furthermore, the wettability of the films could be controlled by varying the roughness structure based on the changes of latex spheres and its spacing. For example, binary colloidal assembly[113-114] could enhance the roughness of the films surface, which will be favorable to the formation of superhydrophobic surface. Shiu *et al.*[115] achieved wettability modulation of films surface by gradually decreasing the latex diameter of close-packed PS nanostructure by oxygen etching, which could effectively modify liquid-solid contact area fractions.

PCs with responsive wettability could be obtained when PCs were made of responsive materials or its surface was coated with responsive polymer. In this case, the change of the surface chemical composition arisen from stimuli leads to the change of the wettability, while roughness structure will amplify the wettability change of the films. For example, Wang *et al.*[116-117] fabricated a PCs with tunable wettability by using latex spheres with amphiphilic materials of PS-b-PMMA-b-PAA. The wettability of the films can be tuned from superhydrophilic to superhydrophobic by varying the assembly temperature. The change of wettability could be mainly attributed to the change of the surface chemical composition driven from phase separation of polymer segments toward minimum interfacial energy when increasing assembly temperature. This phase reversion procedure results in a gradual increase of the water contact angle as the assembly temperature rises. In addition, the change trend of wettability for the PCs can be finely controlled by adjusting the ratio of the soft polymer segment and hard polymer segment,[117] which can modify the phase change temperature of the polymer. Otherwise, the wettability of the film could be flexibly controlled via adjusting the pH of the assembly system[118], in which the variation of pH modified the presence or not of the hydrogen bonding between hydrophilic groups (Figure 9). The stable superhydrophobicity arises from hydrogen bonding association between $SO_3^-Na^+$ of sodium dodecylbenzene sulfonate (NaDBS) and hydrophilic COOH around the latex surface, and the association locks hydrophilic groups of the latex sphere surface into a preferable configuration. Vice versa, the superhydrophilic films were achieved when introducing NH_3H_2O into the latex suspension (pH=12), in this case, hydrogen bonding is suppressed due to deprotonation of COOH to COO^-. Similarly, PCs with photo-responsive wettability have been obtained by Ge *et al.*[119] which were achieved by electrostatic layer-by-layer self-assembly of photoresponsive azobenzene monolayer upon the silica inverse opal. And polypyrrole[120] inverse opal PCs with electrochemical response could be obtained, and its wettability could be electrically tuned from 48.7° to 139.4° when being altered between neutral and oxidation state. In this case, the films show hydrophilicity at the neutral state due to the introduction of hydrophilic Li^+, while the films show hydrophobicity at the oxidation state due to the doping of DBS^- with a hydrophobic long alkyl chain. The typical application of polymer PCs with special wettability is visibly liquid manipulation properties[121] and patterned fabrication.

3.3. Polymer PCs with Stopband Modification

PCs have aroused wide research interest due to their important application of their optic properties. Particularly, the stopband of PCs could be reversibly changed in response to exterior stimulus. The responsive PCs were generally obtained by introducing various functional materials into PCs system, thus, the change of functional material responsive to exterior stimuli can result in the variation of the refractive index or crystal structure, which will lead to the modulation of stopband. The change of stopband of PCs mainly depends on the Bragg law[122]:

$$m\lambda = 2nd\sin\theta \quad (1)$$

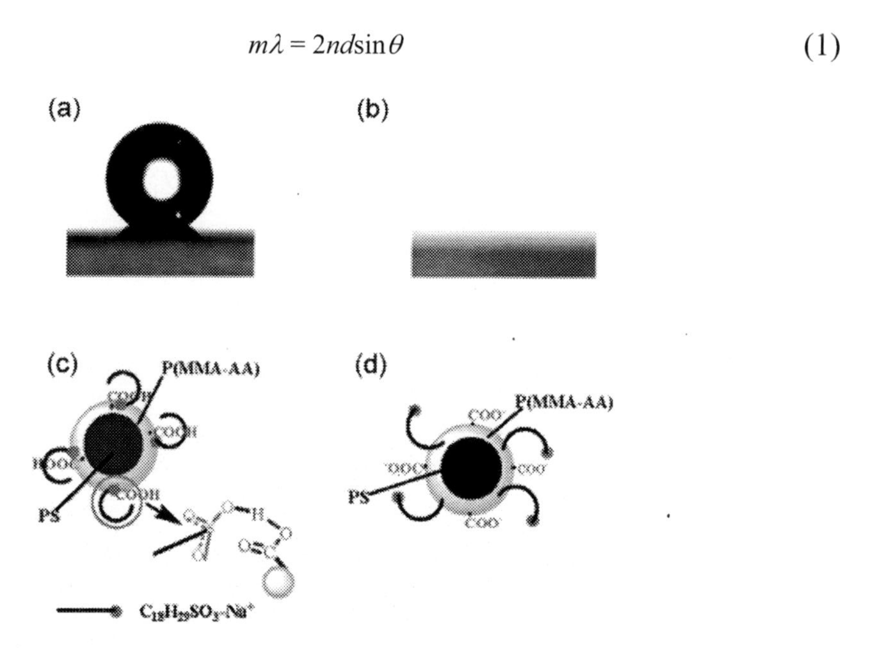

Figure 9. (a and b) Photographs of water droplet shape on the films assembled from suspensions with pH of 6.0 and 12, respectively; (c and d) illustrations of the structure of the latex sphere in the films of parts a and b, respectively. The conformation of hydrogen bonding is noted by the arrow. [118] Copyright 2006 American Chemical Society.

Here, θ is the angle of incidence between the beam and the lattice plane, λ is the wavelength of the reflectance peak maximum (i.e., the stopband position), m is the order of the Bragg diffraction (i.e., m=1), d is the interplanar spacing between (hkl) planes (i.e. the 111 planes), n is the average refractive index of the photonic structure. Therefore, the change of stopband could be attributed to the change of refractive index, interplanar spacing or crystal structure. Bragg diffraction was firstly described for X-rays where atomic lattices caused these interference phenomena.[123]

Asher[124-136] firstly fabricated the responsive PCs by coupling PCs with various functional hydrogel materials, in which, the stopband could be modulated based on the volume phase transition of hydrogel induced by exterior stimuli such as temperature,[124] ionic strength,[129] glucose,[133] light irradiation,[126] magnetic field,[38-39] chemical,[132] etc. Some responsive PCs have been fabricated by other group.[137-139] In this process, the lattice space and refractive index can be effectively modulated. For example, photo-responsive PCs could be obtained when introducing azo-benzene into the polymer structure. The stopband could be reversibly modified by alternatively irradiated by visible and ultraviolet light.[126] While temperature-sensitive PCs could be obtained by introducing PNIPAM into the polymer PCs, the phase change of the PNIPAM molecular will induce the change of the optic properties of the PCs[124] (Figure 10). The colloidal crystal array (CCA) are highly swollen at low temperature,, almost touching, and diffracting weakly, while the particles become compact and diffract nearly all incident light at the Bragg wavelength above the phase-transition temperature. It is found that the temperature change does not affect the lattice spacing, the 1-nm shift of the maximum wavelength of diffraction upon heating from 10 to 40°C results almost entirely form the change in the refractive index of water. The wavelength-tunable diffraction devices could be obtained based on these volume phase transition properties.

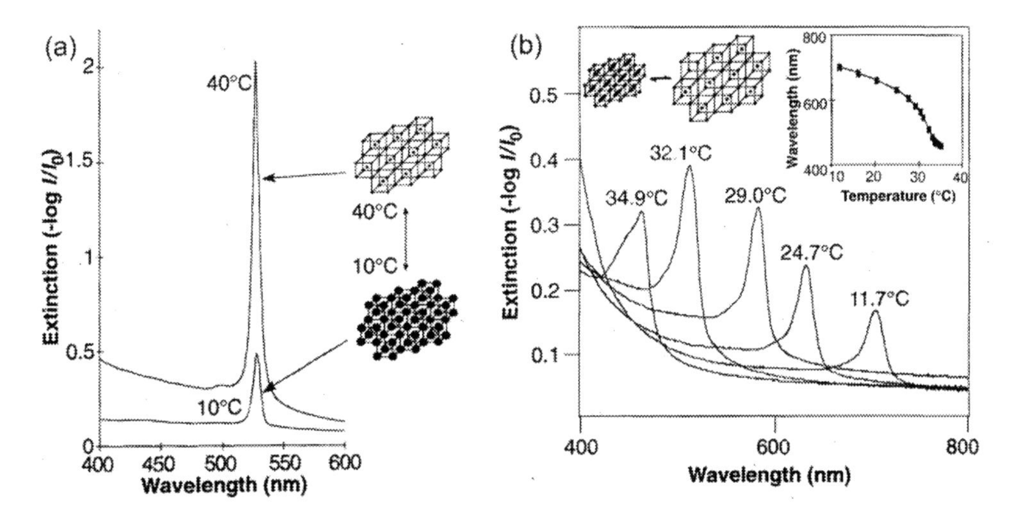

Figure 10. (a) Diffraction from a CCA of PNIPAM spheres at 10 and at 40 °C, insert is temperature switching between a swollen sphere array below the phase –transition temperature and an identical compact sphere array above the transition. (b) temperature tuning of Bragg diffraction from a 125 um thick films of 99 nm PS spheres embedded in a PNIPAM gel. Reproduced from permission of [124].

On the basis of Asher's work, Takeoka[140-142] introduced 2 or 3 kinds of functional materials into the same system and the films could demonstrate multi-stimuli response. For example, the thermo-photo responsive PCs have been fabricated from the PNIPAM hydrogel and exhibit a change in hydrophilicity in response to temperature, whereas the minor monomer, 4-acryloylaminoazobenzene, undergoes a change in the dipole moment upon the photo-isomerization of the azobenzene group.[142] Similarly, the PCs can exhibit various switchable colors over the visible region initiated by both electrochemical reaction and temperature change.[140] The films were fabricated from copolymer of electrolyte hydrogel of methacrylic acid (as a pH-sensitive monomer) and NIPAM (as a thermosensitive monomer). This gel reveals an electrochemically triggered rapid two-state switching between two arbitrary structural colors at constant temperature, and its color could be varied based on the temperature change. The color change with electrochemical response mainly can be attributed that electrolysis of water causes a change in pH of an electrolyte solution in the vicinity of the electrodes, the swelling degree of a pH-sensitive hydrogel on the electrodes can be easily and drastically changed by electrolysis. The responsive PCs could also be obtained when introducing responsive materials into the porous PC structure. For example, Gu and co-workers fabricated the tunable PCs based on filling organic molecules into PCs, whose reflective index can vary with exterior stimuli such as temperature and light irradiation.[143] The stop band of PCs can also be tuned by pH[144-145] or by magnetic field.[146-147]

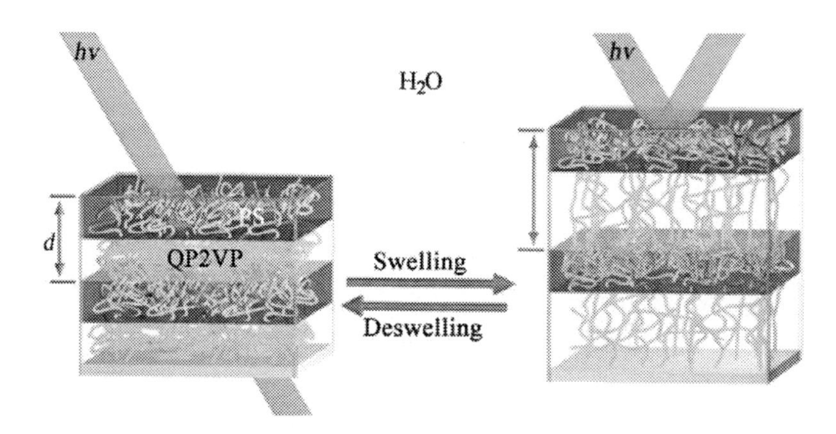

Figure 11.Schematic diagram of the structure of photonic gel film and the tuning mechanism. The photonic gel film was prepared by self-assembly of a diblock copolymer (PS-b-P2VP). Swelling/deswelling of the P2VP gel layers (pink) by aqueous solvents modulates both the domain spacing and the refractive-index contrast, and accordingly shifts the wavelengths of light (hν) reflected by the stop band.

Responsive PCs could be fabricated from microphase separation of functional BCPs. For example, Thomas has fabricated a highly tunable, structural-color reflector based on a hydrophilic/hydrophobic BCP (PS-b-poly(2-vinylpyridine)(P2VP)) that achieved substantial tenability (over 575%) of the primary stop band from ultraviolet (350 nm) to near infrared wavelengths (1600 nm) by changing the salt concentration in the solvent (Figure 11).[131] Recently, they created a full-color pixel by employing the same block copolymer combination with a simple electrochemical cell, the films produce red, green, and blue color through electrochemical stimulation.[148]

4. Applications of Polymer PCs

Polymer PCs showed potential application in high performance optic devices and intelligent optic sensing devices based on its special light manipulation properties. For example, the film could be used as colorful coatings based on its structure color, and specific optic device based on its special light manipulation properties. In the following section, some typical application examples of the polymer PCs will be demonstrated.

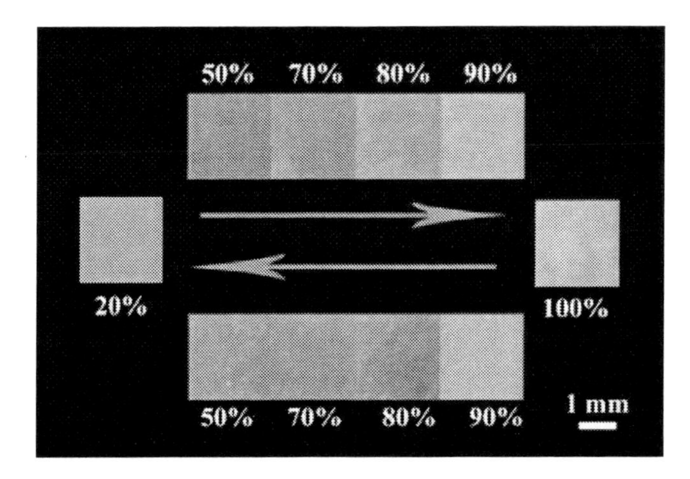

Figure 12. Photographs of as-prepared PC hydrogel responding to relative humidity of 20%, 50%, 70%, 80%, 90% and 100%, respectively. [65] Reproduced by the permission of The Royal Society of Chemistry.

4.1. Optical Sensing Device

The basic principle of PC sensor mainly depends on the change of stopband of PCs responding to exterior stimulus. If the stopband of PCs is in the visible range, the response of the films to environmental change could be observed from visible color change. The PCs with special optical properties could provide a new possibility for the chemical and biological detectors. Asher firstly developed visibly PCs sensing device, the color of the films could be reversibly changed in response to light, temperature, pH, ionic species, mechanical, glucose and solvent when coupling PCs with various functional hydrogel materials. Song *et al.* developed a series of PC sensors[65, 71, 149-150] which could visibly detect environment change, such as humidity[65] or oil kinds,[71, 149] etc. Firstly, they fabricated a colorful humidity sensitive PC hydrogel by infiltrating acrylamide (AAm) solution into poly(St-MMA-AA) PC template and subsequent photo-polymerization. As-prepared PAAm -poly(St-MMA-AA) PC hydrogel successfully combined the humidity sensitivity of PAAm and structure color of the PC template, it could reversibly vary from transparent to violet, blue, cyan, green and red under various humidity conditions (Figure 12), the color change covering the whole visible range. Otherwise, they fabricated oil-sensitive PCs[149] materials based on phenolic resin (PR) inverse opals with both superoleophilicity and superhydrophobicity due to oil adsorption of PR. The macroporous structure of the inverse opal benefited oil adsorption, and optical properties of PCs could provide various optical signal when adsorbing different oils. As a result, a refractive index variation of 0.02 resulted in a shift of stopband of 26 nm, which showed excellent selectivity for different oils. The visible oil monitor [71] could be realized when selecting the PC materials with higher refractive index, such as carbon materials.

Ultrasensive biodetector have been developed based on effective light-manipulation properties. Asher[132] firstly fabricated an interpenetrated CCA glucose sensor by attaching the enzyme glucose oxidase to a Poly colloidal crystal array (PCCA) of PS colloids. Glucose solutions prepared in air cause the interpenetrated CCA to swell and red-shift the diffraction.

This interpenetrated CCA returns to its original diffraction wavelength after removal from glucose. Asher[151] fabricated PC hydrogel with two coupled recognition modules, a creatinine deiminase (CD) enzyme and a 2-nitrophenol titrating group. Creatinine within the gel is rapidly hydrolyzed by the CD enzyme in a reaction which releases OH^-. This elevates the steady-state pH within the hydrogel as compared to the exterior solution. The increased solubility of the phenolate species as compared to that of the neutral phenols causes a hydrogel swelling, which red-shifts the IPCCA diffraction. This PC IPCCA senses physiologically relevant creatinine levels, with a detection limit of 6 *i*M, at physiological pH and salinity. This sensor also determines physiological levels of creatinine in human blood serum samples. The sensing technology platform may be used to fabricate PC sensors for any species for which there exists an enzyme which can catalyze it to release H^+ or OH^-. Song *et al.*[152] verified an ultrasensitive DNA detection using PCs combining fluorescence resonance energy transfer technique (Figure 13). The introduction of PCs into DNA detector system greatly amplified the optical signal; the stopband of PCs inhibited the energy loss of donor molecule. The method can achieve ultrasensitivity up to about 13.5 fm with optimized PCs. Furthermore, the PC-assisted detection method offers a great advantage over conventional techniques for its excellent ability to discriminate single base-pair mismatches, which is of crucial importance for the diagnosis of genetically encoded diseases.

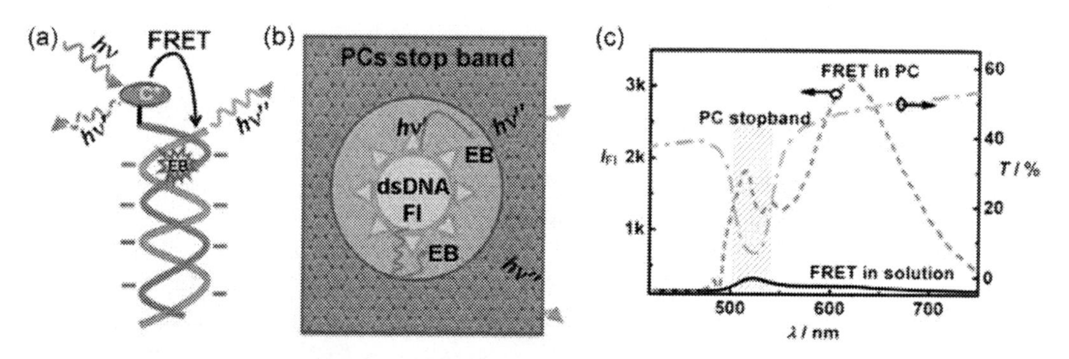

Figure 13. a) DNA sequence detection based on a FRET mechanism. b) Effect of the PC on FRET. c) Emission spectra of dsDNA-Fl/EB in the PC and solution respectively. [152] Copyright Wiley-VCH Verlag GmbH & Co. KGaA. Reproduced with permission.

4.2. PCs Used for Solar Cell

PCs can selectively modify propagation of the light with specific wavelength based on its intrinsic periodic structure. Various photonic devices have been fabricated based on PC structure and light manipulation properties, such as light waveguide that causes light to curve at acute angles,[153-154] dielectric mirrors[155] and so on. In rescent years, scientists have coupled PC structure into dye-sensitive solar cell (DSSC) and investigated its influence on the cell performance.[156-162] Typically, Mallouk *et al.* pioneered to couple TiO_2 PC layer into TiO_2 photo-electrode, which brought about an enhancement of light harvesting efficiency due to light localization properties of PCs.[156-157] Miguez's group carried out series of theorectical and experimental researches to illustrate the light-harvesting enhancement of PCs on the DSSC.[158-160] Huisman *et al.*[161] replaced the conventional TiO_2 photo-anode by its inverse opal. However, the procedure gave rise to an obvious drop of the cell efficiency due to difficult fabrication of PC anode with large area. Song *et al.*[70] developed a high efficient output of DSSC with a PC concentrator (Figure 14). The PC concentrator was developed by combining the advantages of light-focusing of concave-mirror and wavelength-selective filtration of PCs. The as-prepared PC concentrators could effectively improve the output power of the DSSC by more than 5 times by converging the desired light (i.e., the absorbance of the dye) and filtering out harmful IR and UV light, which demonstrated a promising strategy to promote the DSSCs' practical application.

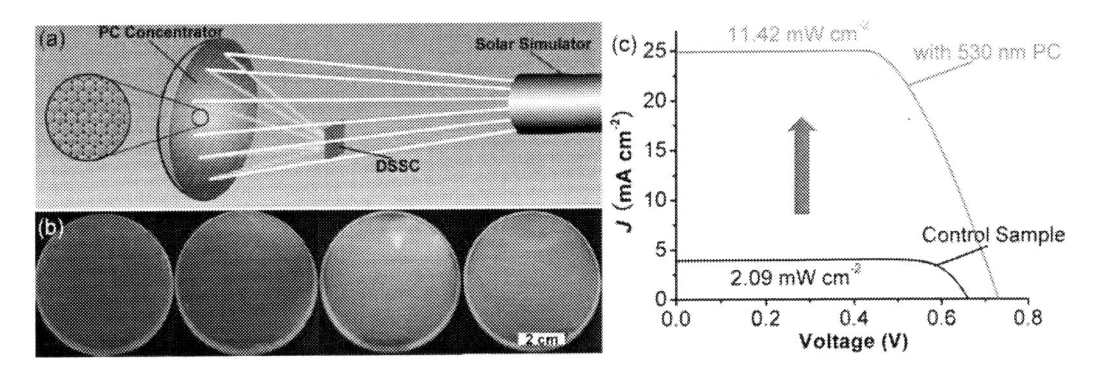

Figure 14. The scheme of the photovoltaic system with a PC concentrator. The insert is a typical SEM image of the PC concentrator; (b) The typical photographs of the PC concentrators with different colours, the stopbands of the PC concentrators from left to right are 448, 475, 530 and 647 nm, respectively. The scale bar is 2 cm. (c) The incident photon-to-current conversion efficiency curve of the DSSC, insert numbers are the P_{max} of DSSC when using PC concentrators with stopband of 530 nm and control sample. [70] Reproduced by permission of The Royal Society of Chemistry.

4.3. PCs Used for Enhanced Fluorescence

PCs have been applied for improving the fluorescence efficiency in many optic devices. The photoluminescence enhancement by PCs has been reported by photonic defects[163] and minimizing surface recombination losses of quantum dots at low temperature.[164] Recently, Klimov *et al.*[165] have reported amplified spontaneous emission in semiconductor nanocrystals uniformly coated on opal PS surfaces by enhancing the optical gain which was achieved by reducing the group velocity at the edge of the photonic stopband. Vos[166] and Cunningham *et al.* [167] have successfully realized a five-fold enhancement of the signal-to-noise ratio by utilizing resonant PC-enhanced fluorescence in a cytokine immunoassay. Song *et al.*[168-169] propose a simple, effective and practicable strategy to enhance fluorescence emission of molecular materials in the solid state by the fabrication of a Bragg mirror of PCs, where, the stopband can inhibit light propagation in a certain direction for a given frequency, therefore, PCs can modify significantly the emission characteristics of embedded optically-active materials (dyes, polymers semiconductors, etc.) as the emission wavelengths of the active materials overlap the stopband. An about 20 times enhancement was obtained when using the PC surface (Figure 15a). They recently amplified the fluorescent contrast by PCs in optical storage by applying PCs to the optical storage system.[170] A 40-fold enhancement of the fluorescence signal and a sevenfold ON/OFF ratio amplification are achieved.

Furthermore, a fluorescence image on the matched PC surface displays both higher brightness and contrast compared with that on glass(Figure 15b). Thus, amplification of fluorescent contrast on the PC surface can provide much better sensitivity and resolution in the bistable photoswitching, which will open a simple and viable way for design and development of high-performance optical memory devices. Song used PCs to enhance the emission density of organic dyes[171] and to achieve nonlinear emission.[172] An high-performance light emitted device (LED) system was obtained when coating PCs upon the surface of LED, which greatly amplify the take-out efficiency of light.[173] Recently, Song [174] synthesized high- photocatalytic performance Ti-Si oxides PCs with hierarchically macro-/mesoporous structure by combination of Polymer PC template and amphiphilic triblock copolymer. It was found that the photodegradation efficiency of i-Ti-Si PCs was 2.1 times higher than that of TiO_2 PCs, and a maximum enhanced factor of 15.6 was achieved in comparison to nanocrystalline TiO_2 films when the energy of slow photon [175-176] was optimized to the abosorption region of TiO_2, which originated from the synergetic effect of slow photon enhancement and high surface area. The slow photon effect of PCs has been applied to improve photo-conduct properties of semiconductor [177] by ozin et al.

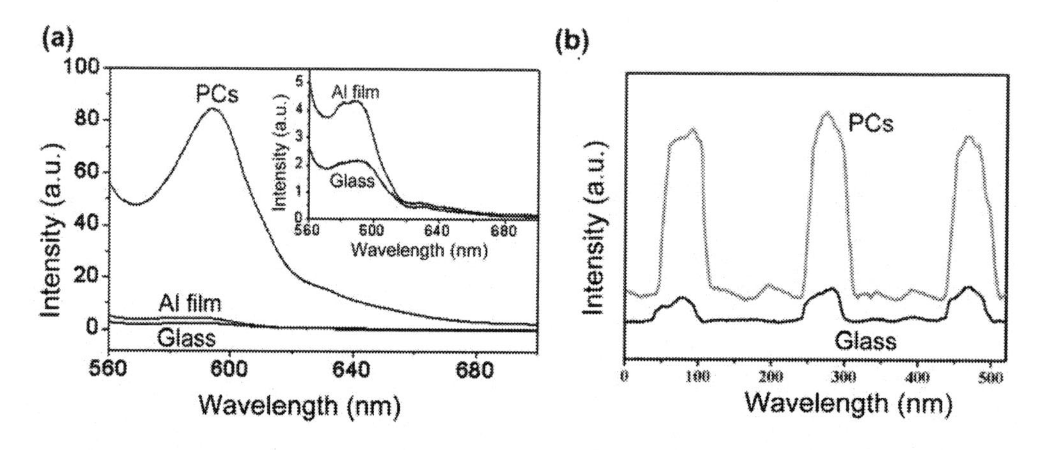

Figure 15. (a) The fluorescence spectra of 20 nm thick RB deposited on yellow PCs, aluminium film and glass substrate (λ_{ex} = 550 nm). Here the fluorescence spectra were collected at the stopband direction of PCs. The inset shows the fluorescence spectra of RB (20 nm) on aluminium film and glass. [168] Reproduced by permission of The Royal Society of Chemistry. (b) Fluorescence intensity associated line profiles from the glass and PC surfaces, respectively. [170] Copyright Wiley-VCH Verlag GmbH & Co. KGaA. Reproduced with permission.

5. Future Outlook

PCs have attracted increasing attention due to their special light manipulation properties. They show promising applications in ultrasensitive detector, fluorescence amplification, optic waveguide in various optic device, photo-catalytic, etc. However, there is still a long way for the practical application of PCs due to the difficult fabrication of PCs with large-scale and high quality. Moreover, it is difficult for the fabrication of full-stopband polymer PCs due to the low refractive index of polymer, which greatly restricts wide application of PCs.

Otherwise, further developing light-manipulation application of PCs in new system, such as detecting minute amount of harmful gas or solid, will be an important research trend for PCs.

References

[1] Yablonovitch, E. (1987). Inhibited Spontaneous Emission in Solid-State Physics and Electronics. *Phys. Rev. Lett*, **58**(20), 2059-2062.

[2] John, S. (1987). Strong localization of photons in certain disordered dielectric superlattices. *Phys. Rev. Lett*, **58**(23), 2486-2489.

[3] López, C. (2003). Materials aspects of photonic crystals. *Adv. Mater*, **15(20)**, 1679-1704.

[4] Xia, YN., Gates, B., Yin, Y. D. & Lu, Y. (2000). Monodispersed colloidal spheres: Old materials with new applications. *Adv. Mater*, **12**(10), 693-713.

[5] Xia, Y. N., Gates, B. & Li, Z. Y. (2001). Self-assembly approaches to three-dimensional photonic crystals. *Adv. Mater*, **13**(6), 409-413.

[6] Norris, D. J. & Vlasov, Y. A. (2001). Chemical approaches to three-dimensional semiconductor photonic crystals. *Adv. Mater*, **13**(6), 371-376.

[7] Manoharan, V. N., Imhof, A., Thorne, J. D. & Pine, D. J. (2001). Photonic crystals from emulsion templates. *Adv. Mater*,2001**,** **13**(6), 447-450.

[8] Flaugh, P. L., Odonnell, S. E. & Asher, S. A. (1984). Development of a New Optical Wavelength Rejection Filter - Demonstration of Its Utility in Raman-Spectroscopy. *Appl. Spectrosc*, **38**(6), 847-850.

[9] Texter, J. (2003). Polymer colloids in photonic materials. *C. R. Chim.*, **6**(11-12), 1425-1433.

[10] Edrington, A. C., Urbas, A. M., DeRege, P., Chen, C. X., Swager, T. M., Hadjichristidis, N., Xenidou, M., Fetters, L. J., Joannopoulos, J. D. & Fink, Y. (2001). Thomas, E. L., Polymer-based photonic crystals. *Adv. Mater*, **13**(6), 421-425.

[11] Campbell, M., Sharp, D. N., Harrison, M. T., Denning, R. G. & Turberfield, A. J. (2000). Fabrication of photonic crystals for the visible spectrum by holographic lithography. *Nature*, **404**(6773), 53-56.

[12] Tondiglia, V. P., Natarajan, L. V., Sutherland, R. L., Tomlin, D. & Bunning, T. J. (2002). Holographic formation of electro-optical polymer-liquid crystal photonic crystals. *Adv. Mater*, **14**(3), 187-191.

[13] Míguez, H., Meseguer, F., López, C., López-Tejeira, F. & Sánchez-Dehesa, J. (2001). Synthesis and photonic bandgap characterization of polymer inverse opals. *Adv. Mater*, **13**(6), 393-396.

[14] Stöber, W., Fink, A. & Bohn, E. (1968). Controlled growth of monodisperse silica spheres in the micron size range. *J. Colloid Interf. Sci.*, **26**(1), 62-69.

[15] Holland, B. T., Blanford, C. F. & Stein, A. (1998). Synthesis of macroporous minerals with highly ordered three-dimensional arrays of spheroidal voids. *Science*, **281**(5376), 538-540.

[16] Zhang, J. H., Sun, Z. Q. & Yang, B. (2009). Self-assembly of photonic crystals from polymer colloids. *Current Opinion in Colloid & Interface Science*, **14**(2), 103-114.

[17] Denkov, N. D., Velev, O. D., Kralchevsky, P. A., Ivanov, I. B., Yoshimura, H. & Nagayama, K. (1992). Mechanism of Formation of 2-Dimensional Crystals from Latex-Particles on Substrates. *Langmuir*, **8**(12), 3183-3190.

[18] Denkov, N. D., Velev, O. D., Kralchevsky, P. A., Ivanov, I. B., Yoshimura, H. & Nagayama, K. (1993). 2-Dimensional Crystallization. *Nature*, **361**(6407), 26-26.

[19] Dimitrov, A. S. & Nagayama, K. (1996). Continuous convective assembling of fine particles into two-dimensional arrays on solid surfaces. *Langmuir*, **12**(5), 1303-1311.

[20] Yamaki, M., Higo, J. & Nagayama, K. (1995). Size-Dependent Separation of Colloidal Particles in 2-Dimensional Convective Self-Assembly. *Langmuir*, **11**(8), 2975-2978.

[21] Jiang, P., Bertone, J. F., Hwang, K. S. & Colvin, V. L. (1999). Single-crystal colloidal multilayers of controlled thickness. *Chem. Mater*, **11**(8), 2132-2140.

[22] Yang, S. M., Míguez, H. & Ozin, G. A. (2002). Opal circuits of light - Planarized microphotonic crystal chips. *Adv. Funct. Mater*, 2002, **12**(6-7), 425-431.

[23] Vlasov, Y. A., Bo, X. Z., Sturm, J. C. & Norris, D. J. (2001). On-chip natural assembly of silicon photonic bandgap crystals. *Nature*, **414**(6861), 289-293.

[24] Wong, S., Kitaev, V. & Ozin, G. A. (2003). Colloidal crystal films: Advances in universality and perfection. *J. Am. Chem. Soc.*, **125**(50), 15589-15598.

[25] Kitaev, V. & Ozin, G. A. (2003). Self-assembled surface patterns of binary colloidal crystals. *Adv. Mater*, **15**(1), 75-78.

[26] Gu, Z. Z., Fujishima, A. & Sato, O. (2002). Fabrication of high-quality opal films with controllable thickness. *Chem. Mater*, **14**(2), 760-765.

[27] Fudouzi, H. (2004). Fabricating high-quality opal films with uniform structure over a large area. *J. Colloid Interf. Sci.*, **275**(1), 277-283.

[28] Meng, Q. B., Fu, C. H., Einaga, Y., Gu, Z. Z., Fujishima, A. & Sato, O. (2002). Assembly of highly ordered three-dimensional porous structure with nanocrystalline TiO2 semiconductors. *Chem. Mater*, **14**(1), 83-88.

[29] Zheng, Z. Y., Liu, X. Z., Luo, Y. H., Cheng, B. Y., Zhang, D. Z., Meng, Q. B. & Wang, Y. R. (2007). Pressure controlled self-assembly of high quality three-dimensional colloidal photonic crystals. *Appl. Phys. Lett*, **90**(5), 051910.

[30] Zeng, F., Sun, Z. W., Wang, C. Y., Ren, B. Y., Liu, X. X. & Tong, Z. (2002). Fabrication of inverse opal via ordered highly charged colloidal spheres. *Langmuir*, **18**(24), 9116-9120.

[31] Yeh, S. R., Seul, M. & Shraiman, B. I. (1997). Assembly of ordered colloidal aggregates by electric-field-induced fluid flow. *Nature*, **386**(6620), 57-59.

[32] Trau, M., Saville, D. A. & Aksay, I. A. (1996). Field-induced layering of colloidal crystals. *Science*, **272**(5262), 706-709.

[33] Lumsdon, S. O., Kaler, E. W. & Velev, O. D. (2004). Two-dimensional crystallization of microspheres by a coplanar AC electric field. *Langmuir*, **20**(6), 2108-2116.

[34] Braun, P. V. & Wiltzius, P. (1999). Microporous materials - Electrochemically grown photonic crystals. *Nature*, **402**(6762), 603-604.

[35] Deutsch, M., Vlasov, Y. A. & Norris, D. J. (2000). Conjugated-polymer photonic crystals. *Adv. Mater*, 2000, **12**(16), 1176-1180.

[36] Bibette, J. (1993). Monodisperse Ferrofluid Emulsions. *J. Magn. Magn. Mater*, **122**(1-3), 37-41.

[37] Calderon, F. L., Stora, T., Monval, O. M., Poulin, P. & Bibette, J. (1994). Direct Measurement of Colloidal Forces. *Phys. Rev. Lett*, **72**(18), 2959-2962.

[38] Xu, X., Friedman, G., Humfeld, K. D., Majetich, S. A. & Asher, S. A. (2001). Superparamagnetic Photonic Crystals. *Adv. Mater*, 2001, **13**(22), 1681-1684.

[39] Xu, X., Friedman, G., Humfeld, K. D., Majetich, S. A. & Asher, S. A. (2002). Synthesis and Utilization of Monodisperse Superparamagnetic Colloidal Particles for Magnetically Controllable Photonic Crystals. *Chem. Mater*, **14**(3), 1249-1256.

[40] Ge, J. P., Hu, Y. X., Biasini, M., Beyermann, W. P. & Yin, Y. D. (2007). Superparamagnetic magnetite colloidal nanocrystal clusters. *Angew. Chem., Int. Ed*, **46**(23), 4342-4345.

[41] Ge, J. P. & Yin, Y. D. (2008). Magnetically tunable colloidal photonic structures in alkanol solutions. *Adv. Mater*, **20**(18), 3485-3491.

[42] Ge, J. P., He, L., Goebl, J. & Yin, Y. D. (2009). Assembly of Magnetically Tunable Photonic Crystals in Nonpolar Solvents. *J. Am. Chem. Soc.*, **131**(10), 3484-3486.

[43] Ge, J. P. & Yin, Y. D. (2008). Magnetically responsive colloidal photonic crystals. *J. Mater. Chem.*, **18**(42), 5041-5045.

[44] van Blaaderen, A., Ruel, R. & Wiltzius, P. (1997). Template-directed colloidal crystallization. *Nature*, **385**(6614), 321-324.

[45] Amos, R. M., Rarity, J. G., Tapster, P. R., Shepherd, T. J. & Kitson, S. C. (2000). Fabrication of large-area face-centered-cubic hard-sphere colloidal crystals by shear alignment. *Phys. Rev. E*, **61**(3), 2929-2935.

[46] Cheng, Z. D., Russell, W. B. & Chaikin, P. M. (1999). Controlled growth of hard-sphere colloidal crystals. *Nature*, **401**(6756), 893-895.

[47] Lin, K. H., Crocker, J. C., Prasad, V., Schofield, A., Weitz, D. A., Lubensky, T. C. & Yodh, A. G. (2000). Entropically driven colloidal crystallization on patterned surfaces. *Phys. Rev. Lett*, **85**(8), 1770-1773.

[48] Velikov, K. P., Christova, C. G., Dullens, R. P. A. & van Blaaderen, A. (2002). Layer-by-layer growth of binary colloidal crystals. *Science*, **296**(5565), 106-109.

[49] Gates, B., Qin, D. & Xia, Y. N. (1999). Assembly of nanoparticles into opaline structures over large areas. *Adv. Mater*, **11**(6), 466-469.

[50] Yin, Y. D., Lu, Y., Gates, B. & Xia, Y. N. (2001). Template-assisted self-assembly: A practical route to complex aggregates of monodispersed colloids with well-defined sizes, shapes, and structures. *J. Am. Chem. Soc.*, **123**(36), 8718-8729.

[51] Yin, Y. D. & Xia, Y. N. (2003). Self-assembly of spherical colloids into helical chains with well-controlled handedness. *J. Am. Chem. Soc.*, **125**(8), 2048-2049.

[52] Ozin, G. A. & Yang, S. M. (2001). The race for the photonic chip: Colloidal crystal assembly in silicon wafers. *Adv. Funct. Mater*, 2001, **11**(2), 95-104.

[53] Urbas, A., Sharp, R., Fink, Y., Thomas, E. L., Xenidou, M. & Fetters, L. J. (2000). Tunable block copolymer/homopolymer photonic crystals. *Adv. Mater*, **12**(11), 812-814.

[54] Winey, K. I., Thomas, E. L. & Fetters, L. J. (1991). Swelling a Lamellar Diblock Copolymer with Homopolymer - Influences of Homopolymer Concentration and Molecular-Weight. *Macromolecules*, **24**(23), 6182-6188.

[55] Joannopoulos, J. D., Meade, R. D. & Winn, J. N. (1995). *Photonic crystals: molding the flow of light*. Princeton University Press: Princeton, N.J.

[56] Bockstaller, M. R., Mickiewicz, R. A. & Thomas, E. L. (2005). Block copolymer nanocomposites: Perspectives for tailored functional materials. *Adv. Mater*, **17**(11), 1331-1349.

[57] Hajduk, D. A., Harper, P. E., Gruner, S. M., Honeker, C. C., Kim, G., Thomas, E. L. & Fetters, L. J. (1994). The Gyroid - a New Equilibrium Morphology in Weakly Segregated Diblock Copolymers. *Macromolecules*, **27**(15), 4063-4075.

[58] Fink, Y., Winn, J. N., Fan, S. H., Chen, C. P., Michel, J., Joannopoulos, J. D. & Thomas, E. L. (1998). A dielectric omnidirectional reflector. *Science*, **282**(5394), 1679-1682.

[59] Urbas, A. M., Maldovan, M., DeRege, P. & Thomas, E. L. (2002). Bicontinuous cubic block copolymer photonic crystals. *Adv. Mater*, **14**(24), 1850-1853.

[60] Mao, G. P., Wang, J. G., Ober, C. K., Brehmer, M., O'Rourke, M. J. & Thomas, E. L. (1998). Microphase-stabilized ferroelectric liquid crystals (MSFLC): Bistable switching of ferroelectric liquid crystal-coil diblock copolymers. *Chem. Mater*, **10**(6), 1538-1545.

[61] Jiang, P. & McFarland, M. J. (2004). Large-scale fabrication of wafer-size colloidal crystals, macroporous polymers and nanocomposites by spin-coating. *J. Am. Chem. Soc.*, **126**(42), 13778-13786.

[62] Wang, D. Y. & Mohwald, H. (2004). Rapid fabrication of binary colloidal crystals by stepwise spin-coating. *Adv. Mater*, **16**(3), 244-247.

[63] Cui, L. Y., Zhang, Y. Z., Wang, J. X., Ren, Y. B., Song, Y. L. & Jiang, L. (2009). Ultra-Fast Fabrication of Colloidal Photonic Crystals by Spray Coating. *Macromol. Rapid. Comm*, **30**(8), 598-603.

[64] Xia, Y. N., Yin, Y. D., Lu, Y. & McLellan, J. (2003). Template-assisted self-assembly of spherical colloids into complex and controllable structures. *Adv. Funct. Mater*, **13**(12), 907-918.

[65] Tian, E. T., Wang, J. X., Zheng, Y. M., Song, Y. L., Jiang, L. & Zhu, D. B. (2008). Colorful humidity sensitive photonic crystal hydrogel. *J. Mater. Chem.*, **18**(10), 1116-1122.

[66] Colodrero, S., Mihi, A., Haggman, L., Ocana, M., Boschloo, G., Hagfeldt, A. & Míguez, H. (2009). Porous One-Dimensional Photonic Crystals Improve the Power-Conversion Efficiency of Dye-Sensitized Solar Cells. *Adv. Mater*, **21**(7), 764-770.

[67] Suezaki, T., O'Brien, P. G., Chen, J. I. L., Loso, E., Kherani, N. P. & Ozin, G. A. (2009). Tailoring the Electrical Properties of Inverse Silicon Opals - A Step Towards Optically Amplified Silicon Solar Cells. *Adv. Mater*, **21**(5), 559-563.

[68] Chassagneux, Y., Colombelli, R., Maineult, W., Barbieri, S., Beere, H. E., Ritchie, D. A., Khanna, S. P., Linfield, E. H. & Davies, A. G. (2009). Electrically pumped photonic-crystal terahertz lasers controlled by boundary conditions. *Nature*, **457**(7226), 174-178.

[69] Xie, Z. Y., Sun, L. G., Han, G. Z. & Gu, Z. Z. (2008). Optical Switching of a Birefringent Photonic Crystal. *Adv. Mater*, **20**(19), 3601-3604.

[70] Zhang, Y. Z., Wang, J. X., Zhao, Y., Zhai, J., Jiang, L., Song, Y. L. & Zhu, D. B. (2008). Photonic crystal concentrator for efficient output of dye-sensitized solar cells. *J. Mater. Chem.*, **18**(23), 2650-2652.

[71] Li, H. L., Chang, L. X., Wang, J. X., Yang, L. M. & Song, Y. L. (2008). A colorful oil-sensitive carbon inverse opal. *J. Mater. Chem.*, **18**(42), 5098-5103.

[72] Wang, J. X., Liang, J., Wu, H. M., Yuan, W. F., Wen, Y. Q., Song, Y. L. & Jiang, L. (2008). A facile method of shielding from UV damage by polymer photonic crystals. *Polymer International*, **57**(3), 509-514.

[73] Fudouzi, H. & Xia, Y. N. (2003). Photonic papers and inks: Color writing with colorless materials. *Adv. Mater*, **15**(11), 892-896.
[74] Arsenault, A. C., Puzzo, D. P., Manners, I. & Ozin, G. A. (2007). Photonic-crystal full-colour displays. *Nat. Photon*, **1**(8), 468-472.
[75] Puzzo, D. P., Arsenault, A. C., Manners, I. & Ozin, G. A. (2009). Electroactive Inverse Opal: A Single Material for All Colors. *Angew. Chem., Int. Ed.*, **48**(5), 943-947.
[76] Ozin, G. A. & Arsenault, A. C. (2008). P-Ink and Elast-Ink from lab to market. *Materials Today*, **11**(7-8), 44-51.
[77] Zeng, Y., He, M. & Harrison, D. J. (2008). Microfluidic self-patterning of large-scale crystalline nanoarrays for high-throughput continuous DNA fractionation. *Angew. Chem., Int. Ed.*, **47**(34), 6388-6391.
[78] Scrimgeour, J., Sharp, D. N., Blanford, C. F., Roche, O. M., Denning, R. G. & Turberfield, A. J. (2006). Three-dimensional optical lithography for photonic microstructures. *Adv. Mater*, **18**(12), 1557-1560.
[79] George, M. C., Mohroz, A., Piech, M., Bell, N. S., Lewis, J. A. & Braun, P. V. (2009). Direct Laser Writing of Photoresponsive Colloids for Microscale Patterning of 3D Porous Structures. *Adv. Mater*, **21**(1), 66-70.
[80] Gu, Z. Z., Fujishima, A. & Sato, O. (2002). Patterning of a colloidal crystal film on a modified hydrophilic and hydrophobic surface. *Angew. Chem., Int. Ed.*, **41**(12), 2068-2070.
[81] Dziomkina, N. V., Hempenius, M. A. & Vancso, G. J. (2005). Symmetry control of polymer colloidal monolayers and crystals by electrophoretic deposition onto patterned surfaces. *Adv. Mater*, **17**(2), 237-240.
[82] Arsenault, A. C., Miguez, H., Kitaev, V., Ozin, G. A. & Manners, I. (2003). A polychromic, fast response metallopolymer gel photonic crystal with solvent and redox tunability: A step towards photonic ink (P-Ink). *Adv. Mater*, **15**(6), 503-507.
[83] de Gans, B. J., Duineveld, P. C. & Schubert, U. S.(2004). Inkjet printing of polymers: State of the art and future developments. *Adv. Mater*, **16**(3), 203-213.
[84] Tekin, E., Smith, P. J. & Schubert, U. S. (2008). Inkjet printing as a deposition and patterning tool for polymers and inorganic particles. *Soft Matter*, **4**(4), 703-713.
[85] Berggren, M., Nilsson, D. & Robinson, N. D. (2007). Organic materials for printed electronics. *Nat. Mater*, **6**(1), 3-5.
[86] Park, J. U., Hardy, M., Kang, S. J., Barton, K., Adair, K., Mukhopadhyay, D. K., Lee, C. Y., Strano, M. S., Alleyne, A. G., Georgiadis, J. G., Ferreira, P. M. & Rogers, J. A. (2007). High-resolution electrohydrodynamic jet printing. *Nat. Mater*, **6**(10), 782-789.
[87] Fan, H. Y., Lu, Y. F., Stump, A., Reed, S. T., Baer, T., Schunk, R., Perez-Luna, V., López, G. P. & Brinker, C. J. (2000). Rapid prototyping of patterned functional nanostructures. *Nature*, **405**(6782), 56-60.
[88] Ko, H. Y., Park, J., Shin, H. & Moon, J. (2004). Rapid self-assembly of monodisperse colloidal spheres in an ink-jet printed droplet. *Chem. Mater*, **16**(22), 4212-4215.
[89] Wang, D., Park, M., Park, J. & Moon, J. (2005). Optical properties of single droplet of photonic crystal assembled by ink-jet printing. *Appl. Phys. Lett*, **86**(24), 241114.
[90] Park, J., Moon, J., Shin, H., Wang, D. & Park, M. (2006). Direct-write fabrication of colloidal photonic crystal microarrays by ink-jet printing. *J. Colloid Interf. Sci.*, **298**(2), 713-719.

[91] Park, J. & Moon, J. (2006). Control of colloidal particle deposit patterns within picoliter droplets ejected by ink-jet printing. *Langmuir*, **22**(8), 3506-3513.

[92] Wang, D., Park, M., Park, J. & Moon, J. (2006). Reflectance spectroscopy of single photonic crystal island fabricated by ink-jet printing. *Mater. Res. Soc. Symp. Proc.*, **901***E*, 0901.

[93] Cui, L. Y., Li, Y. F., Wang, J. X., Tian, E. T., Zhang, X. Y., Zhang, Y. Z. & Song, Y. L., L., J. (2009). Fabrication of large-area patterned photonic crystals by ink-jet printing. *J. Mater. Chem.*, 19, 5499-5502.

[94] Kim, H., Ge, J., Kim, J., Choi, S., Lee, H., Lee, H., Park, W., Yin, Y. & Kwon, S. (2009). Structural colour printing using a magnetically tunable and lithographically fixable photonic crystal. *Nat. Photon.*, **3**(9), 534-540.

[95] Ge, J. P., Goebl, J., He, L., Lu, Z. D. & Yin, Y. D. (2009). Rewritable Photonic Paper with Hygroscopic Salt Solution as Ink. *Adv. Mater*, **21**(42), 4259-4264.

[96] Bertone, J. F., Jiang, P., Hwang, K. S., Mittleman, D. M. & Colvin, V. L. (1999). Thickness dependence of the optical properties of ordered silica-air and air-polymer photonic crystals. *Phys. Rev. Lett*, **83**(2), 300-303.

[97] Gates, B., Park, S. H. & Xia, Y. N. (2000). Tuning the photonic bandgap properties of crystalline arrays of polystyrene beads by annealing at elevated temperatures. *Adv. Mater*, **12**(9), 653-656.

[98] Mayoral, R., Requena, J., Moya, J. S., López, C., Cintas, A., Míguez, H., Meseguer, F., Vázquez, L., Holgado, M. & Blanco, A. (1997). 3D long-range ordering in an SiO2 submicrometer-sphere sintered superstructure. *Adv. Mater*, **9**(3), 257-260.

[99] Míguez, H., Tetreault, N., Hatton, B., Yang, S. M., Perovic, D. & Ozin, G. A. (2002). Mechanical stability enhancement by pore size and connectivity control in colloidal crystals by layer-by-layer growth of oxide. *Chem. Commun*, (22), 2736-2737.

[100] Ruhl, T. & Hellmann, G. P. (2001). Colloidal crystals in latex films: Rubbery opals. *Macromol. Chem. Phys.*, **202**(18), 3502-3505.

[101] Wang, J. X., Wen, Y. Q., Ge, H. L., Sun, Z. W., Zheng, Y. M., Song, Y. L. & Jiang, L. (2006). Simple fabrication of full color colloidal crystal films with tough mechanical strength. *Macromol. Chem. Phys.*, **207**(6), 596-604.

[102] Chen, X., Wang, L. H., Wen, Y. Q., Zhang, Y. Q., Wang, J. X., Song, Y. L., Jiang, L. & Zhu, D. B. (2008). Fabrication of closed-cell polyimide inverse opal photonic crystals with excellent mechanical properties and thermal stability. *J. Mater. Chem.*, **18**(19), 2262-2267.

[103] He, X., Thomann, Y., Leyrer, R. J. & Rieger, J. (2006). Iridescent colors from films made of polymeric core-shell particles. *Polym. Bull.*, **57**(5), 785-796.

[104] McGrath, J. G., Bock, R. D., Cathcart, J. M. & Lyon, L. A. (2007). Self-assembly of "paint-on" colloidal crystals using poly(styrene-co-N-isopropylacrylamide) spheres. *Chem. Mater,* **19**(7), 1584-1591.

[105] Lyon, L. A., Debord, J. D., Debord, S. B., Jones, C. D., McGrath, J. G. & Serpe, M. J. (2004). Microgel colloidal crystals. *J. Phys. Chem. B*, **108**(50), 19099-19108.

[106] You, B., Wen, N. G., Shi, L., Wu, L. M. & Zi, J. (2009). Facile fabrication of a three-dimensional colloidal crystal film with large-area and robust mechanical properties. *J. Mater. Chem.*, **19**(22), 3594-3597.

[107] Tian, E. T., Cui, L. Y., Wang, J. X., Song, Y. L. & Jiang, L. (2009). Tough photonic crystals fabricated by photo-crosslinkage of latex spheres. *Macromol. Rapid. Comm.*, **30**(7), 509-514.

[108] Wang, J. X., Zhang, Y. Z., Zhao, T. Y., Song, Y. L. & Jiang, L. (2010). Recent progress on the fabrication of colloidial crystals. *Sci. China Ser. B-Chem.*, In Press.

[109] Gu, Z. Z., Uetsuka, H., Takahashi, K., Nakajima, R., Onishi, H., Fujishima, A. & Sato, O. (2003). Structural color and the lotus effect. *Angew. Chem., Int. Ed.*, (8), 894-897.

[110] Sato, O., Kubo, S. & Gu, Z. Z. (2009). Structural color films with lotus effects, superhydrophilicity, and tunable stop-bands. *Acc. Chem. Res.*, **42**(1), 1-10.

[111] Gu, Z. Z., Fujishima, A. & Sato, O. (2004). Biomimetic titanium dioxide film with structural color and extremely stable hydrophilicity. *Appl. Phys. Lett*, **85**(21), 5067-5069.

[112] Ge, H. L., Song, Y. L., Jiang, L. & Zhu, D. B. (2006). One-step preparation of polystyrene colloidal crystal films with structural colors and high hydrophobicity. *Thin Solid Films*, **515**(4), 1539-1543.

[113] Zhang, G., Wang, D. Y., Gu, Z. Z. & Mohwald, H. (2005). Fabrication of Superhydrophobic Surfaces from Binary Colloidal Assembly. *Langmuir*, **21**(20), 9143-9148.

[114] Du, C. G., Cui, L. Y. Z., You Zhuan, Zhao, T. Y., Wang, J. X., Song, Y. L. & Jiang, L. (2009). Fabrication of Colloidal Crystals with Hierarchical Structure and Its Water Adhesion Properties. *Journal of Nanoscience and Nanotechnology*, Accepted.

[115] Shiu, J. Y., Kuo, C. W., Chen, P. & Mou, C. Y. (2004). Fabrication of tunable superhydrophobic surfaces by nanosphere lithography. *Chem. Mater*, **16**(4), 561-564.

[116] Wang, J. X., Wen, Y. Q., Feng, X. J., Song, Y. L. & Jiang, L. (2006). Control over the wettability of colloidal crystal films by assembly temperature. *Macromol. Rapid. Comm.*, **27**(3), 188-192.

[117] Wang, J. X., Wen, Y. Q., Hu, J. P., Song, Y. L. & Jiang, L. (2007). Fine control of the wettability transition temperature of colloidal-crystal films: from superhydrophilic to superhydrophobic. *Adv. Funct. Mater*, **17**(2), 219-225.

[118] Wang, J. X., Hu, J. P., Wen, Y. Q., Song, Y. L. & Jiang, L. (2006). Hydrogen-bonding-driven wettability change of colloidal crystal films: From superhydrophobicity to superhydrophilicity. *Chem. Mater*, **18**(21), 4984-4986.

[119] Ge, H. L., Wang, G. J., He, Y. N., Wang, X. G., Song, Y. L., Jiang, L. & Zhua, D. B. (2006). Photoswitched wettability on inverse opal modified by a self-assembled azobenzene monolayer. *ChemPhysChem.*, **7**(3), 575-578.

[120] Xu, L., Wang, J. X., Song, Y. L. & Jiang, L. (2008). Electrically tunable polypyrrole inverse opals with switchable stopband, conductivity, and wettability. *Chem. Mater*, **20** (11), 3554-3556.

[121] Dorvee, J. R., Derfus, A. M., Bhatia, S. N. & Sailor, M. J. (2004). Manipulation of liquid droplets using amphiphilic, magnetic one-dimensional photonic crystal chaperones. *Nat. Mater*, **3**(12), 896-899.

[122] Richel, A., Johnson, N. P. & McComb, D. W. (2000). Observation of Bragg reflection in photonic crystals synthesized from air spheres in a titania matrix. *Appl. Phys. Lett*, **76** (14), 1816-1818.

[123] James, R. W. (1962). *The optical principles of the diffraction of x-rays*. Ox Bow Press: Woodbridge, Connecticut.

[124] Weissman, J. M., Sunkara, H. B., Tse, A. S. & Asher, S. A. (1996). Thermally Switchable Periodicities and Diffraction from Mesoscopically Ordered Materials. *Science*, **274**(5289), 959-963.

[125] Reese, C. E., Baltusavich, M. E., Keim, J. P. & Asher, S. A. (2001). Development of an intelligent polymerized crystalline colloidal array colorimetric reagent. *Anal. Chem.*, **73**(21), 5038-5042.

[126] Reese, C. E., Mikhonin, A. V., Kamenjicki, M., Tikhonov, A. & Asher, S. A. (2004). Nanogel nanosecond photonic crystal optical switching. *J. Am. Chem. Soc.*, **126**(5), 1493-1496.

[127] Walker, J. P. & Asher, S. A. (2005). Acetylcholinesterase-based organophosphate nerve agent sensing photonic crystal. *Anal. Chem.*, **77**(6), 1596-1600.

[128] Asher, S. A., Sharma, A. C., Goponenko, A. V. & Ward, M. M. (2003). Photonic crystal aqueous metal cation sensing materials. *Anal. Chem.*, **75**(7), 1676-1683.

[129] Reese, C. E. & Asher, S. A. (2003). Photonic crystal optrode sensor for detection of Pb2+ in high ionic strength environments. *Anal. Chem.*, **75**(15), 3915-3918.

[130] Holtz, J. H., Holtz, J. S. W., Munro, C. H. & Asher, S. A. (1998). Intelligent polymerized crystalline colloidal arrays: Novel chemical sensor materials. *Anal. Chem.*, **70**(4), 780-791.

[131] Kang, Y., Walish, J. J., Gorishnyy, T. & Thomas, E. L. (2007). Broad-wavelength-range chemically tunable block-copolymer photonic gels. *Nat. Mater*, **6**(12), 957-960.

[132] Holtz, J. H. & Asher, S. A. (1997). Polymerized colloidal crystal hydrogel films as intelligent chemical sensing materials. *Nature*, **389**(6653), 829-832.

[133] Ben-Moshe, M., Alexeev, V. L. & Asher, S. A. (2006). Fast responsive crystalline colloidal array photonic crystal glucose sensors. *Anal. Chem.*, **78**(14), 5149-5157.

[134] Alexeev, V. L., Sharma, A. C., Goponenko, A. V., Das, S., Lednev, I. K., Wilcox, C. S., Finegold, D. N. & Asher, S. A. (2003). High ionic strength glucose-sensing photonic crystal. *Anal. Chem.*, **75**(10), 2316-2323.

[135] Asher, S. A., Alexeev, V. L., Goponenko, A. V., Sharma, A. C., Lednev, I. K., Wilcox, C. S. & Finegold, D. N. (2003). Photonic crystal carbohydrate sensors: Low ionic strength sugar sensing. *J. Am. Chem. Soc.*, **125**(11), 3322-3329.

[136] Lee, Y. J., Pruzinsky, S. A. & Braun, P. V. (2004). Glucose-sensitive inverse opal hydrogels: Analysis of optical diffraction response. *Langmuir*, **20**(8), 3096-3106.

[137] Kang, J. H., Moon, J. H., Lee, S. K., Park, S. G., Jang, S. G., Yang, S. & Yang, S. M. (2008). Thermoresponsive Hydrogel Photonic Crystals by Three-Dimensional Holographic Lithography. *Adv. Mater*, **20**(16), 3061-3065.

[138] Debord, J. D., Eustis, S., Debord, S. B., Lofye, M. T. & Lyon, L. A. (2002). Color-tunable colloidal crystals from soft hydrogel nanoparticles. *Adv. Mater*, **14**(9), 658-662.

[139] Hu, Z. B., Lu, X. H. & Gao, J. (2001). Hydrogel opals. *Adv. Mater*, **13**(22), 1708-1712.

[140] 1Matsubara, K., Watanabe, M. & Takeoka, Y. (2007). A thermally adjustable multicolor photochromic hydrogel. *Angew. Chem., Int. Ed.*, **46**(10), 1688-1692.

[141] Takeoka, Y. & Watanabe, M. (2003). Tuning structural color changes of porous thermosensitive gels through quantitative adjustment of the cross-linker in pre-gel solutions. *Langmuir*, **19**(22), 9104-9106.

[142] Ueno, K., Matsubara, K., Watanabe, M. & Takeoka, Y. (2007). An electro- and thermochromic hydrogel as a full-color indicator. *Adv. Mater*, **19**(19), 2807-2812.

[143] Gu, Z. Z., Hayami, S., Meng, Q. B., Iyoda, T., Fujishima, A. & Sato, O. (2000). Control of photonic band structure by molecular aggregates. *J. Am. Chem. Soc.*, **122**(43), 10730-10731.

[144] Gu, Z. Z., Fujishima, A. & Sato, O. (2000). Photochemically Tunable Colloidal Crystals. *J. Am. Chem. Soc.*, **122**(49), 12387-12388.

[145] Gu, Z. Z., Iyoda, T., Fujishima, A. & Sato, O. (2001). Photo-reversible regulation of optical stop bands. *Adv. Mater*, **13**(17), 1295-1298.

[146] Kubo, S., Gu, Z. Z., Takahashi, K., Ohko, Y., Sato, O. & Fujishima, A. (2002). Control of the optical band structure of liquid crystal infiltrated inverse opal by a photoinduced nematic-isotropic phase transition. *J. Am. Chem. Soc.*, **124**(37), 10950-10951.

[147] Kubo, S., Gu, Z. Z., Takahashi, K., Fujishima, A., Segawa, H. & Sato, O. (2005). Control of the optical properties of liquid crystal-infiltrated inverse opal structures using photo irradiation and/or an electric field. *Chem. Mater*, **17**(9), 2298-2309.

[148] Walish, J. J., Kang, Y., Mickiewicz, R. A. & Thomas, E. L. (2009). Bioinspired Electrochemically Tunable Block Copolymer Full Color Pixels. *Adv. Mater.* **21**(30), 3078-3081.

[149] Li, H. L., Wang, J. X., Yang, L. M. & Song, Y. L. (2008). Superoleophilic and Superhydrophobic Inverse Opals for Oil Sensors. *Adv. Funct. Mater*, **18**(20), 3258-3264.

[150] Li, M. Z., Wang, J. X., Feng, L., Wang, B. B., Jia, X. R., Jiang, L., Song, Y. L. & Zhu, D. B. (2006). Fabrication of tunable colloid crystals from amine-terminated polyamidoamine dendrimers. *Colloids and Surfaces a-Physicochemical and Engineering Aspects*, **290**(1-3), 233-238.

[151] Sharma, A. C., Jana, T., Kesavamoorthy, R., Shi, L. J., Virji, M. A., Finegold, D. N. & Asher, S. A. (2004). A general photonic crystal sensing motif: Creatinine in bodily fluids. *J. Am. Chem. Soc.*, **126**(9), 2971-2977.

[152] Li, M. Z., He, F., Liao, Q., Liu, J., Xu, L., Jiang, L., Song, Y. L., Wang, S. & Zhu, D. B. (2008). Ultrasensitive DNA detection using photonic crystals. *Angew. Chem., Int. Ed.*, **47**(38), 7258-7262.

[153] Mekis, A., Chen, J. C., Kurland, I., Fan, S., Villeneuve, P. R. & Joannopoulos, J. D. (1996). High Transmission through Sharp Bends in Photonic Crystal Waveguides. *Phys. Rev. Lett*, **77**(18), 3787-3790.

[154] Qiu, M. & He, S. L. (2000). Guided modes in a two-dimensional metallic photonic crystal waveguide. *Phys. Lett. A*, **266**(4-6), 425-429.

[155] Solli, D. R., McCormick, C. F. & Hickmann, J. M. (2006). Polarization-dependent reflective dispersion relations of photonic crystals for waveplate mirror construction. *J. Lightwave Technol.*, **24**(10), 3864-3867.

[156] Nishimura, S., Abrams, N., Lewis, B. A., Halaoui, L. I., Mallouk, T. E., Benkstein, K. D., van de Lagemaat, J. & Frank, A. J. (2003). Standing wave enhancement of red absorbance and photocurrent in dye-sensitized titanium dioxide photoelectrodes coupled to photonic crystals. *J. Am. Chem. Soc.*, **125**(20), 6306-6310.

[157] Halaoui, L. I., Abrams, N. M. & Mallouk, T. E. (2005). Increasing the conversion efficiency of dye-sensitized TiO2 photoelectrochemical cells by coupling to photonic crystals. *J. Phys. Chem. B*, **109**(13), 6334-6342.

[158] Mihi, A. & Míguez, H. (2005). Origin of light-harvesting enhancement in colloidal-photonic-crystal-based dye-sensitized solar cells. *J. Phys. Chem. B*, **109**(33), 15968-15976.

[159] Mihi, A., López-Alcaraz, F. J. & Míguez, H. (2006). Full spectrum enhancement of the light harvesting efficiency of dye sensitized solar cells by including colloidal photonic crystal multilayers. *Appl. Phys. Lett*, **88**(19), 193110.

[160] Mihi, A., Calvo, M. E., Anta, J. A. & Míguez, H. (2008). Spectral response of opal-based dye-sensitized solar cells. *J. Phys. Chem. C*, **112**(1), 13-17.

[161] Huisman, C. L., Schoonman, J. & Goossens, A. (2005). The application of inverse titania opals in nanostructured solar cells. *Sol. Energy Mater. Sol. Cells*, **85**(1), 115-124.

[162] Somani, P. R., Dionigi, C., Murgia, M., Palles, D., Nozar, P. & Ruani, G. (2005). Solid-state dye PV cells using inverse opal TiO2 films. *Sol. Energy Mater. Sol. Cells*, **87**(1-4), 513-519.

[163] Ye, J. Y., Ishikawa, M., Yamane, Y., Tsurumachi, N. & Nakatsuka, H. (1999). Enhancement of two-photon excited fluorescence using one-dimensional photonic crystals. *Appl. Phys. Lett*, **75**(23), 3605-3607.

[164] Ryu, H. Y., Lee, Y. H., Sellin, R. L. & Bimberg, D. (2001). Over 30-fold enhancement of light extraction from free-standing photonic crystal slabs with InGaAs quantum dots at low temperature. *Appl. Phys. Lett*, **79**(22), 3573-3575.

[165] Maskaly, G. R., Petruska, M. A., Nanda, J., Bezel, I. V., Schaller, R. D., Htoon, H., Pietryga, J. M. & Klimov, V. I. (2006). Amplified spontaneous emission in semiconductor-nanocrystal/synthetic-opal composites: Optical-gain enhancement via a photonic crystal pseudogap. *Adv. Mater*, **18**(3), 343-347.

[166] Bechger, L., Lodahl, P. & Vos, W. L. (2005). Directional fluorescence spectra of laser dye in opal and inverse opal photonic crystals. *J. Phys. Chem. B*, **109**(20), 9980-9988.

[167] Ganesh, N., Zhang, W., Mathias, P. C., Chow, E., Soares, J. A. N. T., Malyarchuk, V., Smith, A. D. & Cunningham, B. T. (2007). Enhanced fluorescence emission from quantum dots on a photonic crystal surface. *Nat. Nanotech*, **2**(8), 515-520.

[168] Zhang, Y. Q., Wang, J. X., Ji, Z. Y., Hu, W. P., Jiang, L., Song, Y. L. & Zhu, D. B. (2007). Solid-state fluorescence enhancement of organic dyes by photonic crystals. *J. Mater. Chem.*, **17**(1), 90-94.

[169] Zhang, Y. Q., Wang, J. X., Chen, X., Liang, J., Jiang, L., Song, Y. L. & Zhu, D. B. (2007). Enhancing fluorescence of tricolor fluorescent powders by silica inverse opals. *Appl. Phys. A-Mater*, **88**(4), 811-811.

[170] Li, H., Wang, J. X., Lin, H., Xu, L., Xu, W., Wang, R. M., Song, Y. L. & Zhu, D. B. (2010). Amplification of Fluorescent Contrast by Photonic Crystals in Optical Storage. *Adv. Mater*, **22**, 1237-1241.

[171] Li, M. Z., Liao, Q., Zhang, J. P., Jiang, L., Song, Y. L., Zhu, D. B., Chen, D., Tang, F. Q. & Wang, X. H. (2007). Energy transfer boosted by photonic crystals with metal film patterns. *Appl. Phys. Lett*, **91**(20), 203516.

[172] Li, M. Z., Xia, A. D., Wang, J. X., Song, Y. L. & Jiang, L. (2007). Coherent control of spontaneous emission by photonic crystals. *Chemical Physics Letters*, **444**(4-6), 287-291.

[173] Li, M. Z., Liao, Q., Liu, Y., Li, Z. Y., Wang, J. X., Jiang, L. & Song, Y. L. (2010). A white-lighting LED system with a highly efficient thin luminous film. *Appl. Phys. A-Mater*, **98**(1), 85-90.

[174] Liu, J., Li, M. Z., Wang, J. X., Song, Y. L., Jiang, L., Murakami, T. & Fujishima, A. (2009). Hierarchically Macro-/Mesoporous Ti-Si Oxides Photonic Crystal with Highly Efficient Photocatalytic Capability. *Environmental Science & Technology*, **43**(24), 9425-9431.

[175] Chen, J. I. L., Loso, E., Ebrahim, N. & Ozin, G. A. (2008). Synergy of slow photon and chemically amplified photochemistry in platinum nanocluster-loaded inverse titania opals. *J. Am. Chem. Soc.*, **130**(16), 5420-5421.

[176] Chen, J. I. L. & Ozin, G. A. (2008). Tracing the Effect of Slow Photons in Photoisomerization of Azobenzene. *Adv. Mater*, **20**(24), 4784-+.

[177] O'Brien, P. G., Kherani, N. P., Zukotynski, S., Ozin, G. A., Vekris, E., Tetreault, N., Chutinan, A., John, S., Mihi, A. & Miguez, H. (2007). Enhanced photoconductivity in thin-film semiconductors optically coupled to photonic crystals (vol 19, pg 4117, 2007). *Adv. Mater*, **19**(24), 4326-4326.

In: Photonic Crystals
Editor: Venla E. Laine, pp. 31-66
ISBN: 978-1-61668-953-7

Chapter 2

ACHIEVING COMPLETE BAND GAPS USING LOW REFRACTIVE INDEX MATERIAL

***Dahe Liu*[1, 2,*]*, Tianrui Zhai*[2] *and Zhaona Wang*[2]
[1]Key Laboratory of Nondestructive Test (Ministry of Education), Nanchang Hangkong University, Nachang 330063, China.
[2]Applied Optics Beijing Area Major Laboratory, Department of Physics, Beijing Normal University, Beijing 100875, China

1. Introduction

Photonic crystals (PCs) were introduced by E. Yablonovitch [1] and S. John [2] in 1987. In 1990, Ho *et al.* demonstrated theoretically that a diamond structure possesses complete band gaps (CBGs) [3]. Since then, great interest was focused on fabricating three-dimensional (3D) photonic crystals (PCs) in order to obtain CBGs. Several methods were reported, and CBGs were achieved in the range of microwave or submicrowave. Theoretical analysis showed that although CBGs can be obtained by diamond structure, a strict condition should be satisfied, i.e., the modulation of the refractive index of the material used should be larger than 2.0 [4,5]. So, attention was paid to finding the materials with high refractive index. Some scientists tried to fill the templates with high refractive index materials to increase the modulation of the refractive index [6-8], and CBGs were obtained. However, the CBGs achieved in 3D PCs were mostly in microwave or infrared regions [7,9-11]. Holography is a cheap, rapid, convenient, and effective technique for fabricating 3D structures. In 1997, holographic technique was introduced for fabricating the face centered cubic (fcc) structure [12]. Campbell *et al.* actually fabricated the fcc structure with holographic lithography [13]. Several authors reported their works on this topic [4,5,9,14]. Toader *et al.* also showed theoretically a five-beam "umbrella" configuration in the synthesis of a diamond photonic crystal [15]. Because both the value and modulation of the refractive index of the holographic recording materials are commonly low, there would be no CBGs in PCs made directly by holography. For example, the epoxy photoresist generally used for holographic lithography

* E-mail address: dhliu@bnu.edu.cn. (Corresponding author: Dahe Liu)

[13,16-18] has n=1.6, which is a little bit too low. They may, however, be used as templates for the production of inverse replica structures by, for example, filling the void with high refractive index and burning out or dissolving the photoresist [13], and a good work was done by Meisel *et al* [19,20]. However, to find materials with large refractive index is not easy, and the special techniques needed are very complicated and expensive. It limits the applications of PCs, especially for industrial productions.

Although some efforts had been made [20], CBGs in the visible range had not yet been achieved by using the materials with low refractive index. Therefore, it is a big challenge to fabricate 3D PCs possessing CBGs in the visible range by using materials with low refractive index, though it is greatly beneficial for the future PC industry. As a first step for achieving this, it is important to obtain very wide band gaps. In the previous investigations [21,22], it was evidently shown that the anisotropy of a photonic band gap in a two-dimensional photonic crystal is dependent on the symmetry of the structure, and as the order of the symmetry increases, it becomes easier to obtain a complete band gap. One would naturally ask whether or not such a method is applicable to 3D PCs. In view of this question, we proposed a series holographic method for fabricating some special structures by using materials with low refractive index, and the features of the band gap in such structures were then studied experimentally and theoretically.

2. Complex Diamond Structure [23]

1) Experimental Method and the Samples

It is known that a cell of diamond structure consists of two cells of fcc structures, and the two cells have a distance of one-quarter of the diagonal length of the cell along the diagonal line. According to this, the diamond structure here was implemented by holography through two exposures: an fcc structure was recorded in the first exposure, then, after the recording material was translated one-quarter of the diagonal length along the diagonal line of the fcc structure, a second exposure was made to record another fcc structure. In this way, a PC with diamond structure was obtained.

Figure 1 shows schematically the optical layout in our experiments. Four beams split from a laser beam were converged to a small area. The central beam was set along the normal direction of the surface of the plate, while the other three outer beams were set around the central beam symmetrically with an angle of 38.9° with respect to the central one. The laser used was a diode pumped laser working at 457.9 nm with linewidth of 200 kHz (Melles Griot model 85-BLT-605). The polarization state of each beam was controlled to achieve the best interference result [24,25]. The recording material was mounted on a one-dimensional translation stage driven by a stepping motor with a precision of 0.05 μm/step. The translation stage was mounted on a rotary stage. The holographic recording material used was dichromated gelatin (DCG) with refractive index n=1.52. The maximum value of its refractive index modulation Δn can reach around 0.1, which is a very small value for obtaining wide band gaps. The thickness T of the material was 36 μm. The DCG emulsion was coated on an optical glass plate with flatness of $\lambda/10$ and without any doping.

It is known that there are several directions with high symmetry in the first Brillouin zone of an fcc structure (see Figure 2). In our experiments, two exposures as mentioned above were made firstly in

Γ-*L* ([111]) direction to get a diamond structure. Then, a second, even a third, diamond structure was implemented by changing the orientation of the recording material by rotating the rotary stage to other symmetric directions, i.e., the direction of Γ-*X* ([100]) and/or the direction of Γ-*K* ([110]) in the first Brillouin zone of the first fcc structure. In this way, photonic crystals with one, two, and even three diamond structures were fabricated. It should be pointed out that when the second or the third diamond structure was recorded, the angle between any two beams should be changed to guarantee that all the beams inside the medium satisfy the relation shown in Figure 1 so that the standard diamond structures can be obtained.

Since the hologram made with DCG is a phase hologram, there is only the distribution of the refractive index but no plastic effect inside the hologram, so a scanning electron microscope (SEM) image cannot be obtained. However, microscopic image can be obtained. To verify the structure in the hologram, the same fcc structure was implemented using the same optical layout but with the photoresist of 2 μm thickness. Figure 3 shows the structure made with the photoresist at [111] plane: Figure 3 (a) shows the SEM image and Figure 3(b) shows the optical microscopic image taken with a charge coupled device (CCD) camera mounted on a 1280× microscope. The pixel size of the CCD is 10 μm. Figure 3(c) shows the optical microscopic image of a diamond structure made with DCG of 36 μm at [111] plane, which was also recorded by the CCD camera mounted on a 1280× microscope as mentioned above.

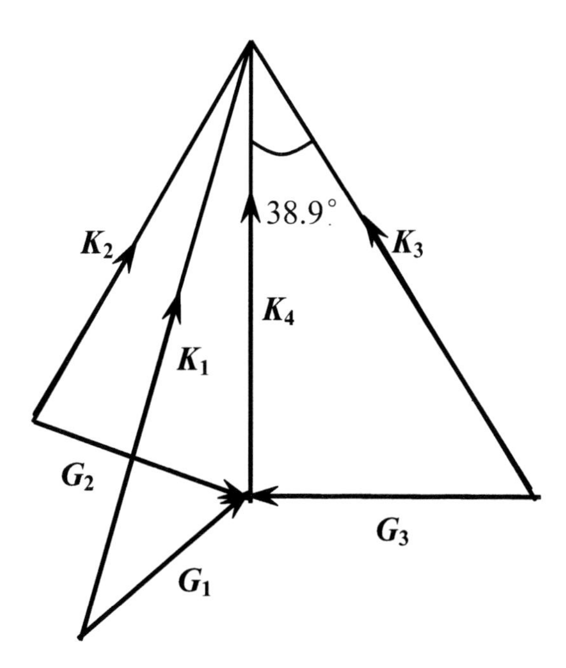

Figure 1. Schematic optical layout for recording an fcc structure.

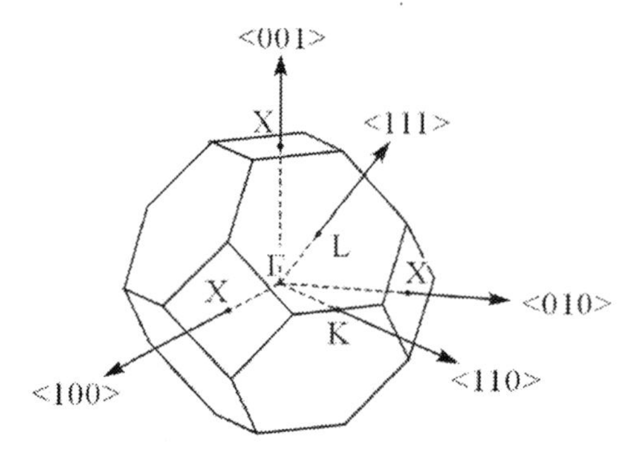

Figure 2. First Brillouin zone of an fcc structure.

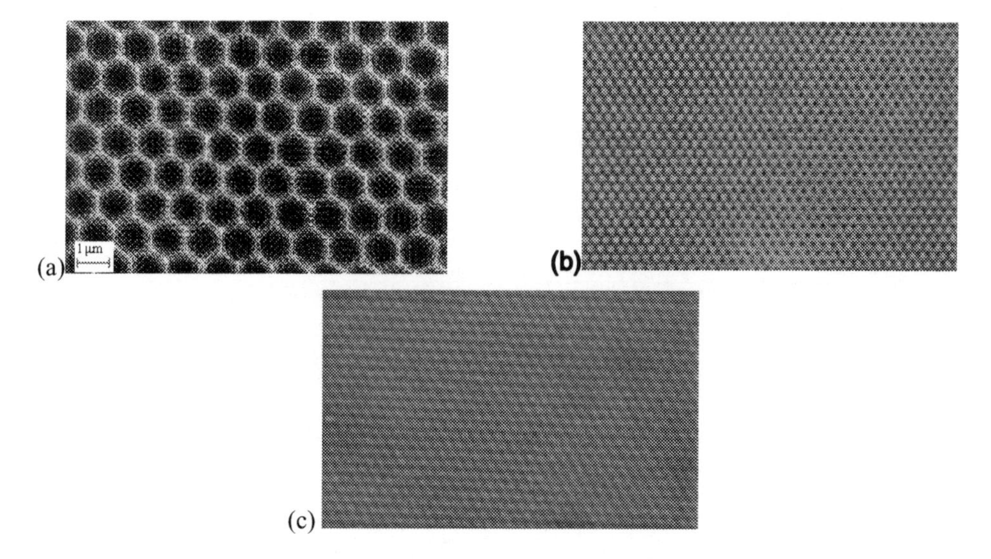

Figure 3. Structure of PCs made in our laboratory. The images (b) and (c) were recorded by a CCD camera mounted on a 1280× microscope. (a) SEM image of an fcc structure made with photoresist of 1 μm thickness. (b) Microscopic image of an fcc structure made with photoresist of 1 μm thickness. (c) Microscopic image of a diamond structure made with DCG of 36 μm.

The theoretical calculated diamond lattice formed by two exposures as mentioned above is shown in Figure 4. It should be pointed out that the shape of the interference results appears like an American football. In this figure, the vertical bar shows a bottom to top gradient corresponding to the outer to inner region of a cross-sectional cut of a football. The gray gradient along the outer surface is related to the value of the vertical bar. Whether a diamond lattice can be obtained depends on how far the two footballs stand apart. If the two footballs are large enough that they overlap, the structure cannot be considered as a diamond lattice. If the two footballs can still be recognized as two, it means that a diamond lattice is formed. In our experiments, the absorption coefficient α of the material was chosen as $\alpha = 1/T$ (Ref. 26) to get optimal interference result. According to previous works, it is well known that for holographic recording materials the relation between the optical density and the exposure is nonlinear. The holographic exposure can be controlled in the region of stronger nonlinear dependence between optical density and exposure as discussed previously [27]. In this way,

two fcc lattices can be recognized, so that the structure can be considered as a diamond lattice.

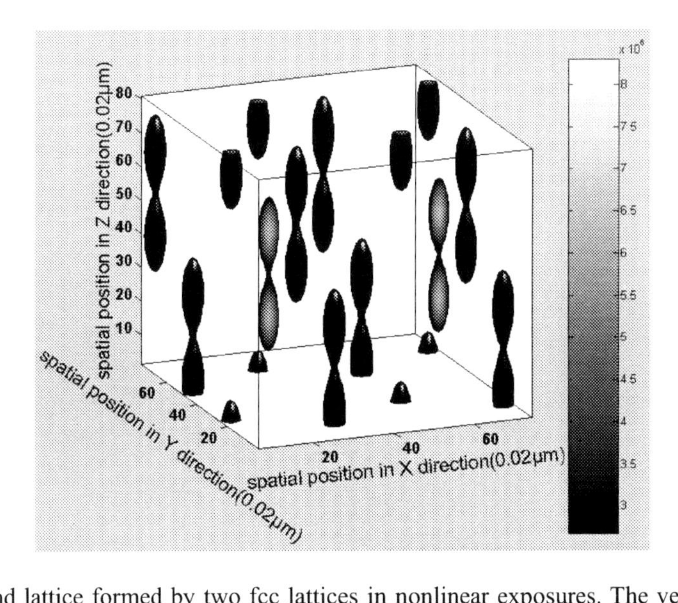

Figure 4. Diamond lattice formed by two fcc lattices in nonlinear exposures. The vertical bar shows a gradient of bottom to top corresponding to the outer to inner region of a cross-sectional cut of a football. The gradient of the outer surface is related to the value of the vertical bar.

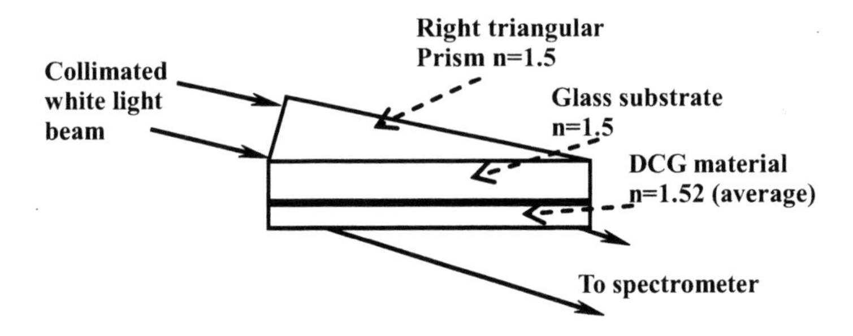

Figure 5. Setup geometry for measuring transmission spectra of photonic crystals.

The transmission spectra of PCs with diamond structure made with DCG were measured. In the measurements, a J-Y 1500 monochromator was employed. To minimize the energy loss from reflection at large incident angle, several right triangular prisms were used. The measuring setup geometry is shown in Figure 5. The measured range of the wavelength was 390–800 nm. The [111] plane of the PC measured was set firstly at an arbitrary orientation. The incident angle of the collimated white light beam was changed from 0° to ±90° (the incident angle is in the plane parallel to the surface of the paper). Then, the PC was rotated to other orientations (the orientation angle is in the plane perpendicular to the surface of the paper), and the incident angle of the collimated white light beam was also changed from 0° to ±90°. The orientation angle and the incident angle are the angles φ and θ in a spherical system indeed. In this way, the measurements give actually 3D results.

2) Measured Spectra and Discussions

The measured transmission spectra of a fabricated triple diamond PC are shown in Figs. 6(a)–6(c), denoting measured results at different orientations of the sample. Each curve in the figures gives the spectrum measured at a certain incident angle and a certain orientation. Different curves give the results at different incident angles and orientations. Thus, whether or not the common gap exists can be determined by the intercept of all the curves. It can be seen that, for the PC with three diamond structures (six fcc structures), the width of the band gaps reached 260 nm, the ratio between the width and the central wavelength of the gaps reached 50 %, and there is a common band gap with a width of about 20 nm at 450–470 nm in the range of 150° of the incident angle. The common gap obtained using DCG with very low refractive index (n=1.52) existed in a wide range, which reached 83 % of the 4π solid angle. Although a complete band gap for all directions was not obtained in our experiments, it is significant to achieve such a wide angle band gap by using a material (DCG) with very low refractive index, because this angular tuning range satisfies most applications in practice, for example, restraining spontaneous radiation with low energy loss in a wide range, wide angle range filter, or reflector with low energy loss.

This interesting result comes from a complex diamond structure. The photonic crystal with triple-diamond structures is an actual multi structure, but not a stack of several same structures. The [111] direction of the first diamond structure is actually the $\Gamma - X$ direction of the second diamond structure and the $\Gamma - K$ direction of the third diamond structure, respectively. When a beam is incident normally on the PC, the beam is in the [111] direction of the first diamond structure. When the incident angle changes, the beam deviates from the [111] direction of the first diamond structure and tends to approach the [111] direction of the second or third diamond structure. Therefore, though the incident angle changes, the light beam remains always around the [111] direction of the other diamond structures. The narrow region in the $K - \omega$ dispersion relation of a diamond lattice may be expanded by other diamond lattices.

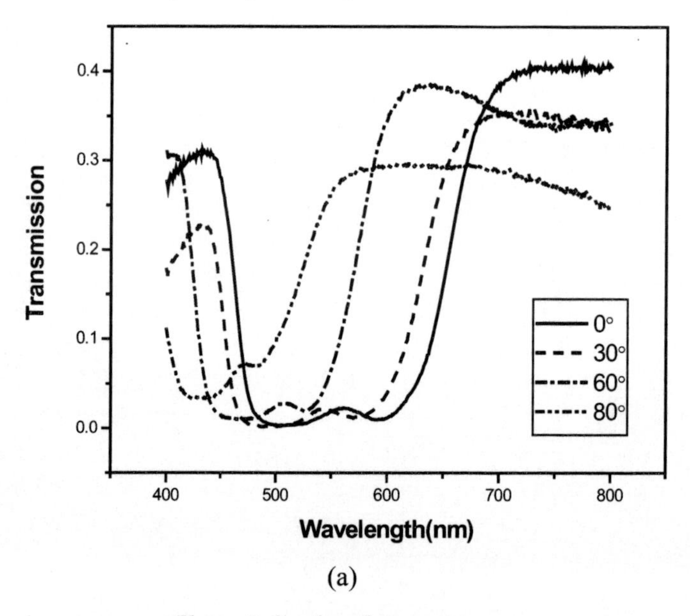

(a)

Figure 6. Continued on next page.

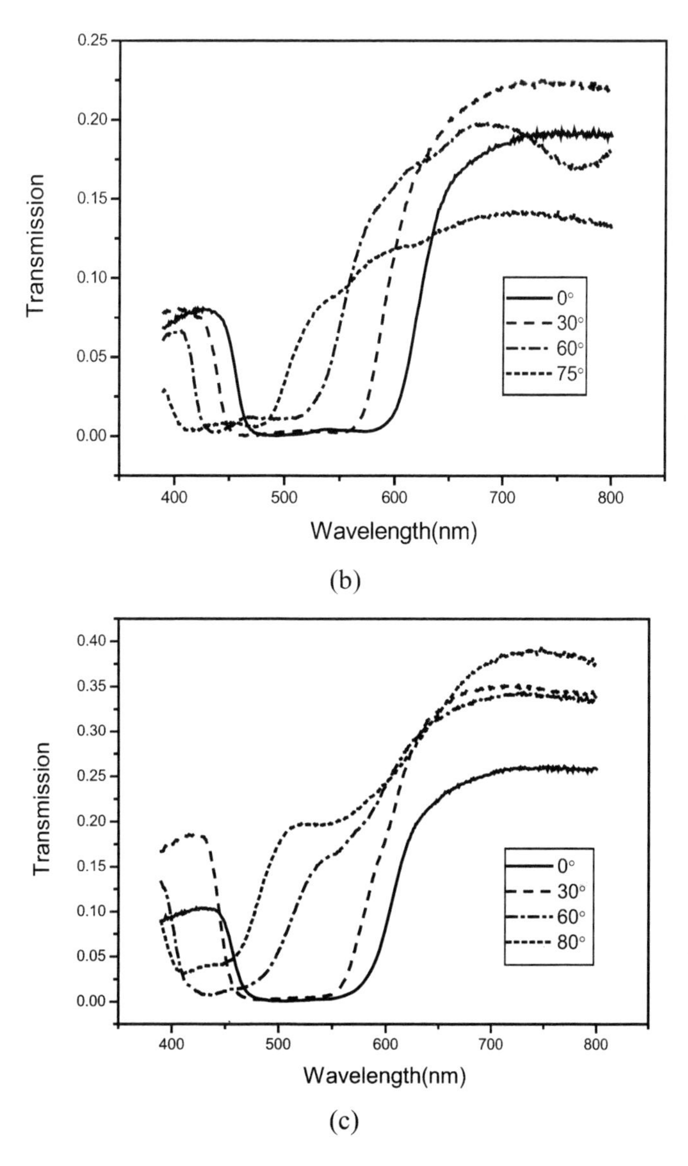

Figure 6. Measured transmission spectra of PC with three diamond lattices recorded by holography in Γ-*L*, Γ-*X*, and Γ-*K* directions of an fcc lattice respectively. (a), (b), and (c) correspond to the measured results. The angles appearing in (a), (b), and (c) are the incident angles, which are in a plane parallel to the surface of the paper. (a) Orientation angle is 0°. (b) Orientation angle is 30°. (c) Orientation angle is 90°.

The advantage of a multi diamond structure can be seen more clearly from a comparison with those having one or two diamond structures. Figures 7 and 8 give the measured transmission spectra of PCs with one and two diamond lattices, respectively, at all orientations. The structure with single diamond lattice was fabricated in $\Gamma-L$ [111] direction, while the structure with two diamond lattices was fabricated in $\Gamma-L$ [111] and $\Gamma-X$ [100] directions. Besides, the method used for measuring the two structures with single and double diamond lattices, respectively, was similar to that mentioned above. It can be found that, for the single diamond structure, the space angle range of the common gap is 40°; for the double diamond structures, it becomes 80°. Comparing Figs. 6–8, it is obvious that the width of the band gaps of a PC can be broadened effectively by increasing the number of diamond lattices. This means that the common gap can be enlarged by means of multi structures. The physical origin of such a phenomenon can be understood from the change of structure symmetry. With the increase of the number of diamond lattices, the symmetries around the center of the structure become higher. This made it easier to obtain a broader response in many directions (common band gap in a wide range of angles), which is similar to the case of two dimensional PCs (see Refs. 19 and 20).

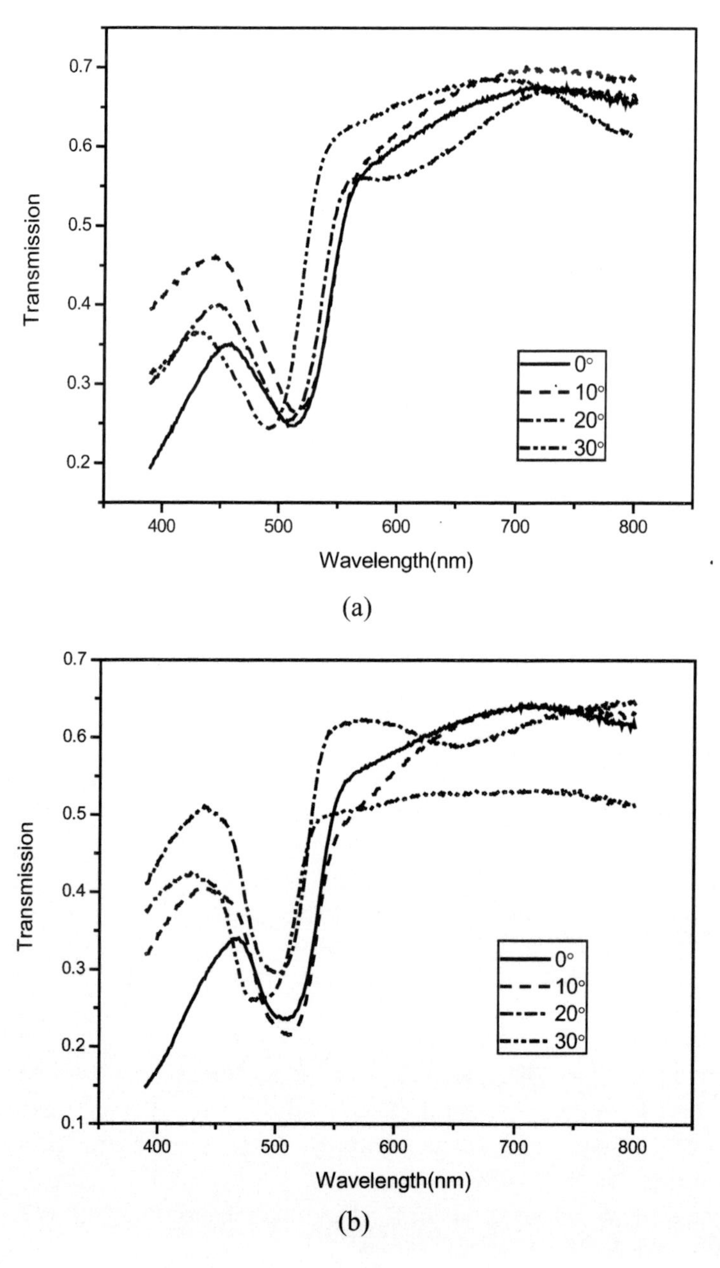

Figure 7. Continued on next page.

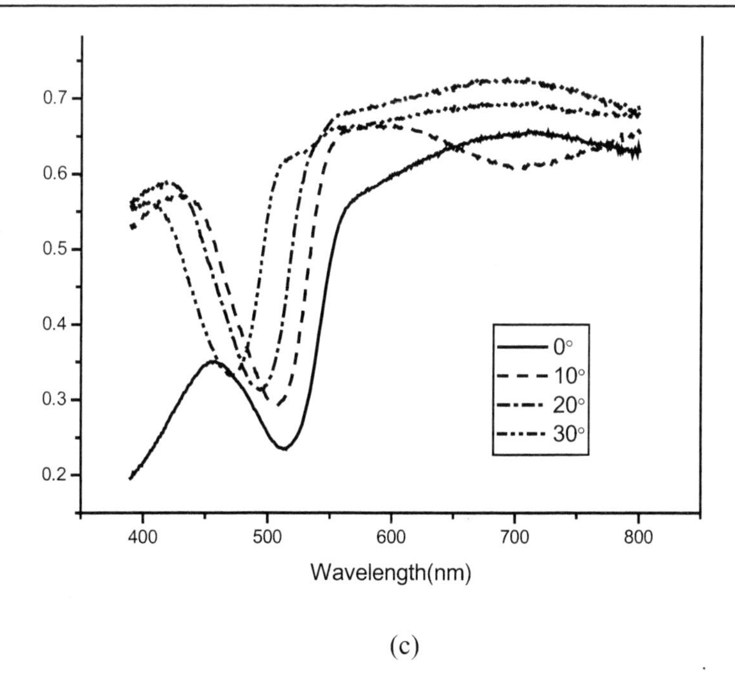

(c)

Figure 7. Measured transmission spectra of PC with single diamond lattice recorded by holography in $\Gamma - L$ direction of an fcc lattice. (a), (b) and (c) correspond to the measured results. The angles appeared in (a), (b) and (c) are the incident angle which is in a plane parallel to the surface of the paper.

(a) Orientation angle is $0°$.

(b) Orientation angle is $30°$.

(c) Orientation angle is $90°$.

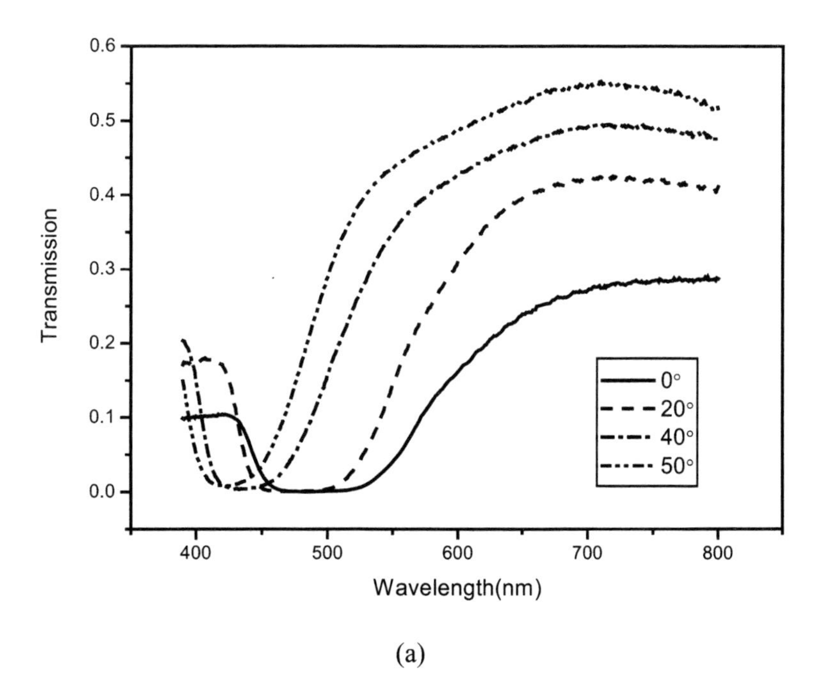

(a)

Figure 8. Continued on next page.

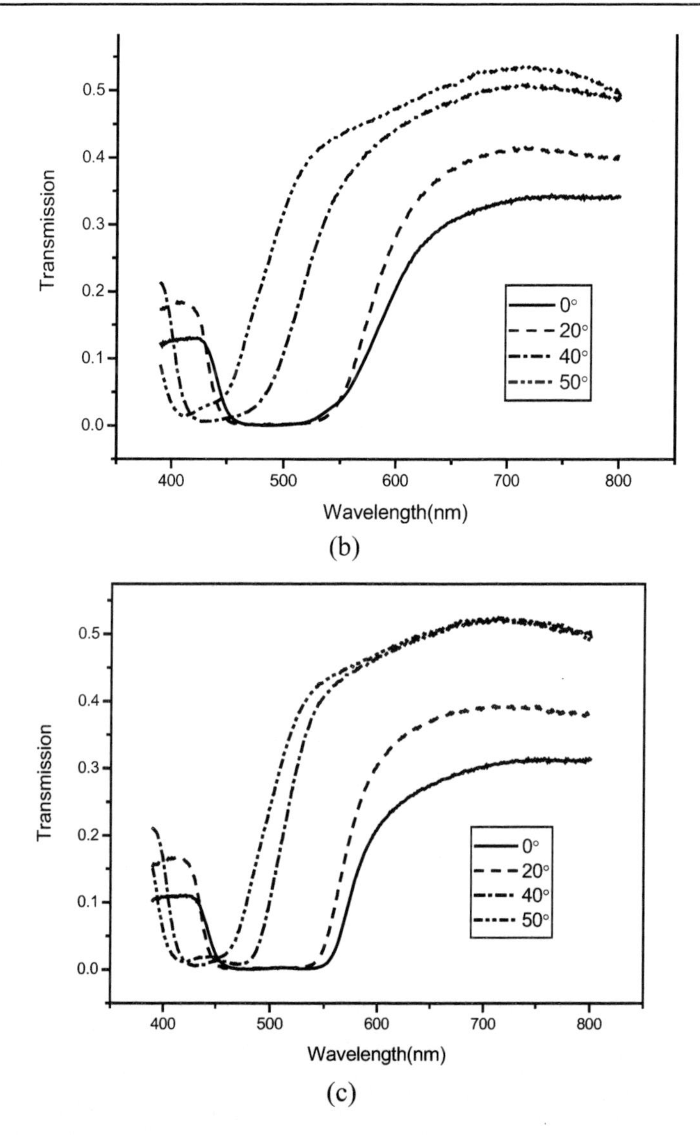

Figure 8. Measured transmission spectra of PC with 2 diamond lattices recorded by holography in $\Gamma - L$ and $\Gamma - X$ directions of an fcc lattice respectively. (a), (b) and (c) correspond to the measured results. The angles appeared in (a), (b) and (c) are the incident angle which is in a plane parallel to the surface of the paper.

(a) Orientation angle is 0°.

(b) Orientation angle is 30°.

(c) Orientation angle is 90°.

3) Conclusion

The width of the band gaps in a PC made by a material with low refractive index can be broadened greatly by multi diamond structures, and a common band gap in the range of 150° of the incident angle can be obtained. This technique will be greatly beneficial in achieving complete band gaps by using materials with low refractive index.

3. Self-simulating Structure [28,29]

As mentioned above, a complex diamond lattice was fabricated using materials with low refractive indices. Although a common gap in the visible range was obtained in a wide range that reached 83 % of the 4π solid angle [23], the CBGs were not yet obtained. However, regardless of whether or not they succeed, the above-mentioned techniques are complicated and difficult to carry out. Now, a novel technique has been developed by which a wide CBG can be achieved easily in the visible range using materials with a low refractive index.

It is obvious that, for a self-simulating spherical system shown in Figure 9, when a light beam passes through the center of the sphere, it will be symmetric in all directions with respect to the center. Therefore, if a PC has this self-simulating spherical structure, it will possess complete band gaps. It should be emphasized that achieving CBGs using this structure does not pose any special requirement on the refractive index of the material used. This type of self-simulating spherical structure can be implemented easily via holography. This includes three steps: 1) preparation of the recording material by coating the emulsion on a soft substrate; 2) fabrication of a 1D PC in rectangular coordinates by using this recording material; 3) to make this PC into a sphere. In this way, a 1D PC in rectangular coordinates was transformed to a spherical 1D PC system.

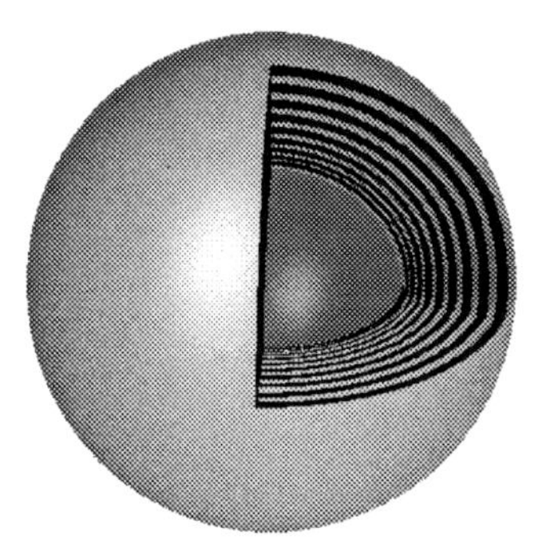

Figure 9. Self-simulating spherical structure.

Figure 10 shows the experimental set-up geometry. Dichromated gelatin (DCG) was still used as recording material. The recording material was prepared by coating a thick layer of 36 mm dichromated gelatin (DCG) onto polyethylene terephthalate (PET). The refractive index of the DCG is 1.52, and the refractive index of the PET is 1.64. Figure 6 shows the optical layout. The laser used was a diode-pumped laser working at 457.9 nm with a line width of 200 KHz (Melles Griot model 85-BLT 605). The monochrometer used was a J–Y 1500. The light beam for measuring was a narrow beam with a diameter of 3 mm. By controlling the wavelength of the laser and the angle between the two beams incident on both sides of the recording material, the period of the PC could be changed for working in the visible, IR, or UV range. Because of the influences of the absorption by the recording material [30], and some factors in processing after exposure of the recording material [31], the structure of the PC in rectangular coordinates is not strictly periodical. Figure 11 shows the structure and distribution of the modulation Δn of the refractive index inside the PC.

When the PC in rectangular coordinates is transformed to a sphere, its structure is just the self-simulating spherical system shown in Figure 9.

Several self-simulating spherical structures with different diameters were fabricated. All of them showed identical characteristics under the same experimental conditions. The measured transmission and reflection spectra of a self-simulating spherical structure with the radius of 20 mm were demonstrated as follows. During the measurements, the angles θ and φ of the spherical system were changed from $0°$ through $180°$ and from $0°$ through $360°$, respectively. Figs. 12 and 13 give only the measured results for φ changes from $0°$ through $360°$ while θ was set at a fixed value of $90°$. When the value of θ is changed in the range of $0-180°$, the measured transmission and reflection spectra are quite consistent. For comparison, the theoretical results calculated by transfer matrix method are also shown.

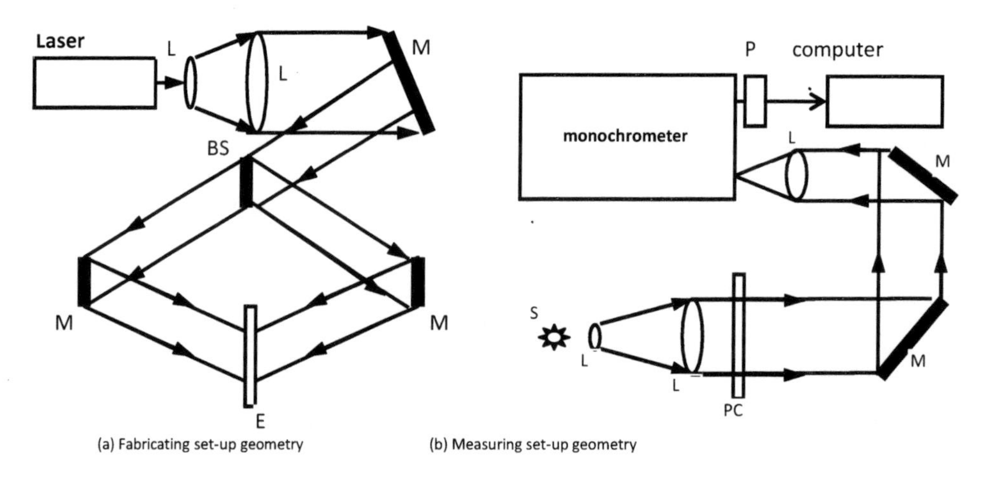

Figure 10. Optical layout for fabricating and measuring PC. L is lens, M is mirror, BS is beam splitter, E is recording material, S is white light source, PC is photonic crystal, P is photomultiplier tube.

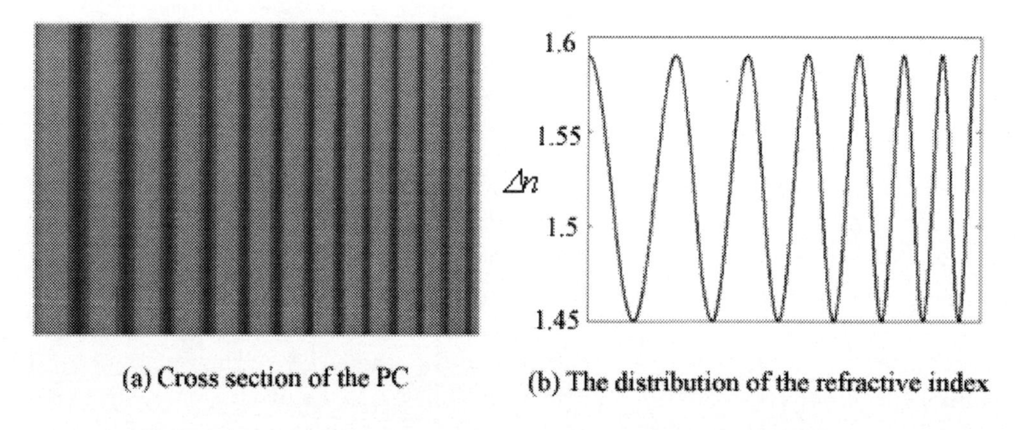

Figure 11. Structure of the fabricated PC in rectangular coordinates.

From Figure 12, it can be seen clearly that, for different values of w in the range of 0–360°, the measured transmission spectra almost completely overlapped with little frequency shift. The width of the CBG reaches 120 nm although the refractive index of DCG is pretty low. The reason is that a structure with non-strict period can broaden the band gaps effectively [32]. Figure 13 shows that the reflection spectra are just complementary to the transmission spectra.

In practical experiments and applications, frequently, the incident beam may not pass through the center of the self-simulating spherical structure strictly. It may affect the characteristics of the band gaps. Figure 13 show the measured transmission spectra when the incident beam deviates from the center of the self-simulating spherical structure. It can be seen that, even though the deviation is as large as 15 mm from the center of the sphere with a radius of 20 mm, there is still a complete band gap with a width of 30 nm. This deviation corresponds to the condition that a light beam incidents on the structure with an incident angle of 49°.

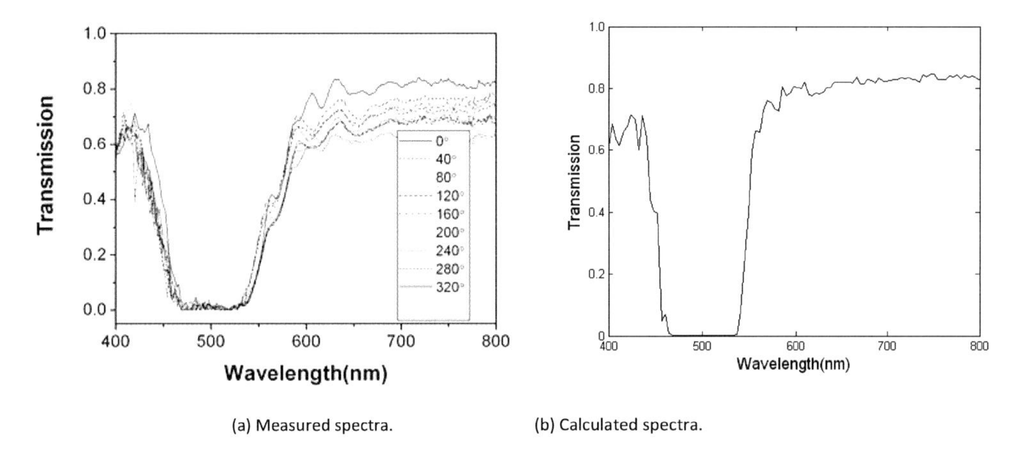

(a) Measured spectra. (b) Calculated spectra.

Figure 12. Measured and calculated transmission spectra of a self-simulating spherical structure for different value of φ at $\theta = 90°$.

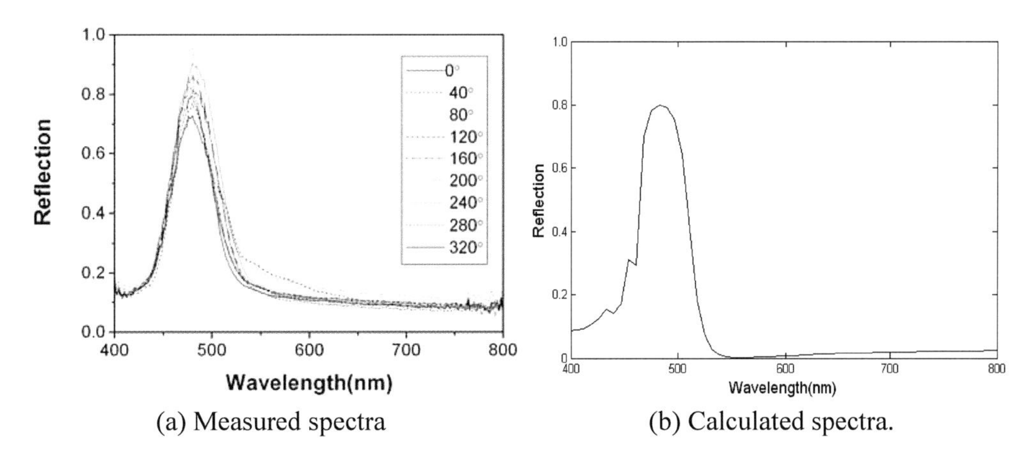

(a) Measured spectra (b) Calculated spectra.

Figure 13. Measured and calculated reflection spectra of a self-simulating spherical structure for different value of φ at $\theta = 90°$

Self-simulating spherical structures with radii of 5 mm, 7.5 mm, 10 mm, 20 mm, and 25 mm were fabricated and measured in our laboratory, and similar results to those shown in Figs. 12 and 13 were obtained. The measured transmission and reflection spectra revealed that a true complete band gap actually exists for this kind of structure. The fact that the light was forbidden from passing through the structure is really caused by the band gaps, rather than induced by other physical mechanisms, for example, total inner reflection and absorption. An actual fabricated Self-simulating spherical structure is shown in Figure 14.

In summary, a self-simulating spherical structure possesses complete band gaps. This kind of structure can be implemented easily in terms of holography. Achieving complete band gaps with this structure can use common holographic recording materials with low refractive indices rather than high refractive index materials. Even though the working condition has a relative large deviation from ideal conditions, there is still a complete band gap.

4. A D_{nv} Point Group Structure Based on Heterostructure [33]

Although the self-simulating spherical structures can obtain CBG, the incident beam must pass through the center of the sphere. however, when light incidents on the surface of the sphere with a incident angle larger than 49°, the CBG will disappear. This disadvantage limits the application of the structure, such as the use of wide incident beam in practical applications. So, improvement of the self-simulating spherical structures is needed.

It was demonstrated that PC heterostructure can improve effectively the characteristics of PCs, and can provide a necessary variation in material properties to turn a photonic crystal raw material into a functional device [34] by means of self-organizing [35-37], lithography with electron beam [38,39] or autocloning [40]. PC heterostructures, like their semiconductor counterparts which are made by combining at least two materials that have distinct band structures, can be fabricated by changing the lattice constant, hole size, or even lattice geometry in the crystal. This can be done either abruptly or gradually [34].

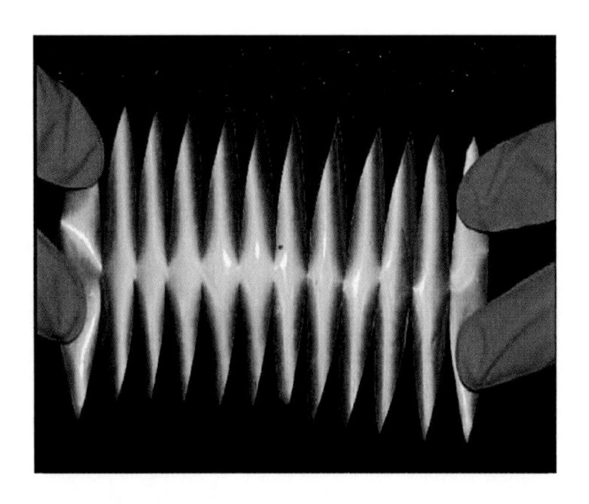

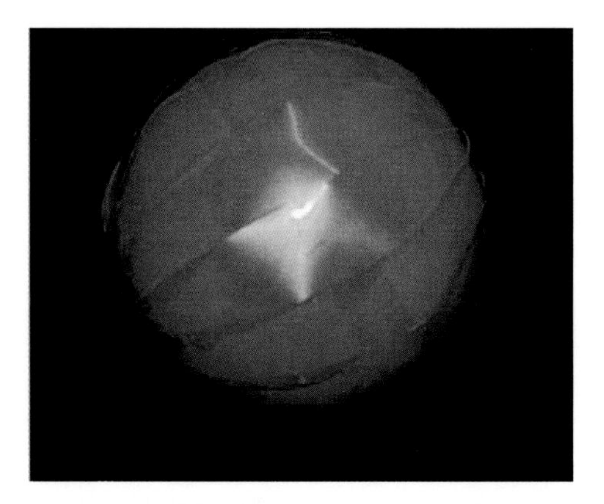

Figure 14. Photographs of cut piece of 1D PC in rectangular coordinates and PC in self-simulating spherical structure

Now, a new technique is proposed: we first developed a technique to fabricate a heterostructure with gradual refractive index distribution and gradual period, and obtained 2D ODBG. Further, a self-simulating sphere structure was implemented based on this kind of heterostructure, and a genuine CBG in discretionary condition can be achieved.

Figure 15 shows the set-up geometry for fabricating our heterostructure. The laser used was a DPSS laser working at 457.9 nm with line width of 200 kHz (Melles Griot model 85-BLT-605). The holographic recording material was dichromated gelatin (DCG). The DCG emulsion of 36 μm thickness was coated on an optical glass plate with flatness of λ/10 and without any doping. The plate was mounted on a 1D translation stage driven by a stepping motor with a precision of 0.05 μm/step.

For an elementary reflection hologram made by two beams with 180°angle, because of the absorption by the recording material the distribution of refractive index inside the medium is a hyperbolic cosine function [30] (see Figure 16), but neither strictly sinusoidal nor strictly period. Besides, swelling of the DCG emulsion exists during the processing after exposure, and the swelling is not uniform [31]. So, we introduced a linear expansion parameter M_z, and the thickness swelling of jth layer at position z can be expressed as [31]

$$M_{z,j} = M_1 - (M_1 - M_N)j/N + r_j/a \tag{1}$$

where $M_{z,j}$ is the expansion parameter for jth period, r_j is a random number in the range [0,1], N is the number of layers, parameter a controls the amplitude of the random fluctuation. The linear random expansion is shown in Figure 17 [31]. So, the final result actually likes a chirped grating with a variable period.

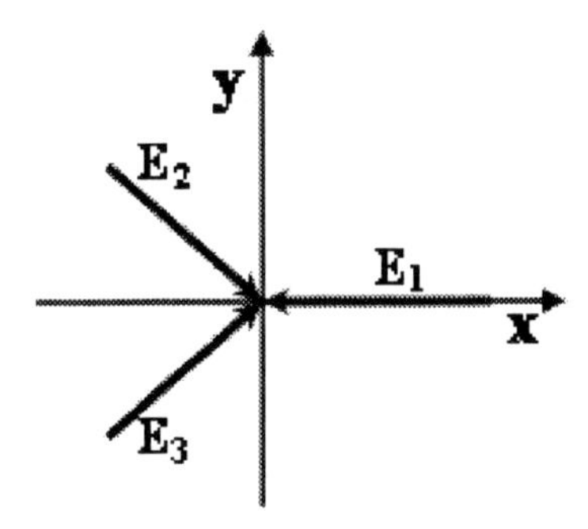

Figure 15. Set-up geometry of triangular lattice. The angles between E1 and E2, E2 and E3, E1 and E3 are all 120°.

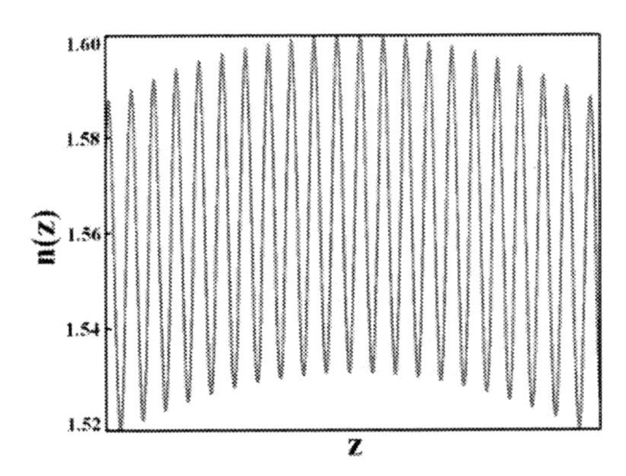

Figure 16. Schematic diagram of distribution of refractive index inside an element reflection hologram. T is the thickness of the material.

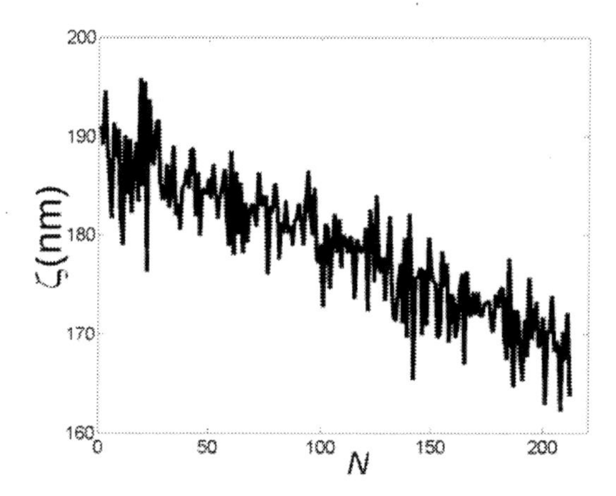

Figure 17. Linear random expansion series (Ordinate) of holographic recording material during the processing after exposure. Abscissa represents period numbers.

The results in Figures 16 and 17 show that what we obtained for an elementary reflection hologram is actually a heterostructure, and the interfaces of the heterostructure are gradual surfaces rather than step functions. In Figure 16, T is the thickness of the material. Also, the period is gradually changed rather than a constant. Therefore, the PC made by the method shown in Figure 15 is a PC heterostructure.

In our experiments, two exposures were taken for fabricating triangular lattices using the set-up geometry shown in Figure 15. After the first exposure, the recording material was translated a distance of d/3 along the $\Gamma - K$ direction of the first Brillouin zone, and then the second exposure was made. In this way, the lattice-point shape is not circular, but elongated and became elliptical. Figure 18 shows the transmission spectra of the triangular lattice heterostructure at different incident angles. It can be seen clearly that there is an absolute band gap in the wavelength range of 430 nm-500 nm at different incident angles, the width (FWHM) of the gap is about 50 nm (445-495 nm). It is an ODBG in deed. The oscillation at the bottom of the band gap is induced by the nonconforming periodicity.

The above experimental results could be explained theoretically as follows.

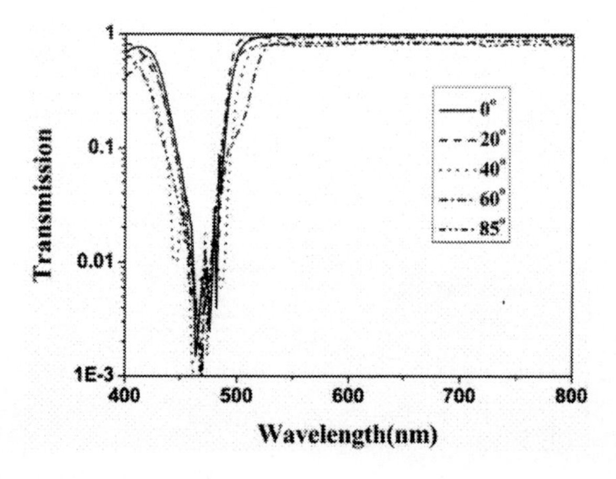

Figure 18. Transmission spectra of triangular lattice heterostructure made by two exposures with a translation.

For the dielectric columniation triangular lattice, the first band and the second band of TM wave at K of its first Brillouin zone is degenerated because of high symmetry. It results in that there is no common gap for TE and TM waves in $\Gamma - K$ direction. In our experiments, the method of two exposures with a translation was used, it made the lattice point shape to be non-circular (see the left iconograph at the bottom in Figure 19). So, the symmetry point group of this kind of triangular lattice is debased to C_{2v} from C_{6v}. It releases the band degeneration of TM wave at K of the first Brillouin zone, and this property is shown clearly in Figure 19 (pointed by the black arrows 1-4). The non-circular lattice-point shape of the triangular lattice induces two extra symmetric points K1 and M1 (see the right iconograph at the bottom in Figure 19)

The gradual period of the heterostructure is an important factor for obtaining ODBG. In this kind of structure the lattice constant is changed gradually and linearly. According to our experimental conditions, in our fabricated structure, there are 150 layers, they can be treated as 150 sub-structures, the minimum lattice constant is d=234.8 nm. During the processing of DCG after exposure a few air holes may be formed by dehydration, considering the refractive index of DCG be 1.52 before exposure, we choose the minimum refractive index $n_0 = 1$. The distribution of refractive index is proportional to light intensity and can be expressed by

$$n = n_0 + \frac{I}{I_{max} - I_{min}} \Delta n \quad , \tag{2}$$

$$I = \sum_{s=0,\ a/3} E_0^2 \left[1 + 4cos\left(\frac{\sqrt{3}}{2}ky\right) cos\left(\frac{3}{2}k(x+s)\right) + 4cos^2\left(\frac{\sqrt{3}}{2}ky\right) \right] \tag{3}$$

where, s is displacement quantity along $\Gamma - K$ direction of the first Brillouin zone. The permittivity is

$$\varepsilon = n_0^2 + 2n_0 \Delta n \cdot \frac{I}{I_{max} - I_{min}} + \left(\frac{I}{I_{max} - I_{min}} \Delta n \right)^2 \tag{4}$$

where, $\Delta n = 0.5$, $I_{max} - I_{min} = 9$.

Because the heterostructure is a gradual period structure consist of 150 sub-structures, the upper band edge of the first sub-structure and the bottom band edge of the last (N=150) sub-structure determine the width of the total band gap of the heterostructure. In our calculations, the structure with the minimum period d (the first sub-structure) and that with the maximum period 1.15 d (the Nth sub-structure) were taken for calculations so that the upper and the bottom band edges can be obtained. Figure 19 gives the results calculated by plane wave expansion method. The dark solid line and the dark dots represent the bands of the first sub-structure for TM and TE waves respectively. The upper edges of the two bands form the high frequency edges of the heterostructure for TM and TE waves. The light dashed line and the light dots represent the bands of the Nth sub-structure for TM and TE waves respectively, The bottom edges of the two bands form the low frequency edgės of the heterostructure for TM and TE waves. The total band gap of the heterostructure is the overlap of the band gaps of the 150 sub-structures. In any high symmetric direction of the first Brillouin zone, the directional band gap of the two sub-structures form the upper and the bottom band edges of the total band gap of the heterostructure in this direction respectively (as shown in light grey zone). The dark grey zone is the common gaps of all directional gaps so that an ODBG of the heterostructure is formed. The calculated position of the band gap locates at $3.8\times10^{15}\sim4.1\times10^{15}$ s^{-1} (460nm~496nm), they (and the width) are well consistent with the experimental results.

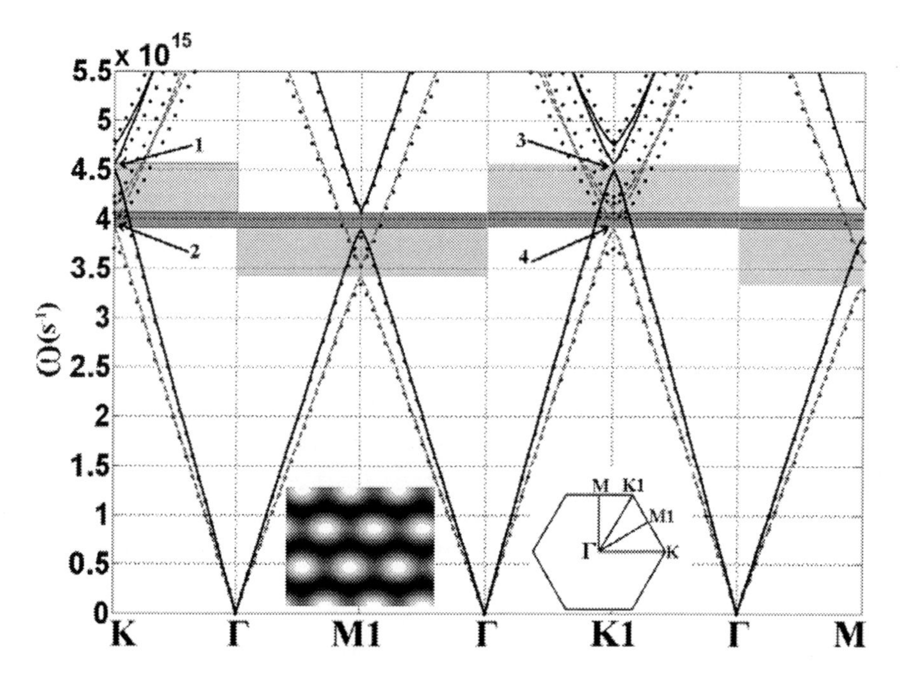

Figure 19. Calculated dispersion relation of gradual triangular lattice heterostructure. The left iconograph at the bottom shows the lattice geometry. The right iconograph at the bottom shows the first Brillouin zone of the triangular lattice.

One of the characters of this kind of triangular lattice is that the bands of TE and TM waves are almost superimposed, and along every direction band gaps are close to each other. This character is very helpful for achieving ODBG through gradual period heterostructures, when the layers of the heterostructure is more enough, and the expansion of the recording material reaches a certain range, a wider ODBG will be obtained by use of narrow directional band gaps. Obviously, it will be helpful for further works to achieve 3-D complete band gaps.

Based on the principle in Ref. [28] this kind of heterostructure can be transformed into a self-simulating spherical structure in the following steps: Firstly, preparing holographic recording material by coating 36 μm thick DCG emulsion on soft substrate polyethylene terephthalate (PET); Then, fabricating gradual index distribution and period heterostructure using the recording material; Finally, transforming the heterostructure into self-simulating sphere. This self-simulating sphere structure can be expressed by D_{nv} point group. Figure 20 shows a D_{nv} self-simulating sphere structure based on the gradual index distribution and gradual period heterostructure.

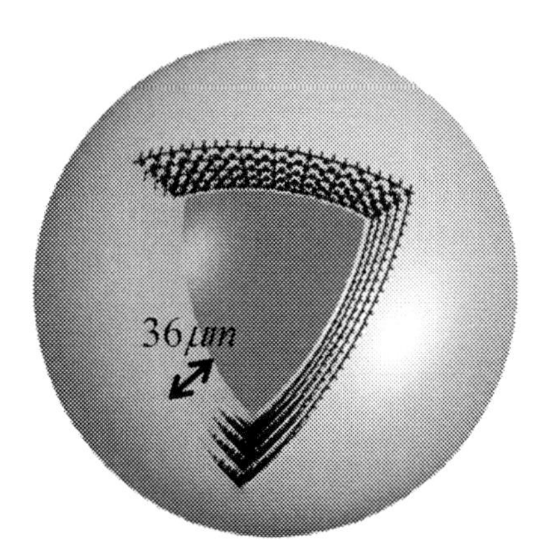

(a) Schematic diagram of D_{nV} self-simulating sphere structure.

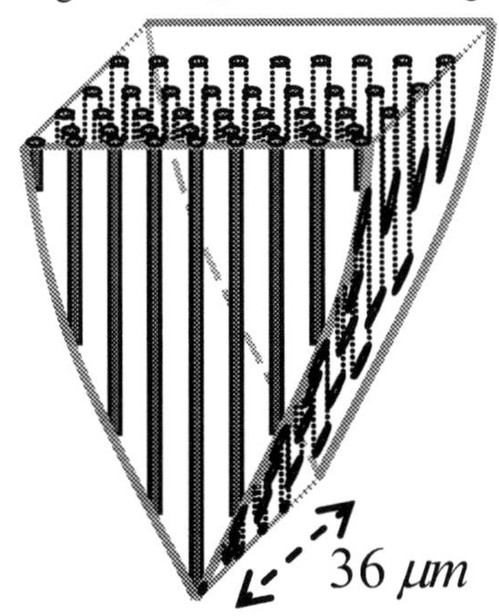

(b) The schematic diagram of the cut slice of the sphere shown in (a)

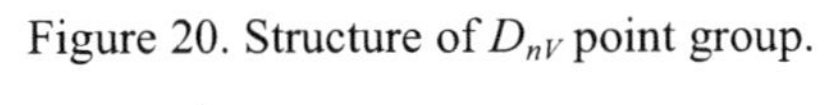

Figure 20. Structure of D_{nV} point group.

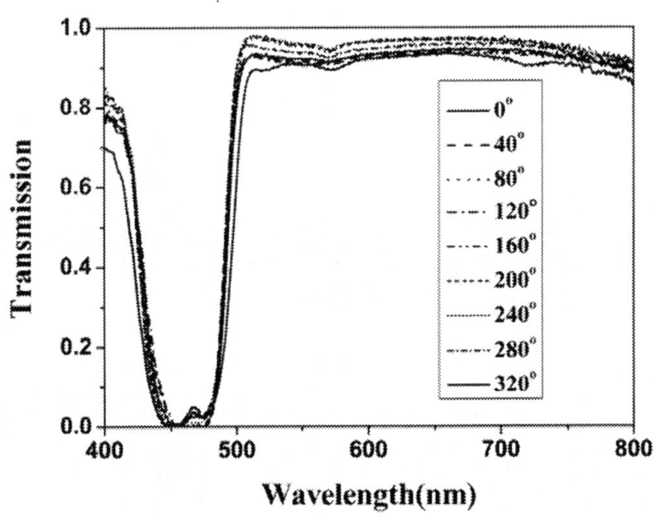

Figure 21. Measured transmission spectra of a D_{12V} self-simulating sphere structure for different value of φ at $\theta = 90°$.

The transmission spectra of a D_{12v} self-simulating sphere structure was measured and shown in Figure 21. In the measurements, a parallel beam with a diameter of 1 *mm* was used as the incident light, and was incident on the surface of the sphere passing through the center of the sphere. It is formed by a collimated beam passing through a 1 *mm* diaphragm. During the measurements, the angles θ and φ of the spherical system were changed from $0°$ through $180°$ and from $0°$ through $360°$ respectively. In our experiments, several D_{12v} self-simulating sphere structures with different diameters were fabricated. All of them showed identical characteristics under the same experimental conditions. It can be seen obviously that, for incident beams from different φ at $\theta = 90°$, the position of the band gap kept almost unchanged, and its width reaches about 70 nm.

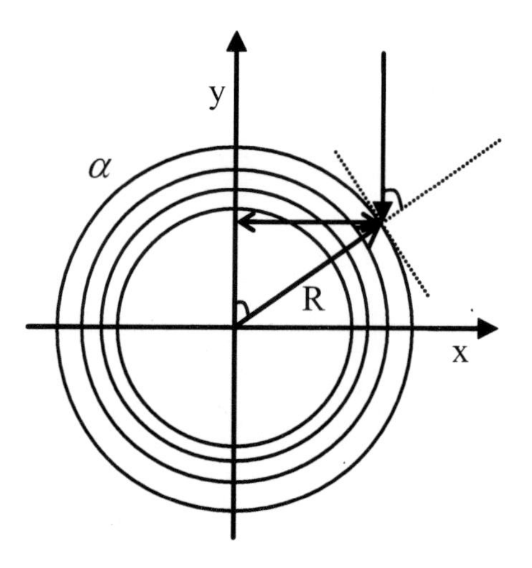

Figure 22. Schematic diagram when the incident light deviates from the center of the sphere. The corresponding incident angle is α = arcsin (l/R).

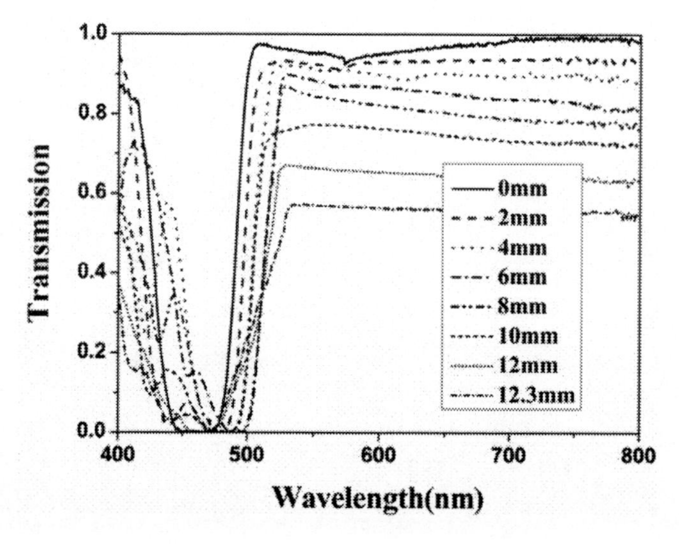

Figure 23. Measured transmission spectra of a D_{12v} self-simulating sphere structure for different values of φ at $\theta = 90°$ when the incident beam shifts different distances from the center. The radius of the sphere is 12.5 mm.

Besides, when the incident light is translated from the center of the sphere for different shifts *l* (see Figure 22). the transmission spectra are measured (see Figure 23). In the measurements, the radius of the sphere was 12.5 *mm*. It can be found that, even the shift is as large as 12.3 mm from the center of the sphere, there still is a CBG with a width of 40 nm. This shift corresponds to the condition that a light beam incidents on the surface of the sphere with an incident angle larger than 80°. In our experiments, two D_{12v} self-simulating sphere structures were superimposed orthogonally to improve further the property of the band gap.

In summary, a D_{nv} point group structure based on 2-D gradual index distribution and gradual period heterostructure and self-simulating sphere possesses genuine complete band gap in a discretionary condition. The D_{nv} self-simulating sphere structure can be implemented easily by holography using low refractive index materials.

5. Theoretical Investigation [41]

The band gaps of PC were described with various methods, such as plane wave expansion method (PWEM) [3,42,43], transfer matrix method (TMM) [44,45], multi-scattering theory [46], tight-binding formulation [47] and finite-difference-time-domain (FDTD) method [48,49]. However, only numerical solutions were obtained by these methods, in which their physical images are not intuitionistic or not clear enough. Therefore, a concise analytical solution (AS) will be valuable for analyzing more thoroughly the properties of the band gaps. Besides, these methods were mostly used for calculating the problems of PCs with step distribution of the refractive index. Recently, the present authors obtained very wide 3-D band gaps [23], even complete band gaps [28,33], using holographic technology and a low refractive index material basing on the principle of multi-beam interference [12,13], and has been used widely, only that the distribution of refractive index inside the recording material (such as photopolymers [50], liquid [51,52], or dichromated gelatin [23,28,33]) is gradual function rather than step function. Liu et al. [53] derived an analytical solution for 1-D PC with a sinusoidal distribution of refractive index. Samokhvalove et al. [54] made an approximation analysis of 2-D PC with rectangular dielectric rods. Nusinsky et al. [55] also made analytical calculation of rectangular PC. Analytical solutions presented by [54] and [55] are valid only for rectangular, step index photonic crystals. While the solution presented in this paper is valid for different structure having gradual distribution of refractive index.

Now, the theoretical analysis is investigated for the band gap characteristics of the PC made by multi beam interference. An analytical solution can be obtained

(1) Basic Considerations

For multi-beam interference, the intensity distribution is a gradual function. When the distribution of the refractive index inside the recording material is proportional to the interference intensity, it should keep in consistency with the gradual distribution according to the property of interference. Therefore, only a few non-zero low order terms exist in the Fourier expansion of the dielectric constant. Since the wave function is generally a coupled multivariate linear equation group, a clear and simple AS can be obtained by approximation according to the spectral property.

For simplicity, an example of 2-D triangular lattice formed by interference of three coherent beams is shown in Figure 15. In Figure 15, the angle between and is , and points just to the bisector of this angle. Since polarization and phase have no effect on the final result, they were neglected in the following analysis.

The interference intensity in the x-y plane is

$$I = \left|\sum_{i=1}^{3} E_i\right|^2 = E_0^2\left[1+4\cos(ky\sin\alpha)\cos\left[kx(1+\cos\alpha)\right]+4\cos^2(ky\sin\alpha)\right] \tag{5}$$

Assuming that the permittivity of the material is proportional to the interference intensity, it will be $\varepsilon = n_0^2 + 2n_0\Delta n\frac{I}{I_{max}-I_{min}}$ after exposure, here, n_0 is the minimum refractive index of the material, and Δn is the modulation of the refractive index. Its Fourier transform will be

$$F(\varepsilon)=\left(n_0^2+3C\right)\delta_{m,n}+C\left(\delta_{m-1,n+1}+\delta_{m-1,n}+\delta_{m,n+1}+\delta_{m,n-1}+\delta_{m+1,n}+\delta_{m+1,n-1}\right) \tag{6}$$

where, $C=\frac{2n_0\Delta n}{I_{\max}-I_{\min}}$.

It can be seen from Eq. (6) that there are only direct current (DC) term and first order term . Obviously, the value of DC term n_0^2+3C is much larger (more than 20 times) than that of the first order term. For example, in dichromated gelatin (DCG) [23,28,33], where $n_0=1.52$, $\Delta n=0.07$, for the triangular lattice formed by the inference of three beams with unit amplitude, we have $I_{max}-I_{min}=9$, i.e., $\frac{n_0^2+3C}{C}\approx 101$, and all the first order terms are equal to *C*. So, the coupling among the first order terms could be neglected, and consider only the coupling between the DC term and any one of the first order term. So $\varepsilon_{\vec{G}'-\vec{G}}$ can be taken as $\begin{bmatrix} n_0^2+3C & C \\ C & n_0^2+3C \end{bmatrix}$. For DCG, the relation $n_0+3C >> C$ could be satisfied when $\Delta n \le 0.4$. All the band gaps belong to first Brillouin zone can be obtained by solving a coupled binary linear equation group, and each band gap corresponds to a simple AS.

(2) Analytical Approach (AS)

From Maxwell equations and Bloch'law, for TM wave, we have

$$det\left\{\left(abs\left(\vec{k}+\vec{G}'\right)abs\left(\vec{k}+\vec{G}\right)\right)/\varepsilon_{\vec{G}'-\vec{G}}\right\}=0 \tag{7}$$

for TE wave, we have

$$\det\left\{\left(\mathrm{Re}\left(\vec{k}+\vec{G}'\right)\mathrm{Re}\left(\vec{k}+\vec{G}\right)+\mathrm{Im}\left(\vec{k}+\vec{G}'\right)\mathrm{Im}\left(\vec{k}+\vec{G}\right)\right)\Big/\varepsilon_{\vec{G}'-\vec{G}}\right\}=0 \tag{8}$$

Figure 24 shows schematically a reciprocal lattice of a triangular lattice. The solid line represents the first Brillouin zone, and the dashed line represents the cell in the reciprocal lattice. K_1 and K_2 are two highly symmetric directions of the first Brillouin zone. G_1-G_6 are the representative reciprocal lattice vectors that collect all values of $\vec{G}$ and $\vec{G}'$ in Eqs. (7) and (8) in the cell of the reciprocal lattice. It can be seen that from the geometric relationship in Figure 2, $\left|\vec{k}+\vec{G}_0\right|=\left|\vec{k}+\vec{G}_1\right|$, where, $\vec{k}$ is in K_1 and K_2 direction respectively.

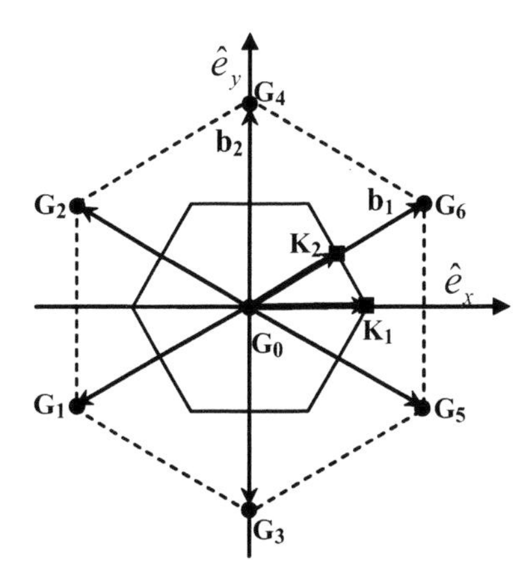

Figure 24. Schematic diagram of the reciprocal lattice of a triangular lattice. The dashed line represents the cell in the reciprocal lattice. The solid line represents the first Brillouin zone

Let a be the lattice constant, so, the vector of its reciprocal lattice are b_1 and b_2 (see Figure 24).

For a triangular lattice $b_1=\frac{2\pi}{a}\left(\hat{e}_x+\tan\left(\frac{\pi}{6}\right)\hat{e}_y\right)$, $b_2=\frac{2\pi}{a\cos(\pi/6)}\hat{e}_y$, then we have $K_1=\frac{4\pi}{3a}\hat{e}_x$, $K_2=\frac{\pi}{a}\hat{e}_x+\frac{\pi}{\sqrt{3}a}\hat{e}_y$.

For TM wave, Eq. (7) can be transformed into

$$\begin{vmatrix} \frac{\left|\vec{k}+\vec{G}_0\right|^2\left(n_0^2+3C\right)}{\left(n_0^2+2C\right)\left(n_0^2+4C\right)}-\frac{\omega^2}{c^2} & \frac{-\left|\vec{k}+\vec{G}_0\right|\left|\vec{k}+\vec{G}_i\right|C}{\left(n_0^2+2C\right)\left(n_0^2+4C\right)} \\ \frac{-\left|\vec{k}+\vec{G}_0\right|\left|\vec{k}+\vec{G}_i\right|C}{\left(n_0^2+2C\right)\left(n_0^2+4C\right)} & \frac{\left|\vec{k}+\vec{G}_0\right|^2\left(n_0^2+3C\right)}{\left(n_0^2+2C\right)\left(n_0^2+4C\right)}-\frac{\omega^2}{c^2} \end{vmatrix}=0 \tag{9}$$

The positions of the top and the bottom edges of the first band gap are at

$$\omega_\pm=\sqrt{\frac{\left(n_0^2+3C\right)\pm C}{\left(n_0^2+2C\right)\left(n_0^2+4C\right)}}\left|\vec{k}+\vec{G}_0\right|c \tag{10}$$

So, $\omega_+ \approx \frac{1}{n_0}\left(1-\frac{2\Delta n}{9n_0}\right)|\vec{k}|c$ and $\omega_- \approx \frac{1}{n_0}\left(1-\frac{4\Delta n}{9n_0}\right)|\vec{k}|c$ can be obtained from Eq.(10) by using the condition $n_0^2 + 3C >> C$ (i,e. $\Delta n \le \frac{9}{34} n_0$).

Therefore, the following conclusions can be obtained:

1) The top and the bottom edges of the first band gap of TM wave are proportional to Δn and $|\vec{k}|$ ($|\vec{k}| \propto \frac{1}{\lambda}$, λ is the light speed in the material), and are inversely proportional to n_0. This is more definite than the common knowledge about "location of the band gap ($\frac{\omega_{i+}+\omega_{i-}}{2}$) is negatively related to the average refractive index", and shows the quantitative dependence of the top and the bottom band edges on material parameters.
2) For a certain material (i.e., n_0 and Δn are fixed), the band gaps will be determined only by the value of $|\vec{k}|$. So, the closer the shape of the first Brillouin zone approaches a circle, the smaller the variation of $|\vec{k}|$ (i.e., $|k_1| \approx |k_2|$). In this case, positions of the directional band gaps in K_1 and K_2 are almost same, obviously, it is more favorable for the appearance of absolute band gaps.
3) The relationship that $\frac{\Delta\omega}{\omega} = 2\frac{\omega_{i+}-\omega_{i-}}{\omega_{i+}+\omega_{i-}} \approx \frac{2\Delta n}{9n_0}|\vec{k}|c$ can be achieved (here $\Delta n \le \frac{9}{34} n_0$), i.e. the bandwidth of the first band gap of TM wave is positively related to Δn and $|\vec{k}|$, and negatively related to n_0. This is more definite than the common knowledge "the higher the ratio of refractive indexes, the wider the band gaps".

Figure 25 shows the relationship between the band width $\Delta\omega/\omega$ and n_0 and Δn. In Figure 25, the color bar gives the normalized bandwidth $\Delta\omega/\omega$, the top line represents the width corresponding to the critical condition $\Delta n = (9/34)n_0$. It shows significantly that one does not have to seek for high refractive index materials for obtaining wider band gap. Band gap could readily be obtained with lower refractive index materials as same as that obtained with higher refractive index material (under the prerequisite condition of $\Delta n = (9/34)n_0$). Large value of the modulation of refractive index needs high refractive index, it is not definitely helpful for obtaining a broad band gap.

These features are obviously different from the relationship between the characteristics of band gap and the material parameters of 1-D PC [56]. But, the basic tendencies of variation of the band gap with material parameter obtained from both considerations are similar.

Also, for TE wave, Eq. (8) could be transformed into

$$\begin{vmatrix} \frac{\left|\vec{k}+\vec{G}_0\right|^2\left(n_0^2+3C\right)}{\left(n_0^2+2C\right)\left(n_0^2+4C\right)}-\frac{\omega^2}{c^2} & \frac{\mathrm{Re}\left[\left(\vec{k}+\vec{G}_0\right)\left(\vec{k}+\vec{G}_i\right)^*\right]C}{-\left(n_0^2+2C\right)\left(n_0^2+4C\right)} \\ \frac{\mathrm{Re}\left[\left(\vec{k}+\vec{G}_0\right)\left(\vec{k}+\vec{G}_i\right)^*\right]C}{-\left(n_0^2+2C\right)\left(n_0^2+4C\right)} & \frac{\left|\vec{k}+\vec{G}_0\right|^2\left(n_0^2+3C\right)}{\left(n_0^2+2C\right)\left(n_0^2+4C\right)}-\frac{\omega^2}{c^2} \end{vmatrix}=0 \tag{11}$$

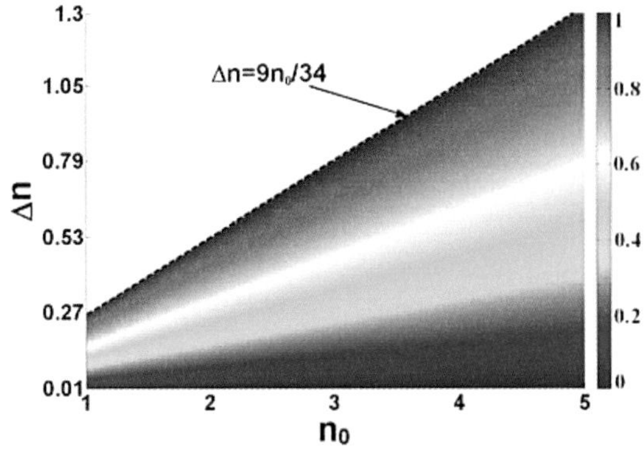

Figure 25. Calculated relationship between band width $\Delta\omega/\omega$ and n_0 and Δn. The color bar gives the band width $\Delta\omega/\omega$, the black dash line represents the width corresponding to $\Delta n=(9/34)n_0$.

The top and the bottom edges of the first band gap are at

$$\omega_{\pm}=c\sqrt{\frac{\left|\vec{k}+\vec{G}_0\right|^2\left(n_0^2+3C\right)\mp\mathrm{Re}\left[\left(\vec{k}+\vec{G}_0\right)\left(\vec{k}+\vec{G}_1\right)^*\right]C}{\left(n_0^2+2C\right)\left(n_0^2+4C\right)}} \tag{12}$$

Let β be the angle between K_1 and K_2 in the direction of K_1 (see Figure 24), then, $\omega_{i\pm}=\sqrt{\frac{\left(n_0^2+3C\right)\pm C\cos 2\beta}{\left(n_0^2+2C\right)\left(n_0^2+4C\right)}}\left|k+G_0\right|c$ which is consistent with Eq.(10). It shows that the influencing factors on the width of the first band gap of TE wave, and the rule of their variations are as same as those for TM wave.

In K_2 direction, $\left(\vec{k}+\vec{G}_1\right)^*=-\left(\vec{k}+\vec{G}_0\right)^*$, then Eq. (8) could be transformed to Eq. (6). This is the reason why same results for TE and TM waves are always obtained in the direction of K_2 calculated by PWEM under the condition of small modulation of refractive index.

(3) Comparison with Plane Wave Expansion Method (PWEM)

It is clear from Eqs. (9) and (11) that the eigenvalue ω^2/c^2 is positively related to the length of $\vec{k}+\vec{G}$. From Figure 24, it can be seen that the lengths of $\left|\vec{k}+\vec{G}_0\right|$ and $\left|\vec{k}+\vec{G}_1\right|$ are minimal in both K_1 and K_2 directions. They correspond respectively to the minimal eigenvalue ω^2/c^2, i.e. the lowest band gaps in two highest symmetric directions.

Figure 26 gives the calculated results using Eqs. (10) and (12). It shows that, $|\vec{k}+\vec{G}_0|$ and $|\vec{k}+\vec{G}_1|$ determine the top and the bottom limit of the first gap, respectively. Each direction of G in Figure 24 corresponds to two band gaps (one for TM wave and the other for TE wave). The top and the bottom edges of the gaps corresponding to TM and TE waves can be obtained by solving Eqs. (9) and (11) when the collection of the values of $\vec{G}$ and $\vec{G}'$ is defined as $\{\vec{G}_0, \vec{G}_i\}$ $(i = 1,2,3,4,5,6)$. The value at any point of the band can be calculated by reducing the value of K while keeping its direction unchanged. The inserts in Figure 4 pointed by the arrows are the zoomed figures of the first band gap in K_1 and K_2 directions respectively, in which the position and the width obtained by PWEM are in good consistency with those obtained by our AS. The bandwidth of TM wave obtained by AS is slightly wider than that obtained by PWEM in K_1 direction, while the width of the gaps of TE wave is slightly narrower. In K_2 direction, those obtained by the two methods coincide almost perfectly. However, it should be noted that in K_1 direction there is no gap for TM wave according to PWEM, but, in AS which is not zero. It indicts that this approximation is more accurate to predict the location of band gaps than the bandwidth of band gaps for TM polarization.

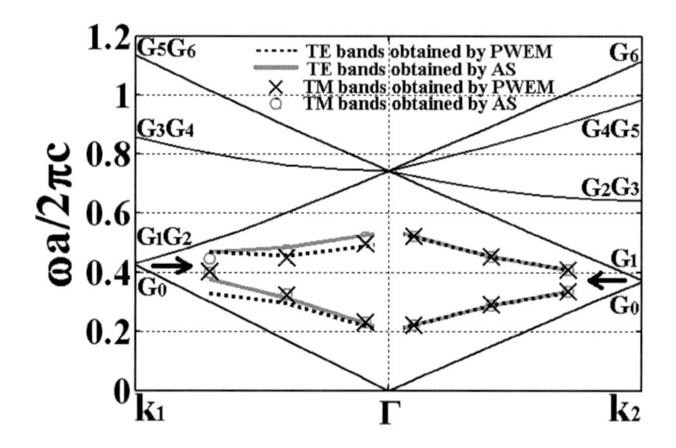

Figure 26. Band gaps calculated by AS and PWEM.

For detail comparison between the two methods, the variation of the top and the bottom band edges with Δn was obtained (see Figure 27) by analyzing Eqs. (10) and (12).

Analytical Solution

In Figure 27, it can be seen that: the top and the bottom edges of TM and TE waves in K_2 direction are in good superposition. In other words, the dielectric constants can be considered as the same for TE and TM waves in K_2 direction when Δn is small. The bandwidth in K_1 and K_2 directions increase with the increase of the refractive index modulation. The space between the bottom edge in K_1 direction and the top edge in K_2 direction decreases with the increase of the refractive index modulation, but they won't intercept when $\Delta n \leq 0.4$. It means that there is no absolute band gap in low refractive index materials with strict period structure for triangular lattice. But, it shows that the bottom edge in K_1 direction and the top edge in K_2 direction will superimpose, and absolute band gap will appear when Δn is large enough. Or, in another way of thinking, absolute band gap could be obtained by optimizing the lattice structure, i.e. using a non-strictly periodical structure under low refractive index modulation[28,33].

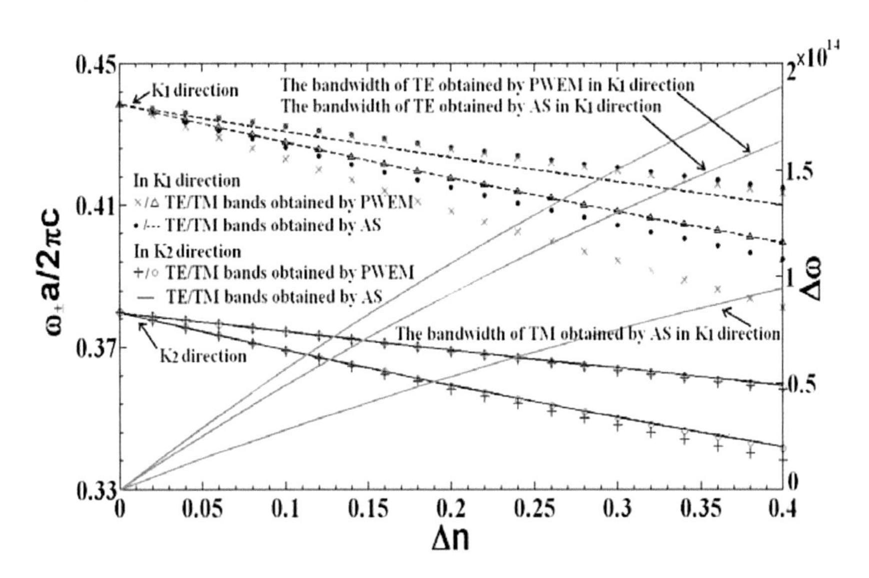

Figure 27. Variation of Position of band edges and bandwidth with modulation of refractive index calculated by AS and PWEM.

Plane Wave Expansion Method

In Figure 27, it can be seen that: the TM mode degenerates in K_1 direction. The top edge of the TE band and the bottom edge of the TM band in K1 direction are in good consistency with those obtained by AS, but, the bandwidth is different from that obtained by AS. While, the band location and the bandwidth of TM and TE waves in K_2 direction obtained by the two methods are all well consistent, and show almost no difference when $\Delta n \leq 0.1$.

(4) Conclusion

Analytical solution was obtained for 2D PCs made by holography. It gives concise and intuitionistic physical image as well as the relationship between the characteristics of the band gaps and the parameters of the materials. The results obtained by the analytical solution are in good consistency with those obtained by plane wave expansion method, but, it should be noted that in some directions the solution may not be accurate enough and may be different from the degenerate case in which predicts by PWEM. This analytical approach is practically the first order approximation of the solution of the PC with step distribution of refractive index. Though the analytical solution is derived for 2D triangular lattice made with a special material (DCG), it is valid for any other kind of 2D lattice made by multi-beam interference or other gradual refractive index materials. And it must be noted that this solution is accurate only for low refractive index modulation materials.

6. Temperature Tunable Random Lasing in Weakly Scattering Structure Formed by Speckle

Recently, the low refractive index material DCG was used to implement a weakly scattering disorder structure formed by speckle, and temperature tunable random lasing using this kind of structure was achieved.

Since the pioneering work of Ambartsumyan et al. [57], random laser has been a subject of intense theoretical and experimental studies [58-65]. In particular, the discovery of intriguingly narrow spectral emission peaks (spikes) in certain random lasers by Cao et al. [60] has given an enormous boost to such a subject. The random laser represents a nonconventional laser whose feedback is mediated by random fluctuations of the dielectric constant in space. In general, there are two kinds of feedback for the random lasing, intensity feedback and field feedback [64,65]. The former is phase insensitive, which is called as incoherent or nonresonant feedback. The latter is phase sensitive, which can be regarded as coherent or resonant feedback.

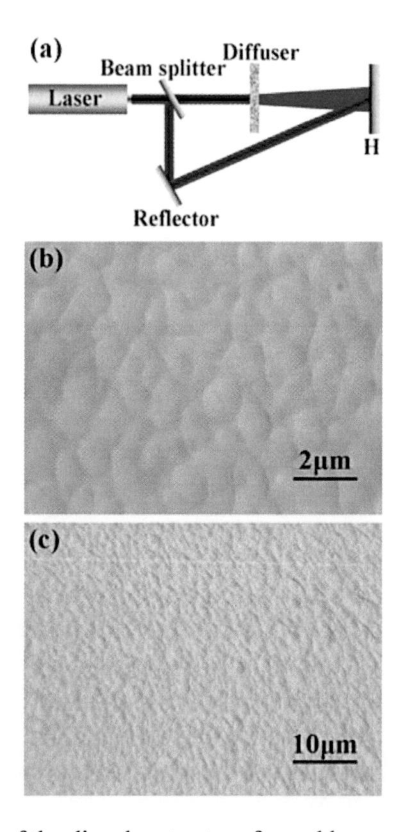

Figure 28. (a) Set-up geometry of the disorder structure formed by speckle. H is holographic recording material. (b) and (c) are the microscopic images of the disordered structure formed by speckle.

The two types of random laser action can not only be found in some strong scattering systems with gain [64-67], they can also be observed in active random media in weakly scattering regime far from Anderson localization. For example, some authors have reported laserlike emission from several weakly scattering samples including conjugated polymer films, semiconductor powders and dye-infiltrated opals [68-72]. In the strongly scattering regime, lasing modes have a nearly one-to-one correspondence with the localized modes of the passive system [64-66,73]. In contrast, the nature of lasing modes in weakly scattering open random systems is still under discussion, although several mechanisms have been proposed to explain such a phenomenon [74-77].

Here we report an experimental observation of random laser in a kind of new random medium in a regime of weak scattering. A disorder structure was formed by speckle. The set-up geometry is shown in Figure 28 (a) schematically. A ground glass was used to generate spatial speckle structure. A reference beam was introduced to implement interference for recording the distribution of the spatial speckle in the holographic recording material. Different distribution of the spatial speckle can be recorded by changing the place of the ground glass or the reference beam. The recording material used was dichromated gelatin (DCG) coated on optical glass. The thickness of the DCG used is $36\,\mu m$. After exposure, the DCG plate was developed in running water for 120 min at $20^{\circ}C$ for developing sufficiently and to remove any residual dichromate, and then soaked into a Rhodamine 6G solution with a concentration of 0.125 mg per milliliter water at the same temperature for 60 min bath enabling the dye molecules to diffuse deep into the emulsion of the gelatin. Then, the DCG plate was dehydrated in turn by soaking it in 50 %, 75 % and 100 % isopropyl alcohol containing the Rhodamine 6G dye with same concentration at $40^{\circ}C$ for 15 min.

After dehydration, the DCG plate was baked at $100^{\circ}C$ for 60 min in an oven. Figure 28 (b) and (c) show the microscopic image of the disordered structure formed by the speckle, and it is a complete disordered structure. The average size of the speckle spot is about $1\,\mu m$. The concentration of the speckle spot is estimated to be the order of magnitude of 10^9 per cm^3. The DCG is a phase type holographic recording material with the refractive index of 1.52 and the modulation of the refractive index less than 0.1. From the transmission measurements, we can estimate the mean free path $l^* > 80\,\mu m$ in the spectral range between 550 nm and 600 nm, which is much larger than the thickness of the system. This means that scattering in the structure is quite weak.

We now turn to the investigations on the photoluminescence (PL) of the above dye-doped random samples under a pump field. Figure 29 (a) shows the set-up geometry for the measuring. A nanosecond (ns) pulsed Nd:YAG laser running at 532 nm with repetition rate of 10 Hz, maximum pulse energy of 1500 mJ , pulse width of 8 ns and a picosecond (ps) pulsed Nd:YAG laser running at 532 nm with repetition rate of 10 Hz, maximum pulse energy of 40 mJ , pulse width of 30 ps were used as the pumping sources respectively. The beam diameters of the ns laser and the ps laser were 13 mm and 3 mm respectively. The laser beam was incident on the surface of the optical glass substrate, then penetrates the substrate and entered the dye doped DCG medium. The detector was put close to the sample to collect the energy as much as possible. By rotating the rotation stage the orientation of the speckle structure can be changed with respect to the laser beam. It results in the variation of the wavelength and the peak number of lasing.

The measured results for the emission spectra under the pump by ps laser are plotted in Figure 29 (b). The blue triangles correspond to the PL spectra of Rhodamine 6G, and the red solid line represents the emission of the dye-doped sample at $\theta = 25^{\circ}$ under the pumping intensity of 120 MW/mm^2. The threshold of the random laser is about 50 MW/mm^2 as shown in the insets. It can be seen that many lasing modes are excited in the gain spectrum of Rhodamine 6G. The widths of these modes are found to be less than 0.4 nm, which exhibits the coherent feedback very well. The phenomenon is very similar to the previous investigations on the random laser in the other weak scattering systems [68-72]. The origin of the phenomenon can also be understood by the previous theory for such a problem [74-77].

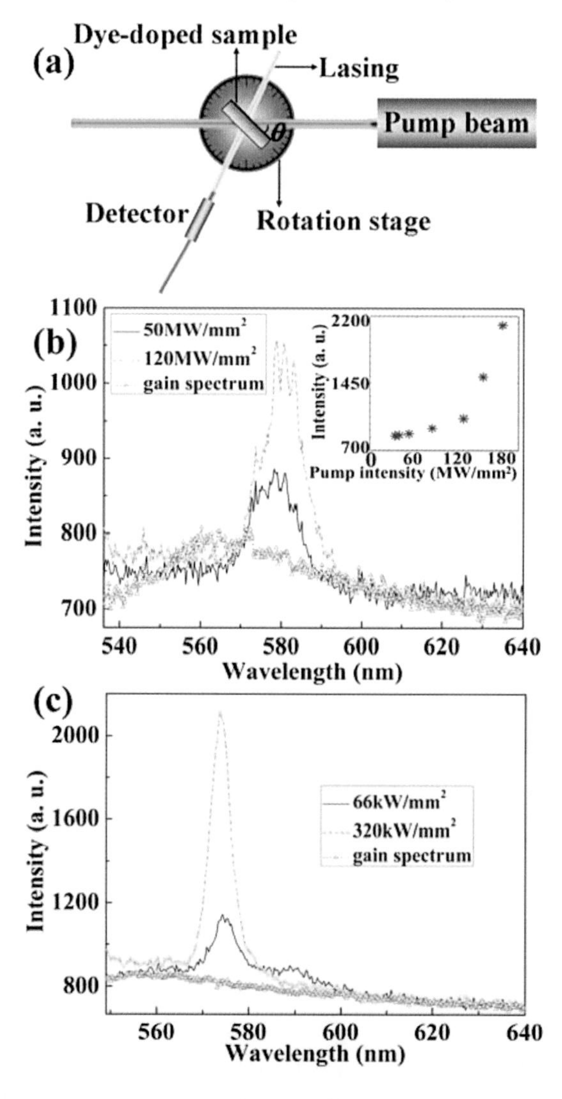

Figure 29. (a) Set-up geometry for measuring random lasing. θ is the angle of the pump beam with respect to the surface of the substrate, it represents the orientation of the sample. (b) Emission pumped by ps laser, and the inset shows the threshold is about 50MW/mm^2 at $\theta = 25°$. (c) Emision pumped by ns laser at $\theta = 30°$. In (b) and (c), the blue triangles are the gain spectrum of rhodamine 6G excited by 532 nm laser beam. The black solid line and the red dashed line represent the emissions with pumped energy around and much higher than threshold respectively.

However, when the dye-doped random sample is pumped by the ns laser, the different phenomenon appears. Figure 29(c) displays the measured results when the pump energy is much higher (320 kW/mm^2) than and around (66 kW/mm^2) the threshold. A single mode is observed and the profile of the mode is fitted by a Gaussian function, and the result demonstrates that this mode shows a perfect Gaussian profile (as shown in Figure 30(a)). Such a character does not change with the change of the pump energy. When $\theta = 30°$, the emission spectra at different pump energy are shown in Figure 30(b). There is only a single mode at 574 nm. The threshold value is found to be 37.7 kW/mm^2, which is much lower than that pumped by ps laser (50 MW/mm^2). The lasing becomes more obvious when the pump energy is higher than 65.9 kW/mm^2.

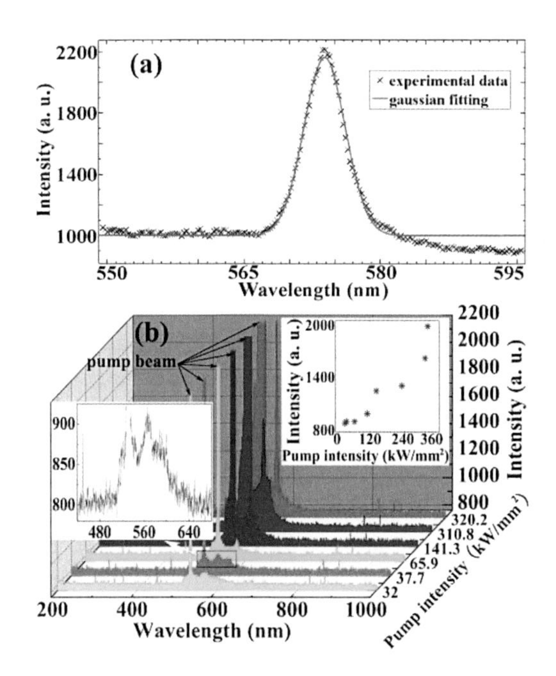

Figure 30. (a) Gaussian function fitted emission peak. (b) Measured emission spectra at different pump energy in the case of $\theta = 30°$, and the inset shows the threshold is about 37.7 kW/mm^2. The left high spikes (marked by black arrows) correspond to 532 nm pump laser beam, and the appeared width does not show its actual line width of the pump laser because of the saturation effect of the CCD detector used.

In fact, our sample is in the shape of a thin film, it can not be considered as a strict isotropic structure. So, the property of the mode also depends on the pump direction of the laser beam. In some directions, the case with two modes can also be found. Figure 31(a) shows the measured results for the emission spectra of the disordered structure at different pump energy using the ns laser. There are two modes of the emission located at 570 nm and 590 nm respectively and the later is much stronger than the former, and the later becomes the dominant mode when the pump energy is strong enough. When the pump energy is around the threshold value, the two modes are in competition (see Figure 31(b)).

In addition, we find that the widths of the modes always keep a few nm for any angle (direction) and intensity of the ns pump beam, which displays the character of intensity feedback. In general, the intensity feedback corresponds to diffusion motion of photons in active random medium, when the photon mean free path is much smaller than the dimension of scattering medium but much longer than the optical wavelength [64,65]. It can be described theoretically by the diffusion equation for the photon energy density in the presence of a uniform and linear gain [57,64,65]. Obviously, it is not such a case for the weak scattering system in the present work, because the mean free path is much larger than the thickness of the sample. However, our experimental results demonstrate that incoherent feedback can also be realized in such a non-diffusion random system by choosing suitable pump sources. In the previous investigations, one has observed the transition between coherent feedback and incoherent feedback by varying the amount of scattering in the gain medium [64,78]. In fact, our present work illustrates that such a transition can also be realized in the same random sample by changing the pump source.

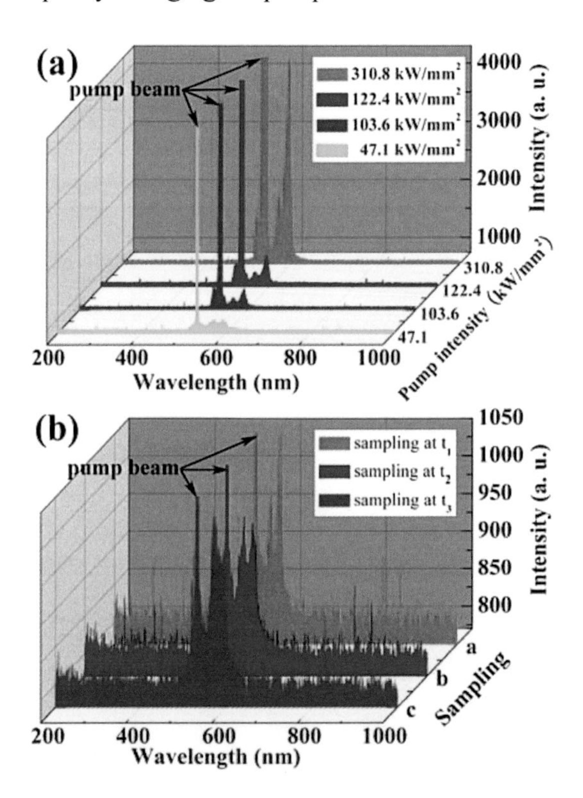

Figure 31. (a) Measured radiation spectra at different pump energy in the case of $\theta = 18°$. (b) Competition of two modes around the threshold value in the case of $\theta = 18°$. The left high spikes (marked by black arrows) correspond to 532 nm pump laser beam, and the appeared width does not show its actual line width of the pump laser because of the saturation effect of the CCD detector used

The phenomena can be understood in terms of the following analysis. For the case of ns laser pumping, the light will go through a path about several meters in the period of one pulse. It means that the light will have much more times of scattering, i.e. has much longer effective interaction length. Therefore, the mode competition has finished and one mode becomes dominant. That is the reason only a single mode is dominant in the emission as long as the pumping energy is higher. However, for the case of ps laser pumping, the light will go through a path about several millimeters. It means that the effective interaction length is much shorter, so the mode competition can not be finished even though the pumping energy is rather high. Besides, for the case pumped by ps laser, because the effective interaction length is very short, i.e., the scattering times is much less. Therefore, much higher pump energy is needed for building up the lasing, it results in the high threshold.

Another interesting phenomenon for the present system is that the emission wavelength for the random laser can be tuned through changing temperature. Figure 32 shows the measured spectra of lasing emission at different temperatures. It can be seen clearly that the emission peaks shift toward to the long wavelength with the increase of the temperature. The lasing will be kept so long as the frequency of the emission is still inside the gain profile when temperature changes. In fact, the temperature tuning for the random laser by infiltrating sintered glass with laser dye dissolved in a liquid crystal had been investigated in previous work [62]. The diffusive feedback was controlled through a change of refractive index of the liquid crystal with temperature. However, such a tuning was employed to turn on and off random lasers. In contrast, the wavelength tuning can be realized in the present system. The phenomenon originates from the special property of the material. With the changes of the temperature, the light path changes due to variation of the distance between the two adjacent speckle spots. It makes the variation of the effective length of interaction, and then the lasing wavelength is changed.

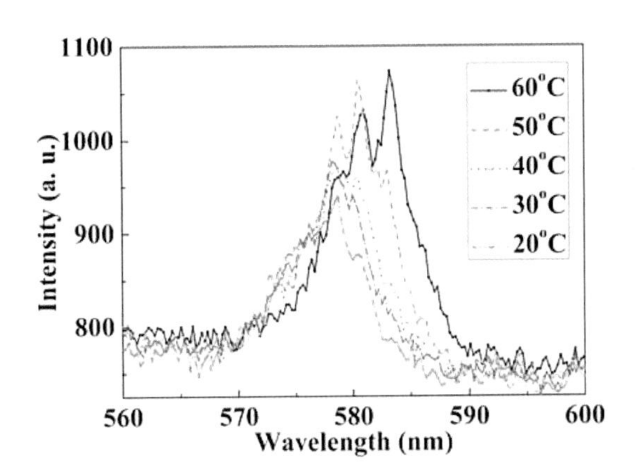

Figure 32. Tunability of wavelength vs. temperature of the random laser.

Finally, it should be pointed out that, the DCG can be coated on a soft substrate, such as polyethylene terephthalate (PET). So, the speckle structure recorded by holography can be formed as a soft film. On the other hand, the holography recorded speckle structure is pretty stable. Hence, the output of this kind of laser has good stability and repeatability. It will be beneficial for actual applications. We did implement the speckle random laser using DCG coated on the PET soft substrate, and the same results mentioned above were obtained.

In summary, we have fabricated a weakly scattering disordered structure formed by speckle using holography. In such a system with gain, we have observed low-threshold random lasing with two kinds of feedback, incoherent and coherent, for the same sample in the case pumped by nanosecond and picosecond lasers, respectively. The wavelength tunability of the random laser with the change of the temperature has been demonstrated. Our results can not only get a deeper understanding on the open question about the properties of the lasing modes in weakly scattering regime, it can also benefit some applications such as remote temperature sensing in hostile environments.

References

[1] John, S. "Strong localization of photons in certain disordered dielectric superlattices," *Phys. Rev. Lett.*, vol. 58, 1987, 2486-2489.

[2] Yablonovitch, E. "Inhibited spontaneous emission in solid-state physics and electronics," *Phys. Rev. Lett.*, vol. 58, 1987, 2059-2062.

[3] Ho, KM; Chan, CT; Soukoulis, CM. *Phys. Rev. Lett*, 1990, 65, 3152-3155.

[4] Chan, TYM; Toader, O; John, S. *Phys. Rev. E*, 2005, 71, 046605.

[5] Sharp, DN; Turberfield, AJ; & RG. Denning, *Phys. Rev. B*, 2003, 68, 205102.

[6] Meisel, DC; Wegener, M; Busch, K. *Phys. Rev. B*, 2004, 70, 165104.

[7] Blanco, A. Emmanuel Chomski, Serguei Grabtchak, Marta Ibisate, Sajeev John, Stephen W. Leonard, Cefe Lopez, Francisco Meseguer, Hernan Miguez, Jessica P. Mondia, Geoffrey A. Ozin, Ovidiu Toader and Henry M. van Driel, *Nature*, 2000, 405, 437-440.

[8] Vlasov, YA; Bo, XZ; Sturm, JC; Norris, DJ. *Nature*, 414, 289-293.

[9] Wu, LJ; Wong, KS. *Appl. Phys. Lett*, 2005, 86, 241102.

[10] Lin, SY ; Fleming, JG; Hetherington, DL; Smith, BK; Biswas, R; Ho, KM; Sigalas, MM; Zubrzycki, W; Kurtz, SR; Jim Bur, *Nature*, 1998, 394, 251- 253.

[11] Noda, S; Tomoda, K; Yamamoto, N; Chutinan, *Science*, 2000, 289, 604-606.

[12] Berger, V; Gauthier-Lafaye, O; Costard, E. *J. Appl. Phys*., 1997, 82, 60-64.

[13] Campbell, M; Sharp, DN; Harrison, MT; RG. Denning, *Nature*, 2000, 404, 53-56.

[14] Zhong, YC ; Zhu, SA; Su, HM; Wang, HZ; Chen, JM; Zeng, ZH; Chen, YL. *Appl. Phys. Lett*, 2005, 87, 061103.

[15] Toader, TYM. Chan and S. John, *Appl. Phys. Lett*, 2006, 89, 101117.

[16] Kondo, T; Matsuo, S; Juodkazis, S. Misawa., *Appl. Phys. Lett*, 2001, 79, 725-727.

[17] Miklyaev, YV; Meisel, DC; Blanco, A; Freymann, GV; Busch, K; Koch, W; Enkrich, C; Deubel, M; Wegener, M. *Appl. Phys. Lett*., 2003, 82, 1284-1286.

[18] Divliansky, TS; Mayer, KS; Holliday, VH. Crespi, *Appl. Phys. Lett*, 2003, 82, 1667-1669.

[19] Meisel, DC; Diem, M; Deubel, M; Willard, FP; Linden, S; Gerthsen, D; Busch, K; Wegener, M. *Adv. Mater*., 2006, 18, 2964-2968.

[20] Cui, LB; Wang, F; Wang, J; Wang, ZN; Liu, DH. *Phys. Lett. A*, 2004, 324, 489-493.

[21] Zoorob, ME; Charlton, MDB; Parker, GJ; Baumberg, JJ; Netti, MC. *Nature* (London), 2000, 404, 740.

[22] Zhang, X; Zhang, ZQ; Chan, CT. *Phys. Rev. B*, 2001, 63, 081105.

[23] Zhi Ren, Zhaona Wang, Tianrui Zhai, Hua Gao, Dahe Liu, Xiangdong Zhang, *Phys. Rev. B*, 2007, 76, 035120.

[24] Su, HM; Zhong, YC; Wang, X; Zheng, XG; Xu, JF; Wang., HZ. *Phys. Rev. E*, 2003, 67, 056619..

[25] Wang, X; Xu, JF; Su, HM; Zeng, ZH; Chen, YL; Wang, HZ; Pang, YK; Tam, WY. *Appl. Phys. Lett*, 2003, 82, 2212-2214.

[26] Dahe Liu, Weiguo Tang, Wunyun Huang, and Zhujian Liang,, *Opt.Eng.*, 1992, 31, 809-812.
[27] Smith ed., HM. *Holographic Recording Materials*, Ch.3, Springer-Verlag, Berlin, 1977.
[28] Zhi Ren, Tianrui Zhai, Zhaona Wang, Jing Zhou, Dahe Liu, *Advanced Materials*, 2008, 20, 2337-2340.
[29] Tianrui Zhai, Zhi Ren, Zhaona Wang, Jing Zhou, Dahe Liu, *IEEE Photon. Tech. Lett.*, 2008, 20, 1066-1068.
[30] Liu, D; Zhou, J. Opt. *Commun*. 1994, 107, 471.
[31] Wang, Z; Liu, D; Zhou, J. Opt. *Lett*, 2006, 31, 3270.
[32] Wang, X; Wang, F; Cui, L; Liu, D. Opt. *Commun*., 2003, 221, 289.
[33] Tianrui Zhai, Zhaona Wang, Rongkuo Zhao, Jing Zhou, Dahe Liu, Xiangdong Zhang, *Appl. Phys. Lett.*, 2008, 93, 210902.
[34] Emanuel Istrate, Edward H. Sargent, *Rev. Mod. Phys.*, 2006, 78, 455-481.
[35] Jiang, P; Ostojic, GN; Narat, R; Mittleman, DM; Colvin, VL. Adv. Mater. (Weinheim, Ger), 2001, 13, 16.
[36] Egen, M; Voss, R; Griesebock, B; Zentel, R; Romanov, S; Torres, CMS. *Chem. Mater*, 2003, 15, 3786-3792.
[37] Wong, S; Kitaev, V; Ozin, GA. *J. Am. Chem. Soc.*, 2003, 125, 15589, 15598.
[38] Song, BS; Noda, S; Asano, T. *Science*, 2003, 300, 1537-1537.
[39] Srinivasan, K; Barclay, PE; Painter, O; Chen, J; Cho, AY; Gmachl, C. *Appl. Phys. Lett*, 2003, 83, 1915.
[40] Kawakami, S; Sato, T; Miura, K; Ohtera, Y; Kawashima, T; Ohkubo, H. *IEEE Photonics Technol. Lett*, 2003, 15, 816-818.
[41] Tianrui Zhai, Zhaona Wang, Rongkuo Zhao, Xiaobin Ren, and Dahe Liu, *IEEE J. Quant. Electron.*, 2009, 45, 1297-1301.
[42] Leung, KM; Liu, YF. "Full vector wave calculation of photonic band structures in face-centered-cubic dielectric media," *Phys. Rev.Lett.*, 1990, vol. 65, 2646-2649.
[43] Zhang, Z; Satpathy, S. "Electromagnetic wave propagation in periodic structures: Bloch wave solution of Maxwell's equations," *Phys.Rev. Lett.*, 1990, vol. 65, 2650-2653.
[44] Pendry, JB; MacKinnon, A. "Calculation of photon dispersion relations," *Phys. Rev. Lett.*, 1992, vol. 69, 2772-2775.
[45] Pendry, J. "Photonic band structures," *J. Mod. Opt.*, vol, 1994, 41, 209-229.
[46] Li, L; Zhang, Z. "Multiple-scattering approach to finite-sized photonic band-gap materials," *Phys. Rev. B*, 1998, vol. 58, 9587-9590.
[47] Lidorikis, E; Sigalas, M; Economou, E; Soukoulis, C. "Tight-binding parametrization for photonic band gap materials," *Phys. Rev. Lett.*, 1998, vol. 81, 1405-1408.
[48] Chan, CT; Datta, S; Yu, QL; Sigalas, M; Ho, KM; Soukoulis, CM. "New structures and algorithms for photonic band gaps," *Physica A*, 1994, vol. 211, 411-419.
[49] Tran, P. "Photonic-band-structure calculation of material possessing Kerr nonlinearity," *Phys. Rev. B*, 1995, vol. 52, 10673-10676.
[50] Jakubiak, R; Bunning, T; Vaia, R; Natarajan, L; Tondiglia, V. "Electrically switchable, one-dimensional polymeric resonators from holographic photopolymerization: A new approach for active photonic bandgap materials," *Adv. Mater.*, 2003, vol. 15, 241-244.
[51] Mach, P; Wiltzius, P; Megens, M; Weitz, D; Lin, K; Lubensky, T; Yodh, A. "Electro-optic response and switchable Bragg diffraction for liquid crystals in colloid-templated materials," *Phys. Rev. E*, 2002, vol. 65, 31720-31723.

[52] Jakubiak, R; Tondiglia, VP; Natarajan, LV; Sutherland, RL; Lloyd, P; Bunning, TJ; Vaia, RA. "Dynamic lasing from all-organic two-dimensional photonic crystals," *Adv. Mater.*, 2005, vol. 17, 2807-2811.
[53] Zheng, J; Ye, Z; Wang, X; Liu, D. "Analytical solution for band-gap structures in photonic crystal with sinusoidal period," *Phys. Lett. A*, 2004, vol. 321, 120-126.
[54] Samokhvalova, K; Chen, C; Qian, B. "Analytical and numerical calculations of the dispersion characteristics of two-dimensional dielectric photonic band gap structures," *J. Appl. Phys.*, 2006, vol. 99, 063104.
[55] Nusinsky, I; Hardy, A. "Approximate analysis of two-dimensional photonic crystals with rectangular geometry. I. E polarization," *J. Opt. Soc. Am. B*, 2008, vol. 25, 1135-1143.
[56] Joannopoulos, JD; Johnson, SG; Winn, JN; Meade, RD. *Photonic Crystals: Molding the Flow of Light*, 2nd ed. Princeton, NJ:Princeton Univ. Press, 2008, 49-52.
[57] Ambartsumyan, RV; Basov, NG; Kryukov, PG; Letokhov, VS. IEEE J. Quantum Electron. QE-2, 442 (1966), V. S. Letokhov, Sov. *Phys. JETP*, 26, 835, 1968.
[58] Lawandy, N; Balachandran, R; Gomes, A; Sauvain, E. *Nature* (London), 368, 436 1994.
[59] Wiersma, D; van Albada, M; Lagendijk, A. *Phys. Rev. Lett*, 1995, 75, 1739.
[60] Cao, H; Zhao, YG; Ong, HC; Ho, ST; Dai, JY; Wu, JY; Chang, RPH. *Appl. Phys. Lett*, 73, 3656, 1998, *Phys. Rev. Lett*, 1999, 82, 2278.
[61] Jiang, XY; Soukoulis, CM. *Phys. Rev. Lett*, 2000, 85, 70.
[62] Wiersma, D; Cavalieri, *S. Nature* (London), 2001, 414, 708-709. A. Rose, Zhengguo Zhu, Conor F. Madigan, Timothy M. Swager and Vladimir Bulovic, *Nature* (London), 2005, 434, 876.
[63] Hanken, E. Türeci, Li Ge, Stefan Rotter, A.Douglas Stone, *Science*, 2008, 320, 643.
[64] Cao, H. *Waves Random Media*, 2003, 13, R1 2003.
[65] Wiersma, DS. *Nature Physics*, 2008, 4, 359.
[66] Milner, V; Genack, AZ. *Phys. Rev. Lett*, 2005, 94, 073901.
[67] Fallert, J; Dietz, R; Sartor, J; Schneider, D; Klingshirn, C; Kalt, H. *Nature Photon*, 2009, 3, 279.
[68] Frolov, SV; Vardeny, ZV; Yoshino, K; Zakhidov, A; Baughman, RH. *Phys. Rev., B*, 1999, 59, R5284.
[69] Ling, Y; Cao, H; Burin, AL; Ratner, MA; Liu, X; Chang, RPH. *Phys. Rev. A*, 2001, 64, 063808.
[70] Mujumdar, S; Ricci, M; Torre, R; Wiersma, D. *Phys. Rev. Lett*, 2004, 93, 053903.
[71] Polson, RC; Vardeny, ZV. *Phys. Rev. B*, 2005, 71, 045205.
[72] Wu, X; Fang, W; Yamilov, A; Chabanov, A; Asatryan, A; Botten, L; Cao, H. *Phys. Rev. A*, 2006, 74, 53812.
[73] Jiang, X; Soukoulis, CM. *Phys. Rev. E*, 2002, 65, 025601.
[74] Apalkov, VM; Raikh, ME; Shapiro, B. *Phys. Rev. Lett*, 2002, 89, 016802.
[75] Florescu, L; John, S. Phys. *Rev. Lett*, 2004, 93, 13602.
[76] Deych, L. *Phys. Rev. Lett*, 2005, 95, 043902.
[77] Vanneste, C; Sebbah, P; Cao, H. *Phys. Rev. Lett*, 2007, 98, 143902.
[78] Cao, H; Xu, JY; Chang, SH; Ho, ST. *Phys. Rev. E*, 2000, 61 1985.

In: Photonic Crystals
Editor: Venla E. Laine, pp. 67-93

ISBN: 978-1-61668-953-7

Chapter 3

PHOTONIC CRYSTALS FOR MICROWAVE APPLICATIONS

Yoshihiro Kokubo
Graduate School of Engineering, University of Hyogo Himeji-shi, Hyogo Japan.

Abstract

This chapter introduces photonic crystals for microwave, millimeter wave and submillimeter wave applications. First, we present a new type of photonic crystal waveguide for millimeter or submillimeter waves which consists of arrayed hollow dielectric rods. The attenuation constant of the waveguide is smaller than that of a Nonradiative Dielectric (NRD) Waveguide.

Second, metallic photonic crystals consisting of metamaterials are designed. Metamaterials are artificially-constructed materials whose permittivity and/or permeability can be negative. A hollow metal rod with a slit in its sidewall is a metamaterial which plays the role of an LC resonator. In addition, an array of such metal rods can behave as a two-dimensional photonic crystal. Conventional metallic photonic crystals have a photonic bandgap for electromagnetic waves with E-polarization (where the electric field is parallel to the axis of the rods). On the other hand, metamaterials consisting of hollow metal rods with sidewall slits have a bandgap for electromagnetic waves with H-polarization (where the electric field is perpendicular to the axis of the rods) because of their negative permeability. Eigenvalues can be calculated using the two-dimensional finite-difference time-domain (FDTD) method, and metallic metamaterial-based photonic crystals can be designed with independently customized bandgaps for E- and H-polarizations.

Third, we analyze propagation losses in metallic photonic crystal waveguides. A conventional metallic waveguide structure whose E-planes are replaced by metal wires can easily confine electromagnetic waves because their electric field direction is parallel to the wires. In this section, the attenuation constants of metallic photonic crystal waveguides, whose E-planes are replaced by arrayed metal rods, are calculated and compared with those of conventional metallic waveguides. Losses are calculated using another FDTD method. The propagation losses are found to decrease when the radii of the rods (relative to the lattice constant) is increased. The bandwidth can be extended by widening the waveguide relative to the lattice constant.

Fourth, we investigate a metallic waveguide in which in-line dielectric rods are inserted. In such a situation, the electromagnetic field is equivalent to that produced by a two-dimensional photonic crystal with a lattice constant in one direction equal to the inter-rod spacing, and in the other direction equal to twice the distance between the rods and the sidewalls. The calculations of the fields are performed using the FDTD method. Our results showed that the frequency range of a single propagation mode in the waveguide can be extended.

Finally, arrayed dielectric rods in a metallic waveguide are known to change the modes and of course the group velocities of electromagnetic waves. In particular, we focus on TE_{10} to TE_{n0} (where n equals 2 or 3) mode converters and the dependence of their behavior on the frequency range. Conventional metallic waveguides only allow propagation at frequencies above the cutoff frequency f_c. The TE_{20} mode, however, is also possible for frequencies higher than $2 f_c$. In this investigation, a mode converter is proposed which passes the TE_{10} mode for frequencies less than twice the cutoff frequency, and converts the TE_{10} to the TE_{20} mode for frequencies higher than twice the cutoff frequency; this is achieved by small variations of the group velocity.

1. Introduction

Photonic crystals (PCs) are periodic optical structures that are designed to control the motion of photons. PCs are generally designed to be used at optical frequencies. However, they also have the potential to be used at microwave or millimeter-wave frequencies.

There are currently no satisfactory waveguides for millimeter- and submillimeter-wave frequencies, because conventional waveguides suffer from dielectric and metallic losses. The few waveguides used at these frequencies suffer from high propagation losses. Nonradiative dielectric (NRD) waveguides [1] are used as waveguides for millimeter-wave frequencies and they do not have bending losses; however, their losses in a straight line are high. A waveguide is required with low propagation losses in these frequency ranges.

PCs consisting of two-dimensional (2D) square lattices of rods have no photonic bandgaps for H-polarization (when the magnetic field is parallel to the rods) unless the rod radius is sufficiently high. Another problem is that the characteristics of conventional photonic crystals change for E-polarization (when the electric field is parallel to the rods) when the characteristics for H-polarization are changed. It would be convenient if the bandgap of a PC for H-polarization is independent of that for E-polarization. Metamaterials are artificial materials that have characteristics that differ from those of natural-occurring materials and which can be controlled by varying the configurations of conventional materials. Metamaterials have negative permeabilities and so electromagnetic waves cannot propagate in them. Using metamaterials, it may be possible to produce PCs that have novel characteristics.

A metallic waveguide structure whose E-planes are replaced by metal wires can easily confine electromagnetic waves because the electric field direction is parallel to the wires. In this section, the attenuation constants of metallic photonic crystal (MPC) waveguides whose E-planes are replaced with metallic rod arrays are calculated and compared with those of conventional metallic waveguides.

Metallic waveguides have several major advantages, including low propagation losses and high power transmissions at microwave frequencies. However, they have the disadvantage that the propagation frequency band is limited above the cutoff frequency, f_c. The usable frequency range is restricted to $f_c < f < 2 f_c$ since the TE_{20} mode can exist only at frequencies higher than $2 f_c$ in rectangular metallic waveguides. A ridge waveguide [2] has a larger propagating frequency range because it has a lower cutoff frequency for the TE_{10} mode. However, it has the disadvantage of a high attenuation constant. PCs have bandgaps in which electromagnetic waves cannot propagate. This characteristic can prevent the TE_{10} mode propagating and permit only the TE_{20} mode to propagate at frequencies higher than $2f_c$ in a metallic waveguide [3].

Power dividers and power combiners may be easily setup using mode converters. For example, a TE_{10}–TE_{30} mode converter easily offers a three-port power divider, and a three-way power combiner can be composed by reversal. A power combiner is useful for application to Gunn diodes in a waveguide array, because it converts the TE_{30} mode to the TE_{10} mode. We have already proposed a mode converter that allows the TE_{10} mode to propagate at low frequencies and efficiently converts the TE_{10} mode into the TE_{20} mode at high frequencies [4]. In this section, a new mode converter is proposed which passes through the TE_{10} mode for the low frequency range and efficiently converts TE_{30} to TE_{10} mode for the high frequency range.

2. A New Type of Photonic Crystal Waveguide for Millimeter-Wave Frequencies

Photonic crystals (PCs), which are photonic bandgap materials, are theoretically predicted to confine and guide light along a channel without bending loss [5]. Waveguides made from PCs are useful optical waveguides since propagation losses may be neglected at optical frequencies due to the small sizes of silicon wafers. However, PC waveguides will have high losses at millimeter-wave frequencies if the same structures are used as those at optical frequencies. We have investigated PC waveguides that have a triangular lattice of hollow dielectric posts and that have lower losses than NRD waveguides at millimeter-wave frequencies.

2.1. Structure of PC Waveguide

We first need to design a PC structure that can reflect electromagnetic waves. A two-dimensional PC structure consisting of a triangular lattice with aligned hollow dielectric posts is suitable for waveguides because it is easier to fabricate a waveguide system using such a structure than with the conventional structure of a dielectric slab containing many holes. Calculations based on the plane wave expansion method predict that the dielectric constant ε_r needs to be greater than 15 for there to be a bandgap in a triangular lattice of hollow dielectric posts for both H-polarization (where the electric field is perpendicular to the posts) and E-polarization (where the electric field is parallel to the posts) at the same frequency when $\omega a/2\pi c < 1$ and the mode number ≤ 10 [6]. In this section, we assume that $\varepsilon_r = 24$ (for example, $LaAlO_3$, $\tan\delta = 3\times10^{-4}$ at 10 GHz). Figure 1 shows the bandgap map of a triangular lattice with hollow dielectric posts. The external radius r_1 was fixed to $a/2$ (where a is the lattice constant) to ensure that the posts are in contact with each other.

The propagation mode in the PC waveguide must be a hybrid mode that has field components in all directions. Since hybrid modes cannot be separated into H- and E-polarizations, a bandgap for either H- or E-polarization prevents electromagnetic waves from propagating in the PC. If we fix the normalized internal radius r_2/a to 0.4, the first bandgap lies between $0.23 < \omega a/2\pi c < 0.37$ for H-polarization.

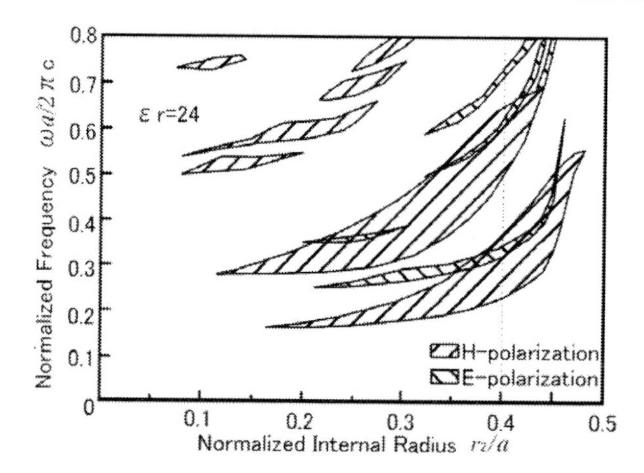

Figure 1. The bandgap map of a triangular lattice with hollow dielectric posts. a represents the lattice constant and r_2 is the internal radius (external radius $r_1 = a/2$)

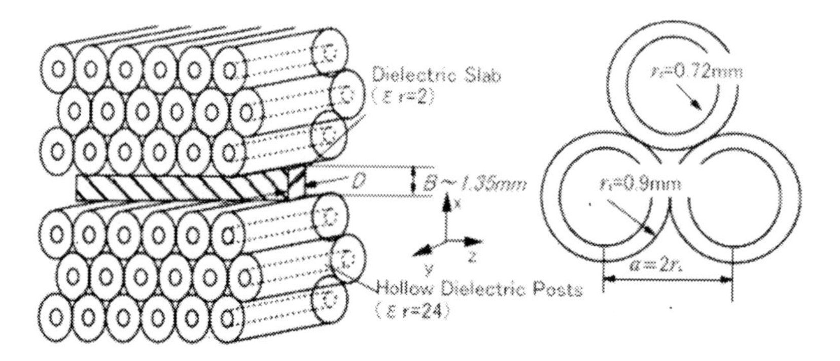

Figure 2. The structure of PC waveguide for 50 GHz. One row of posts has been replaced with a dielectric slab

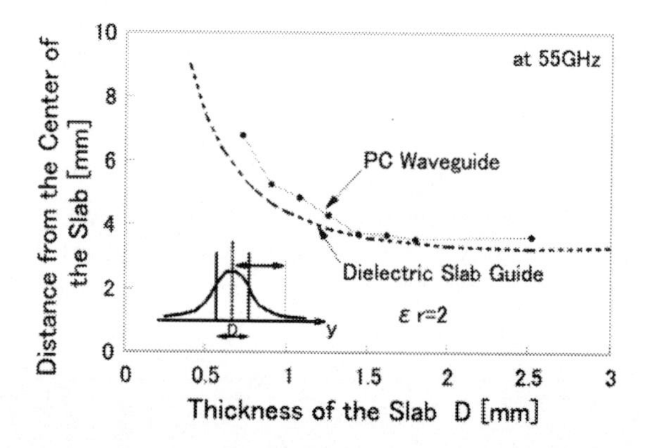

Figure 3. Electric fields in a dielectric slab guide and a PC waveguide. The spreading width is defined as the distance y from the center of the slab to the point at which the intensity of the electric field is one tenth that of the peak

We need to determine the structure of the waveguide for millimeter-wave frequencies after the structure of the PC has been fixed. Of the many structures consisting of dielectric slabs, PCs and metals that we have investigated, the best structure appears to be a slab between PCs. The primary electric field in the slab is in the x direction and only a dominant mode can propagate.

An electromagnetic wave at the center of the slab decays exponentially in the y direction. This spread in the electromagnetic wave depends on the slab width D, making it difficult to determine the length of the posts. We compare the spread of the electric field in the PC waveguide and in a conventional dielectric slab guide when the electric field is parallel to the slab. Figure 3 shows the distance from the center at which the intensity of the electric field is one tenth that of the peak at 50 GHz. The spread in the electric field in the dielectric slab guides is around 4 mm when the slabs are 1.5 mm to 3 mm thick, but it increases drastically when the thickness is less than 1 mm. The electric field in the PC waveguide spreads in a similar manner but it is slightly wider than that in the dielectric slab guide. Since the length of the posts needs to be three to four times longer than the spreading width, we fixed the length of the posts to 28 mm for D greater than 1 mm.

2.2. Propagation Loss

2.2.1. Dielectric and Metallic Losses

Generally, the attenuation constant of a waveguide is given by

$$\alpha = (P_{l,D}+P_{l,M})/2P \tag{1}$$

if the losses are small. $P_{l,D}$ and $P_{l,M}$ are the dielectric and metallic losses, respectively. P represents the travelling wave power. The minimum propagation losses of an NRD waveguide would be 1.7 dB/m (f = 30 GHz), 3.3 dB/m (f = 50 GHz) and 7.8 dB/m (f = 100 GHz) when the dielectric constant of the slab $\varepsilon_r = 2$ and $\tan\delta = 2\times10^{-4}$ when the higher LSM and LSE modes are cutoff. P, $P_{l,D}$ and $P_{l,M}$ are given by

$$P = \frac{1}{2}\int_S (E\times H^*)_z\, dS = \frac{1}{2}\iint_{x,y}(E_x H_y^* - E_y H_x^*)dxdy \tag{2}$$

$$\begin{aligned} P_{l,D} &= \frac{1}{2}\omega\int_V \varepsilon\cdot\tan\delta\cdot|E|^2\, dV \\ &= \frac{1}{2}\omega\varepsilon_0\int_V \varepsilon_r\cdot\tan\delta\cdot\left(|E_x|^2+|E_y|^2+|E_z|^2\right)\cdot dxdydz \end{aligned} \tag{3}$$

$$\begin{aligned} P_{l,M} &= \frac{1}{2\delta_c\sigma}\iint_S |H_t|^2\, dS \\ &= \frac{1}{2\delta_c\sigma}\left(\underset{\substack{y,z \\ \text{at side walls}}}{\iint}\left(|H_y|^2+|H_z|^2\right)dydz + \underset{\substack{x,z \\ \text{at top and bottom walls}}}{\iint}\left(|H_x|^2+|H_z|^2\right)dxdz\right) \end{aligned} \tag{4}$$

where σ represents the conductivity of the guided walls and δ_c is their skin depth.

Losses of PC waveguides are calculated using the three-dimensional FDTD method [7]. In this case, P and $P_{1,D}$ are given by

$$P = \frac{1}{2}\Delta x \cdot \Delta y \cdot \sum_{x=0}^{w}\sum_{y=0}^{h}\Big(E_x(x,y,z)H_y(x,y,z) - E_y(x,y,z)H_x(x,y,z)\Big) \tag{5}$$

$$P_{1,D} = \frac{1}{2}\omega\varepsilon_0 \cdot \Delta x \cdot \Delta y \cdot \Delta z \cdot \sum_{x=0}^{w}\sum_{y=0}^{h}\sum_{z=0}^{unitlength} \varepsilon_r \tan\delta \times \Big(E_x^2(x,y,z) + E_y^2(x,y,z) + E_z^2(x,y,z)\Big) \tag{6}$$

where w is the waveguide width (x direction), h is its height (y direction), and $E(x,y,z)$ and $H(x,y,z)$ are the maximum values of the electric and magnetic fields over time at each point, respectively.

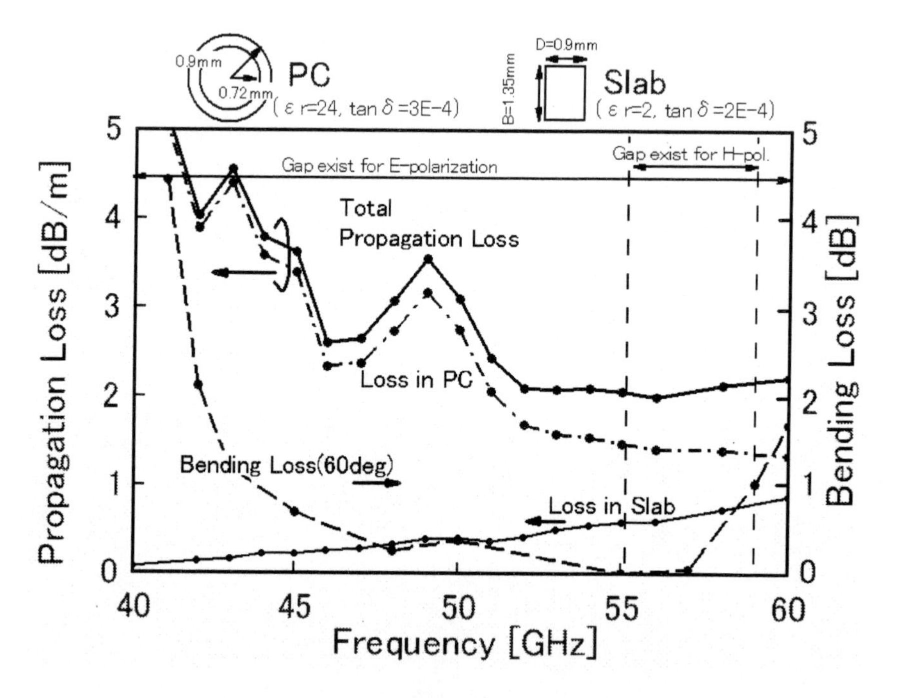

Figure 4. Calculation results for propagation loss.

Because electromagnetic waves in a periodic structure essentially generate reflected waves, it may be difficult to directly determine the net travelling wave power using Eq. (5). However, since the time-averaged power flow must be same at any cross section in the waveguide in the steady state, the net travelling wave power P is given by

$$P = \Delta x \cdot \Delta y \cdot \frac{1}{N}\sum_{t=0}^{T}\sum_{x=0}^{w}\sum_{y=0}^{h}\Big(E_x(x,y,z,t)H_y(x,y,z,t) - E_y(x,y,z,t)H_x(x,y,z,t)\Big) \tag{7}$$

where $E(x,y,z,t)$ and $H(x,y,z,t)$ represent the intensities of the electric and magnetic fields at each time respectively, and T is the period for one cycle and N is the time step of one cycle.

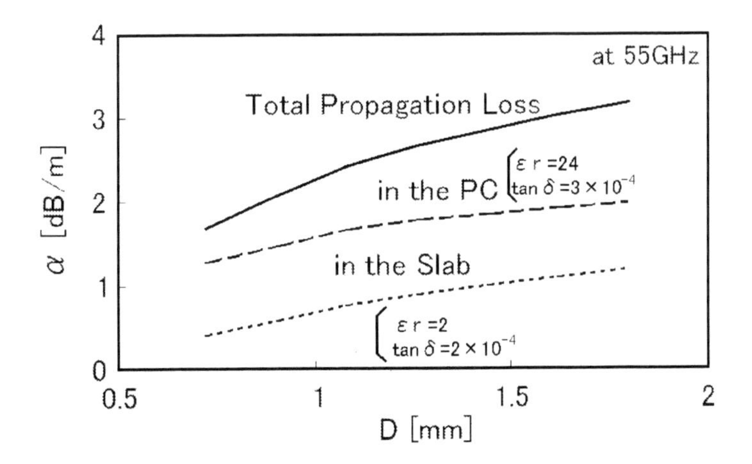

Figure 5. Attenuation constant versus slab width D.

Figure 4 shows the attenuation constants of the PC waveguide around 50 GHz. Dielectric losses in the PC layers are relatively large compared with those in the dielectric slab and have minimum values near 46-47 GHz and 56-57 GHz. Because electromagnetic waves do not penetrate deeply into the PC at frequencies near the middle of the bandgap, the dielectric loss generally tends to be small. A bandgap exists between 38.3 GHz and 61.8 GHz for H-polarization and between 55.1 GHz and 58.1 GHz for E-polarization. Consequently, there are two minimum propagation losses in the PC waveguide.

Attenuation constants for various slab widths D at 50 GHz are calculated and shown in Figure 5 since they depend on the expanse of the propagating wave. As expected, the attenuation constants for both regions are enhanced when D is increased.

2.2.2. Bending Loss

We also calculate the bending loss of the PC waveguide because many studies have found that a conventional PC waveguide suffers from reflection at bending corners. According to calculations using the FDTD method, the bending loss is 1−1.5 dB at a 60-degree bend at frequencies between 45–58 GHz where a bandgap exists for H- and E-polarizations (see Figure 4). The difference between the bending loss and the power reflection may due to scattering because of the limited length of the posts. The length of the posts is determined from calculation results for a straight guide. Therefore, we can calculate the electric field at a bending corner to determine whether the posts are sufficiently long.

Figure 6 shows the electric field E_x at $x = 0$ (center of the slab). E_x decays exponentially to the outside of the slab in straight sections at some times. On the other hand, E_x at a bending corner tends to decrease due to vibrations to the outside of the slab. We set the length of the posts to 28 mm, this may be long enough at a bending corner.

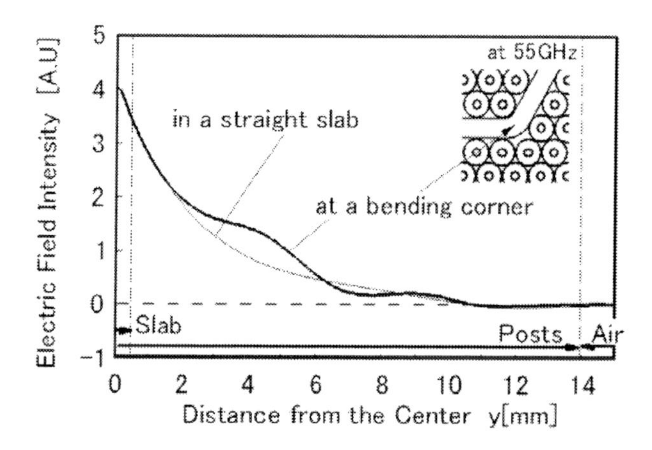

Figure 6. Comparison of electric field (E_x) distributions in a straight slab and at a bending corner (D = 0.90 mm).

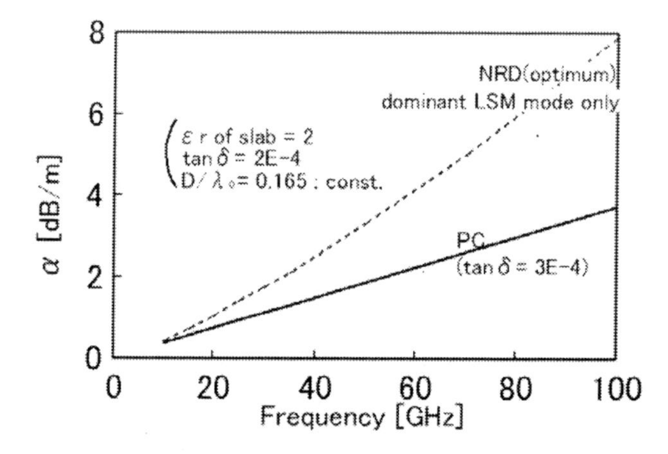

Figure 7. Attenuation constants of a PC waveguide and an NRD waveguide versus frequency.

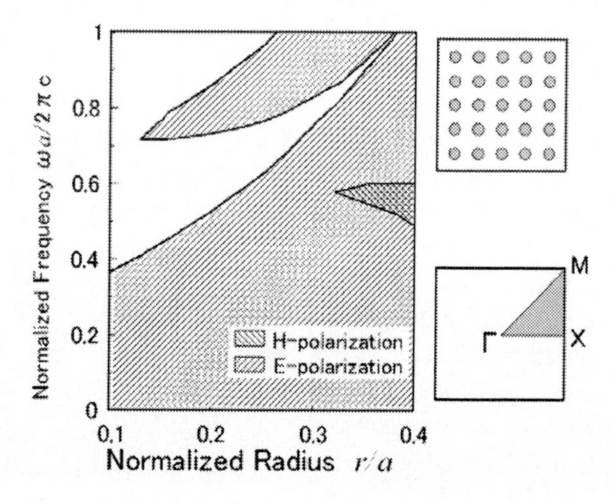

Figure 8. Gap map of arrayed metal column arranged in two dimensional square lattice.

2.2.3. Comparison of Loss Frequency Characteristics of Propagation and NRD Waveguide

Figure 7 shows the minimum attenuation constants of an NRD waveguide at millimeter-wave frequency when $\varepsilon_r = 2$, $\tan\delta = 2\times10^{-4}$ and each dimension is varied with frequency when higher LSM modes cutoff. The attenuation constants increase superlinearly because of metallic losses. Figure 7 also shows the attenuation constants of a PC waveguide given by Eq. (3). If it is assumed that $\varepsilon_r = 2$, $\tan\delta = 2\times10^{-4}$ and $D = 0.165\lambda$ for the slab and $a = 0.33\ \lambda$ for the PC waveguide, the attenuation constants will be 1.1 dB/m ($f = 30$ GHz, $D = 1.65$ mm, $a = 3.3$ mm), 2.1 dB/m ($f = 55$ GHz, $D = 0.9$ mm, $a = 1.8$ mm) and 3.8 dB/m ($f = 100$ GHz, $D = 0.495$ mm, $a = 0.99$ mm). The PC waveguides have lower losses than NRD waveguides in microwave and millimeter-wave frequency ranges.

3. Two-Dimensional Photonic Crystals Using Metamaterials

Photonic crystals arranged in two-dimensional (2D) square lattices have no photonic bandgaps for H-polarization unless they have sufficient column radii as shown in Figure 8. In addition, the characteristics of conventional photonic crystals are changed for E-polarization when the characteristics for H-polarizations are changed.

3.1. Metamaterials

Metamaterials are artificial materials with characteristics that are not available in natural materials and which can be controlled by changing the configurations of conventional materials. One of the metamaterials is the split-ring column structure shown in Figures 9 and 10, which has radius r, lattice constant a, ring thickness t, and ring gap d_g.

After application of a magnetic field H_0 along the axis of the column, resonation occurs from the current in the ring and ring gap. The effective permeability μ, for example, is varied with respect to the frequency, shown in Figure 11. If the frequency dependence of μ follows the Drude model, then a bandgap is generated at $\mu<0$ between the resonant frequency f_0 (lower limit) and the magnetic plasma frequency f_{mp} (upper limit). The resonant frequency f_0 shifts to lower frequency for a larger ring radius, due to the larger inductance. The magnetic plasma frequency f_{mp} is given by the following equation [9].

$$f_{mp} = f_0\sqrt{\frac{1}{F}}$$

Where F is the total cross-sectional area without the column area, and

$$F = 1 - \frac{\pi r^2}{a^2}$$

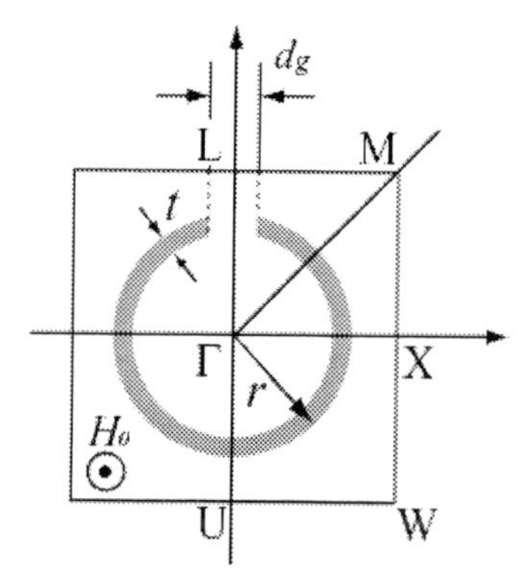

Figure 9. Split-ring structure. [8].

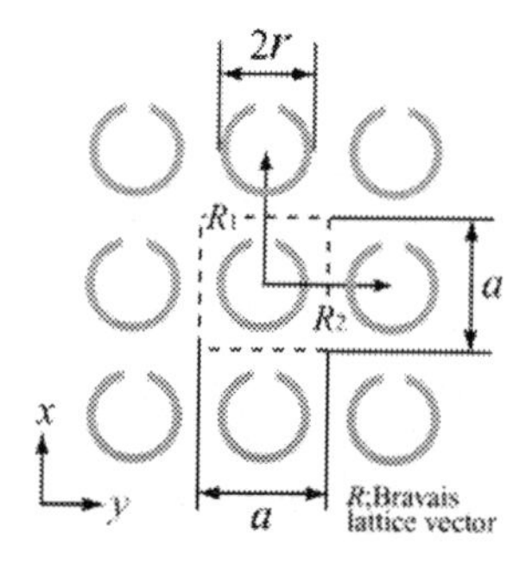

Figure 10. Unit cell of the split-ring structure (region enclosed with the dotted line) [8].

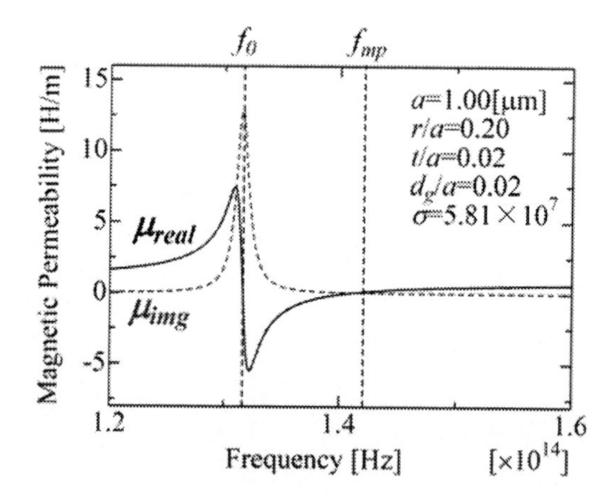

Figure 11. Effective permeability [8].

Accordingly, f_{mp} increases with the radius of the ring. If the electrical resistance is equal to zero, then the effective permeability tends toward positive infinity or negative infinity near f_0. The characteristics of metamaterials are usually regarded as spatial average values, because the size of the metamaterial unit and the distance between units are both very small compared to the wavelength of an electromagnetic wave. However, in this investigation, the spatial period of the metamaterial is approximately equal to the wavelength of an electromagnetic wave and it is therefore regarded as a photonic crystal.

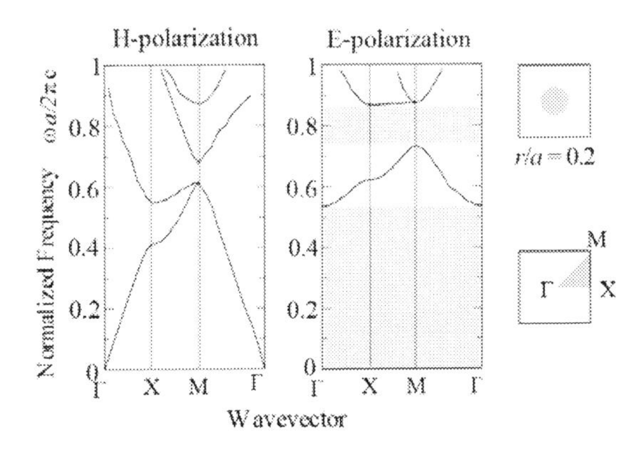

Figure 12. Band diagram of metal columns arranged in a square lattice.

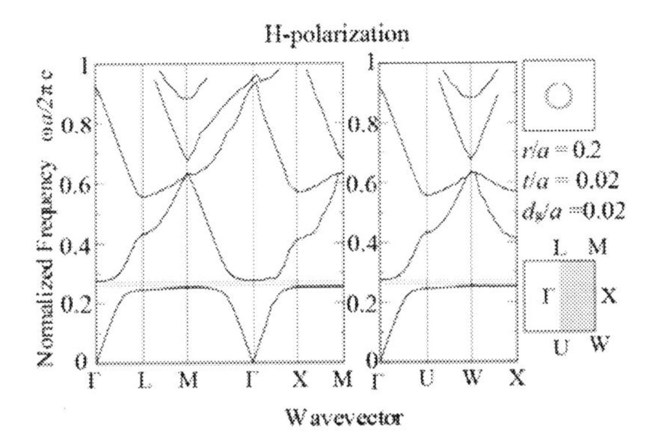

Figure 13. Band diagram for a split-ring type photonic crystal (H-polarization).

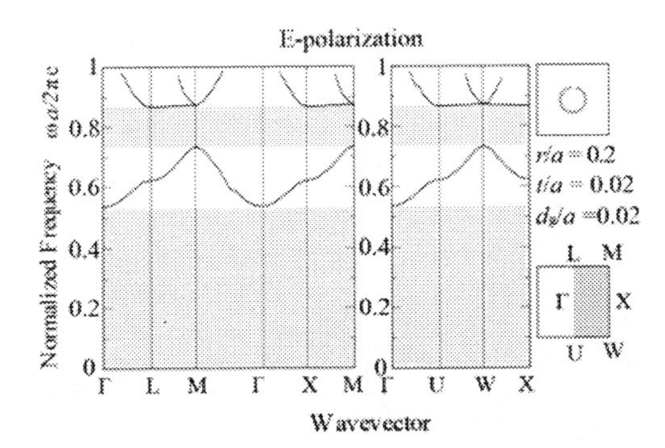

Figure 14. Band diagram for a split-ring type photonic crystal (E-polarization).

3.2. Band Characteristics of Split-Ring Metamaterials

2D metal columns arranged in a square lattice as a photonic crystal have a large photonic bandgap for electromagnetic waves with E-polarization. However, they do not have a bandgap for H-polarization, unless the ring is thick. In addition, bandgaps for E- and H-polarization cannot usually be designed independently. In this section, band diagrams are calculated for a square lattice using the finite difference time domain (FDTD) method [17] for metal split-ring metamaterials. Firstly, a sample of a band diagram for a conventional metal column (r/a=0.20) arranged in a square lattice is shown in Figure 12, where the gray-color area indicates bandgaps. This structure has no bandgap for H-polarization. Band diagrams for photonic crystals comprised of split-ring metamaterials for E- and H-polarization are shown in Figures. 13 and 14 with a normalized radius of rods r/a = 0.20, a normalized thickness t/a = 0.02 and a normalized ring gap d_g/a = 0.02. In case of a conventional metal column, calculations of the wavevector are sufficient only if the area is half quarter of the Brillouin Zone, *i.e.*, ΓXMΓ, due to the symmetry. However, calculations must be performed for the direction along of the critical point ΓLMXWUΓ for this metamaterial. A bandgap exists for H-polarization between $\omega a/2\pi c$=0.255-0.278, as shown in Figure 13, where the frequencies agree with f_0 and f_{mp}. The band diagrams for a conventional metal column and the metamaterial are almost the same for E-polarization, regardless of change in the inner structures. The bandgaps shift toward lower frequency and widen with increasing radius, as shown in Figure 15.

4. Analysis of Propagation Loss of Metallic Photonic Crystal Waveguides [10]

Electromagnetic waves with a frequency lower than some cutoff cannot propagate through a metal wire mesh. This principle is the basis for photonic crystals composed of a three-dimensional periodic structure of metallic wires [11]. The bandgap is calculated for a negative dielectric constant due to surface plasmon polaritons on the metal wire mesh [12]. The propagation modes in a waveguide whose walls are replaced with metal wires have been calculated [13]. A waveguide structure whose E-planes are replaced by metal wires can easily confine electromagnetic waves because their direction is parallel to the wires. Recently, parallel-plate post-wall waveguides [14] and substrate-integrated waveguides (SIWs) [15] have been constructed, consisting of microstrip lines with only one row of metal wires on each side. A SIW is synthesized by forming two rows of metallic holes in a substrate. The field distribution in a SIW is similar to that in a conventional rectangular waveguide. In this letter, the attenuation constants of metallic photonic crystal (MPC) waveguides, whose E-planes are replaced by arrayed metallic rods, are calculated and compared with those of conventional metallic waveguides.

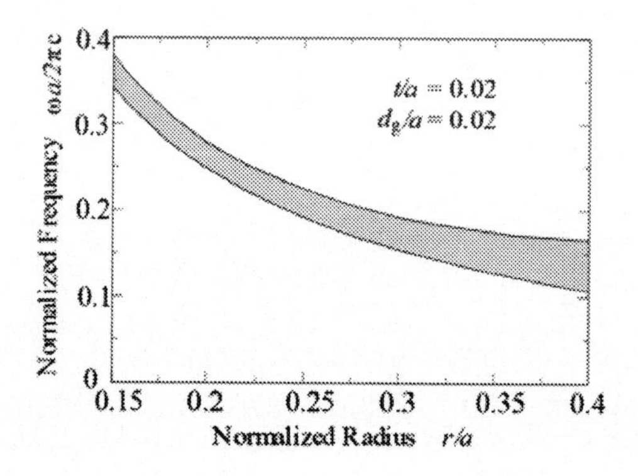

Figure 15. Gap map for various radii of the split-ring structure.

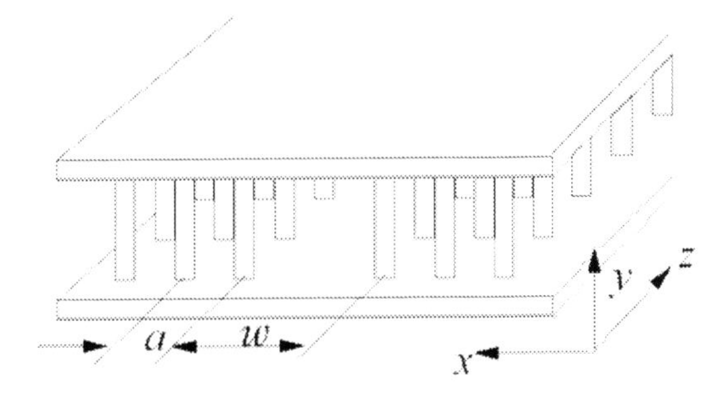

Figure 16. A perspective view of the MPC waveguide. [10]

4.1. Basic Structure

The basic structure of the MPC waveguide is shown in Figure 16 [10]. It is composed of metal rods arranged in a triangular or square lattice. One or more rows are removed and they are sandwiched between parallel metal plates. Consequently, the E-planes have been replaced with MPCs while the H-planes have been left as is. Define *a* as the lattice constant of metal rods and *w* as the gap width between the removed rods. For a triangular lattice, *w* is narrower than that of a square lattice for the same value of *a*.

The field distribution in the waveguide is similar to that in a conventional rectangular waveguide and is uniform along the *y*-axis. Figure 17 shows a bandgap map for E polarization (in which the electric field is parallel to the axis of the metal rods) with varying normalized radius of the rods *r*/*a*. The areas enclosed by the oblique lines represent bandgaps where no propagation modes exist [10]. We fixed the values of *r*/*a* at 0.1, 0.2, and 0.3 for *w*/*a* = 2.5 and removed one row of metal rods from them. The calculation results are shown in Figure 18. The propagation bandwidth is narrower when the normalized radius is reduced because the upper edge of the bandgap decreases. Conversely, a larger normalized radius results in higher modes. Thus, the propagation bandwidth will not become larger than some limiting value.

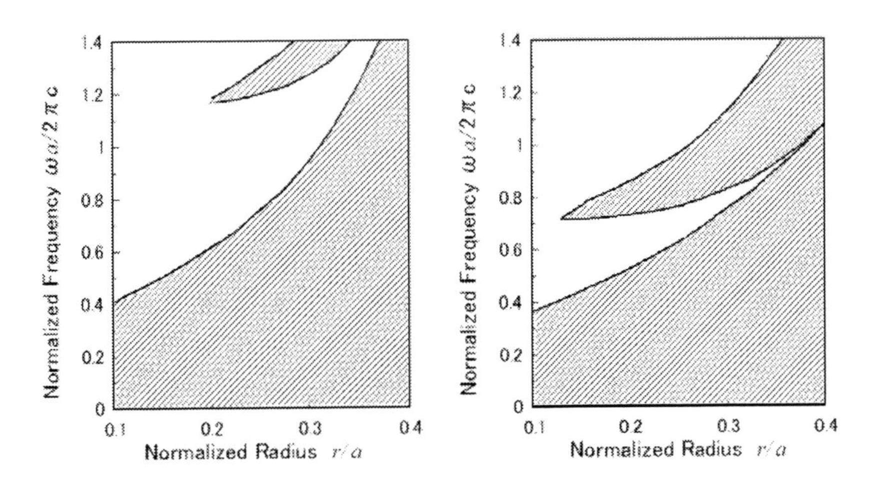

Figure 17. Bandgap map of E-polarized MPC arranged in (a) a triangular lattice, and (b) a square lattice. [10]

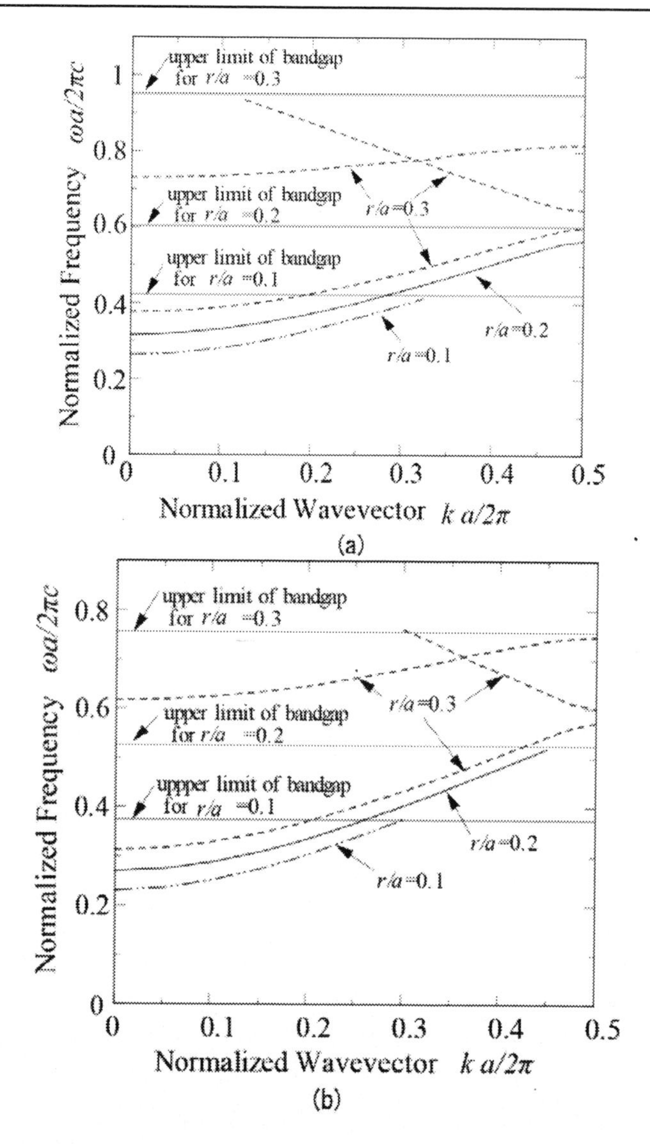

Figure 18. Dispersion of eigenvalues along the guide axis for r/a=0.1, r/a=0.2, r/a=0.3, and w/a =2.5 arranged in (a) a triangular lattice, and (b) a square lattice. [10]

4.2. Attenuation Constant

The attenuation constants of the waveguide with arrayed metal rods were calculated. For comparison, the constants of conventional metallic waveguides were also calculated. The width of both waveguides was set to the same value. However, an electric field only slightly penetrates the row of metal rods, and the width of the MPC waveguide therefore cannot be easily defined. Fixing the cutoff frequency is a better way to compare attenuation constants.

If the losses are small, the attenuation constant of a waveguide is given by Eq. (2) and (5). $P_{l,M}$ are given by Eq. (8) instead of Eq. (4), because the waveguide has metal rods.

$$P_{l,M} = \frac{1}{2\delta_c\sigma}\iint_S |H_t|^2 dS$$

$$= \frac{1}{2\delta_c\sigma}\left(h \cdot \oint_{\text{serface of metal rods}} |H_\theta|^2 d\theta + \iint_{\text{at top and bottom walls}} \left(|H_x|^2 + |H_z|^2\right) dxdz \right) \quad (8)$$

where h represents the height of the waveguide (in the y direction).

The results of the calculation are shown in Figure 19. The standard metallic waveguide is WR-90 (22.4 mm×10.2 mm for which $f_c \approx 6.55$ GHz) with $w/a = 2.5$ and $a = 9.16$ mm.

The attenuation increases drastically above a certain frequency because the electric field penetrates into the MPC array near the edge of the bandgap. The losses at the H-planes are slightly larger than those of conventional waveguides, possibly because the electromagnetic waves repeatedly reflect between the rods during propagation. When the radius of the rods is reduced, the MPC loss grows larger. If the radius of rod is small, the magnetic field penetrates behind the rods and reaches the second row so that the total current flowing at the surface of the rods increases.

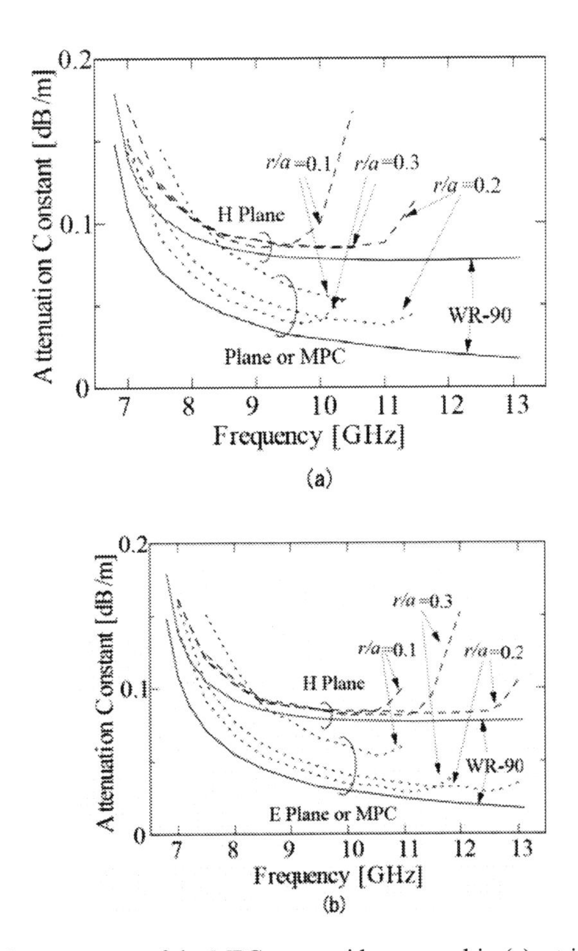

Figure 19. The attenuation constants of the MPC waveguide arranged in (a) a triangular lattice, and (b) a square lattice. [10]

4.3. Varying the Waveguide Width

As shown in Figure 19, the bandwidth of a triangular lattice is narrower than that of a square lattice, because the normalized width w/a of the former waveguide is $\sqrt{3}$ which is narrower than w/a=2 for the latter. Figure 20 shows the eigenvalues of the waveguide with w/a =2, 2.5, 3, and 4 for r/a=0.1. A wider normalized width w/a causes the normalized cutoff frequency to shift down and the bandwidth to widen. If w/a is too large, however, higher modes appear. Increasing w/a for a fixed value of r/a requires reducing r and a while holding the cutoff frequency constant. The attenuation constants of the waveguide for w/a =2, 3, 4 are shown in Figure 21. Widening w/a while simultaneously reducing r and a makes the propagation loss smaller.

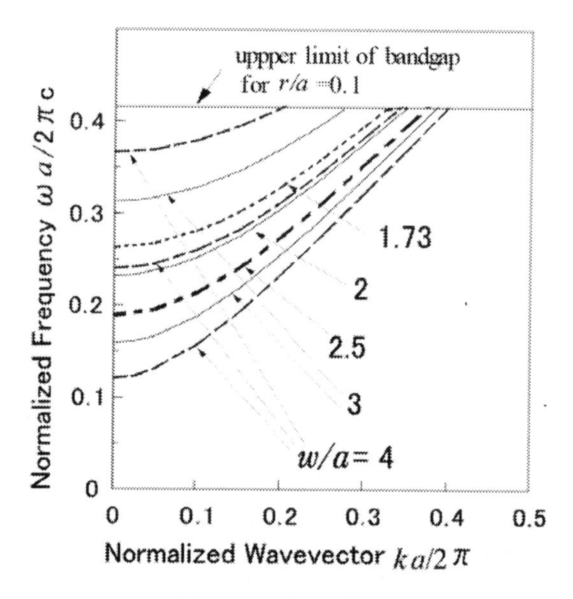

Figure 20. Dispersion of eigenvalues along the guide axis with varying waveguide width for a triangular lattice with r/a = 0.1. [10]

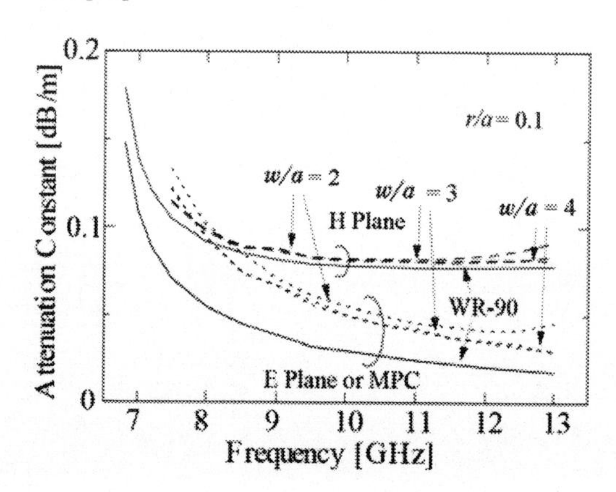

Figure 21. The attenuation constants of the MPC waveguide for various waveguide widths w for a triangular lattice. [10]

5. Wide Band Metallic Waveguide with In-Line Dielectric Rods [3][18]

5.1. Basic Principle

Since the electromagnetic reflection coefficient of a periodic array of dielectric rods has frequency dependence, it is possible that electromagnetic waves will pass through the dual in-line dielectric rods located near the sidewalls at low frequencies, and electromagnetic waves will be reflected between them at high frequencies. As the space between two dielectric arrays is narrower than the space between the metal sidewalls, higher modes are suppressed in the waveguide [19]. We have also suggested a system that has almost the same structure as that in reference [19], but uses a low dielectric constant material to reduce cost [20]. Although an electromagnetic wave is propagated without higher modes above $2f_c$, the total bandwidth of single mode operation is not so wide, because the TE_{10}-like and TE_{20}-like modes are overlapped with each other. If the normalized bandwidth is defined as $W_B = \Delta\omega/\omega_0$ (where $\Delta\omega$ and ω_0 represent bandwidth and center angular frequency, respectively.), then $W_B \approx 0.667$ for the case of conventional rectangular metallic waveguides and $W_B = (\Delta\omega_1+\Delta\omega_2)/\omega_0 \leq 0.75$ for the case of the waveguides reported in references [19] and [20]. Since plural dielectric arrays generally produce plural eigenvalues, the probability of the eigenvalues overlapping with each other may become large. In this paper we have attempted to calculate the eigenvalues for a waveguide with single in-line dielectric rods in order to reduce the probability of overlapping.

The frequency-wavevector (ω-k) characteristics of a conventional rectangular metallic waveguide are shown in Figure 22. The scales of both axes are normalized by the width of the waveguide, w. The eigenvalues of the TE_{10} mode move closer to the asymptote, or so-called air-line ($\omega = ck$), as the frequency increases. The TE_{20} mode can also propagate in the frequency region above $2f_c$. If the TE_{10} mode cannot propagate above $2f_c$, then the single mode propagation frequency range widens, because the TE_{20} mode can only propagate above $2f_c$.

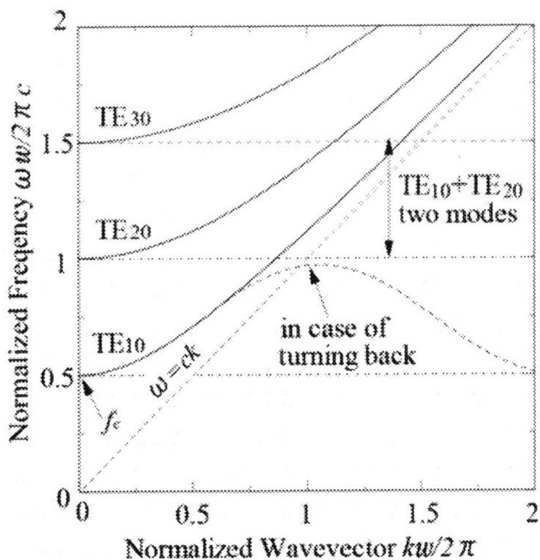

Figure 22. The ω-k diagram for a rectangular waveguide where w represents the width of the waveguide

A stopband occurs for a waveguide with periodic structure or a periodic structure-like photonic crystal [21][22]. If the spatial period is a, and since the eigenvalues turn back as indicated by the dashed line in Figure 22 (at $k = \pi/a$), then the stopband occurs above $\omega > c\pi/a$. If the eigenvalues of the TE_{10} mode turn back at $k = \pi/a$, then single mode propagation is available for the TE_{20} mode above $2f_c$. If we let f be $2f_c$ in ω (= $2\pi f$) = $c\pi/a$, then this means that the width of the waveguide, w, is set to $\lambda_c/2 = (c/f_c)/2 = 2a$. The periodic structure is assumed to be due to the dielectric rods placed at the center of the waveguide, because this will have a strong influence on the TE_{10} mode and will not affect the TE_{20} mode.

5.2. Waveguide Structure

A detailed profile of the wide-band waveguide is shown in Figure 23. An array of circular dielectric rods is set to the center of a metallic waveguide. The dielectric material is assumed to be easily obtainable $LaAlO_3$ (lanthanum aluminate) with a dielectric constant, ε_r, of 24 and a loss tangent, $\tan\delta$, of 3×10^{-4} (at 10 GHz). If the rectangular waveguide is WR-90 (22.9×10.2 mm; $f_c \approx 6.55$ GHz), a is determined as $w/2$ = 22.9 mm/2 = 11.45 mm. The propagation modes for this waveguide are calculated using the supercell approach [16], by applying appropriate periodic Bloch boundary conditions for the unit cell [17]. The supercell profile is also shown in Figure 23.

w is set to $w = \lambda_c/2 = c/f_c/2 = 2.4a$ and $a = 9.54$ mm. The results calculated for $r = 0.054a$ are shown Figure 24 (r represents the radius of the dielectric rod). In this figure, the first band has even symmetry and the second band has odd symmetry, and they are both similar to the TE_{10} and TE_{20} modes in a rectangular metallic waveguide, respectively. The first band is from 6.0 to 13.2 GHz and the second band becomes wider, from 13.2 to 16.8 GHz, and from 17.6 to 21.4 GHz, because the third band is elevated and flattened. The normalized bandwidth of this waveguide $W_B = (\Delta\omega_1+\Delta\omega_2+\Delta\omega_3)/\omega_0 \approx 0.95$.

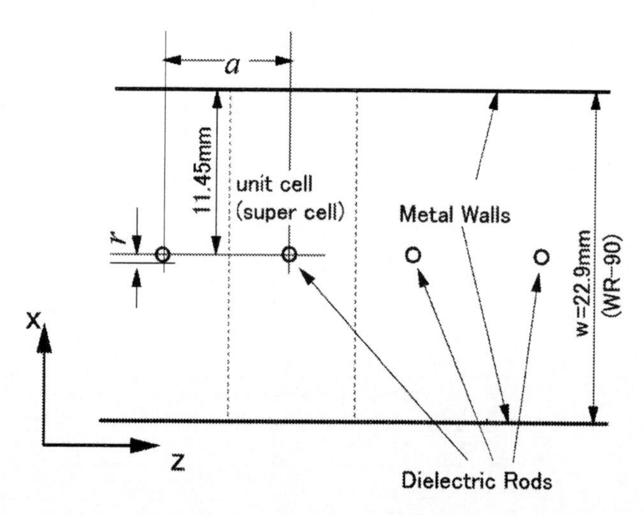

Figure 23. Detailed profile of a WR-90 waveguide in which the dielectric rods are aligned. [3]

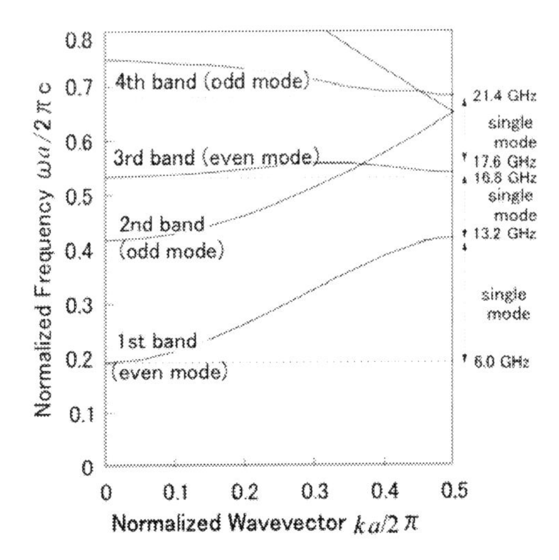

Figure 24. Dispersion of the eigenvalues along the guide axis for ε_r=24, r/a =0.054, and w/a = 2.4. (The right-hand vertical axis shows the single-mode frequency when a is 9.54 mm.) [18]

5.3. Structure of a 90-Degree *H*-Plane Bent Waveguide [18]

When a TE_{10} mode is introduced between 6.0 and 13.2 GHz and a TE_{20} mode at 13.2–16.8 or 17.6–21.4 GHz, single-mode propagation is possible. Dielectric rods are only required at the bent portions because mode conversion does not occur in the straight segments. Reflection, however, will occur at the borders if the dielectric rods are removed along the straight portions. To reduce those reflections, the radius of the rods at the boundaries can be made small.

The 90-degree bent waveguide structure of Ref [23] contains an integral number of dielectric rods. Consequently, rods are located at the corners of the bent portions where they begin or end. In this case, the curvature radius cannot be chosen arbitrarily and one cannot use commercial bent waveguides.

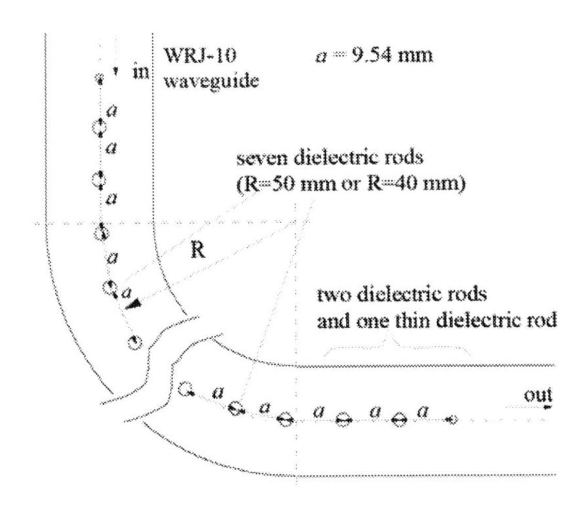

Figure 25. 90-degree *H*-plane bent metallic waveguide.

On the other hand, Figure 24 shows a case in which the radius of curvature R and period a are fixed. Dielectric rods are not necessarily located at the corners of the 90-degree bent portion and three rods are included in the straight portions. The last dielectric rod is thin with $r = 0.36$ mm in order to minimize reflections of the TE_{10}-like mode.

The S parameters of the 90-degree H-plane bent waveguide for R = 40 and 50 mm are plotted as solid lines in Figure 26 using HFSS (high frequency structural simulator) software by Ansoft [24]. These different curvatures (R = 40 mm and R = 50 mm) give similar results. The frequency range is from 8.2 to 10.4 GHz for the first band and from 13.5 to 16.7 and 18.4 to 19.2 GHz for the second band with reflection below –20 dB. The TE_{10} mode transmittances are also shown at the second band. The mode conversions are very small. The TE_{30} cutoff is less than 19.6 GHz.

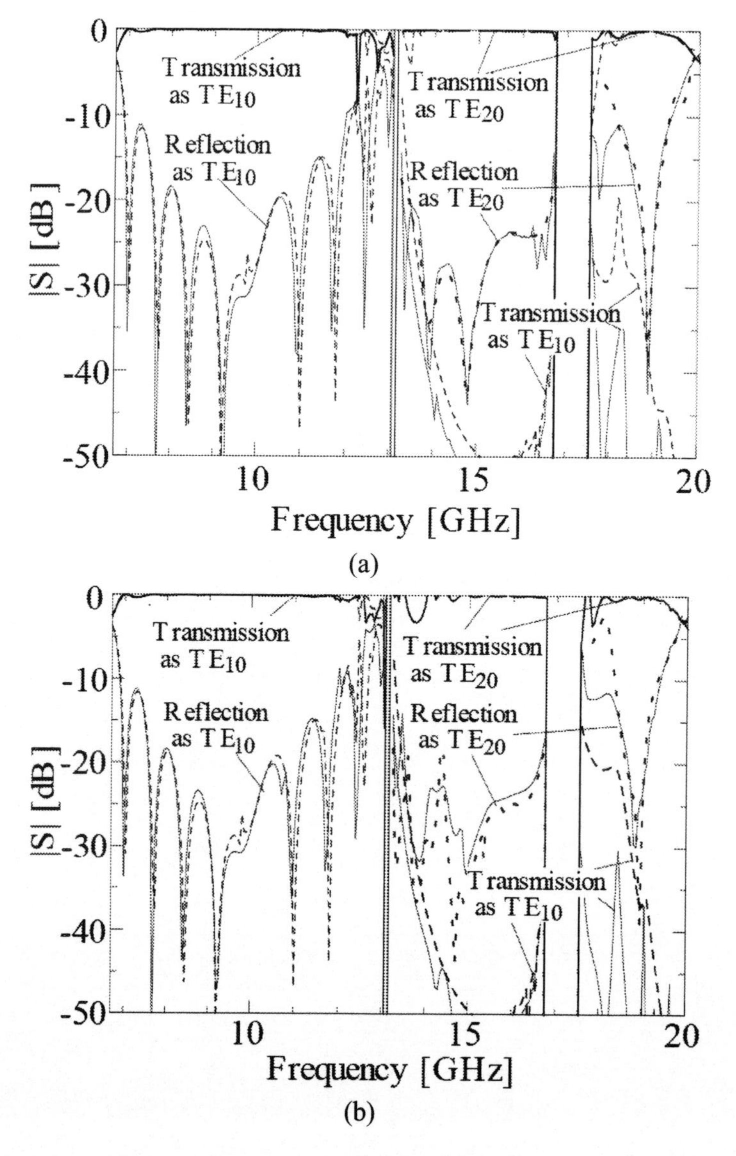

Figure 26. Transmission coefficient |S21| and reflection coefficient |S11| at the corner of a waveguide for (a) R = 40 mm and (b) R = 50 mm. [18]

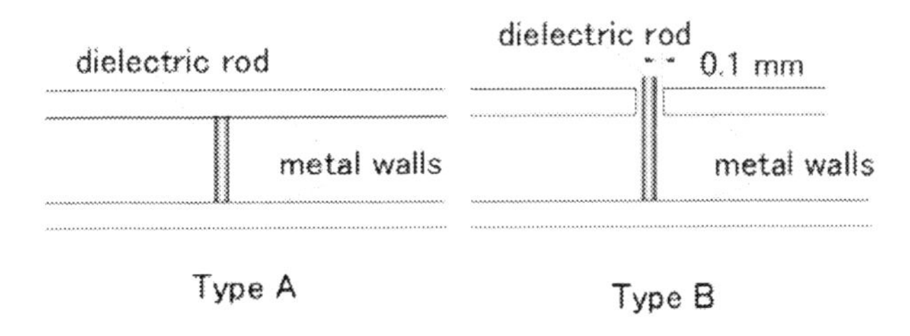

Figure 27 (a) Dielectric rod located in a waveguide that does not have gaps at the top and bottom. (b) Dielectric rod inserted in a hole drilled at the top of the waveguide with a diameter of 0.2 mm that is larger than the diameter of the rod. [18]

5.4. Simple Fabrication Method [18]

It is not trivial to position dielectric rods in a waveguide such as Type A illustrated in Figure 27(a). One must locate the dielectric rods in the waveguide which has no gap at its top or bottom. As a solution, 0.2-mm-diameter holes were drilled into the top of the waveguide and dielectric rods were inserted through them, resulting in Type B in Figure 27(b). The S parameters were calculated using the HFSS software and the results are graphed as dotted lines in Figures. 26(a) and (b). The results for these different structural conditions (solid lines and dotted lines) are similar. The difference between the S parameters of Type A and B is even smaller if the drilled holes are narrower than 0.2 mm.

6. Frequency Range Dependent TE_{30} to TE_{10} Mode Converter

A structure in which two arrays of dielectric rods are setup can convert TE_{30} to TE_{10} mode [25]. We have reported a mode converter which passes through a TE_{10} mode for low frequency range and converts TE_{10} to TE_{20} mode for the high frequency range [4].

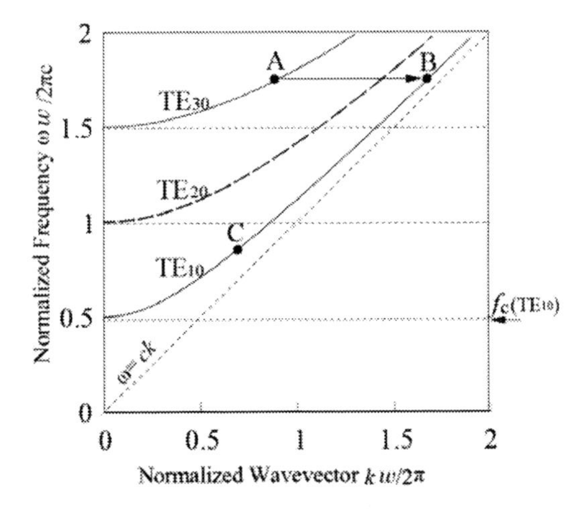

Figure 28. The frequency eigenvalues of a conventional metallic waveguide in a given k wavevector.

Power dividers and power combiners may be easily setup using mode converters. For example, a TE_{10}–TE_{30} mode converter easily offers a three-port power divider, and a three-way power combiner can be composed by reversal. A power combiner is useful for application to Gunn diodes in a waveguide array, because it converts the TE_{30} mode to the TE_{10} mode. In this investigation, a new mode converter is proposed which passes through the TE_{10} mode for the low frequency range and efficiently converts TE_{30} to TE_{10} mode for the high frequency range.

The frequency eigenvalues of a conventional metallic waveguide in a given k wavevector are shown in Figure 28. In this figure, the wavevector k and frequency ω are normalized using the width of the waveguide w. The electromagnetic wave propagates the TE_{10} mode only for $0.5<\omega w/2\pi c<1$, and can propagate TE_{10} and TE_{30} modes for $1.5<\omega w/2\pi c$. If TE_{10} mode is excited by the TE_{30} mode, the group velocity of TE_{30} (A) must be changed to that of TE_{10} (B) for $1.5<\omega w/2\pi c<2$. On the other hand, the group velocity C is not changed for $0.5<\omega w/2\pi c<1$, because this remains in the TE_{10} mode. If the distribution of the transverse electromagnetic field is gradually changed from TE_{30} to TE_{10}, and group velocity A is also gradually changed to B, then the reflection may be reduced for $1.5<\omega w/2\pi c<2$. On the other hand, if the group velocity C is not significantly changed, the reflection may also be suppressed for $0.5<\omega w/2\pi c<1$. Since the mode profile gradually shifts from TE_{30} to TE_{10}, the dielectric rods are replaced from near the center of the waveguide to the sidewall. In other words, the basic setup is shown in Figure 29.

The group velocity is given by $v_g = \frac{1}{\left(dk/d\omega\right)}$. However, it is not simple to determine the group velocity in the waveguide shown in Figure 29. The propagation modes in a waveguide having in-line dielectric rods with period a are calculated using a supercell approach [16] by application of appropriate periodic Bloch conditions at the boundary of the unit cell [17][23]. When the location of the dielectric rods is fixed at a distance d from the sidewall, the group velocity v_g at both of the first and the third bands is changed by varying the radius r. However, the group velocities are also changed at the same time and cannot be changed individually.

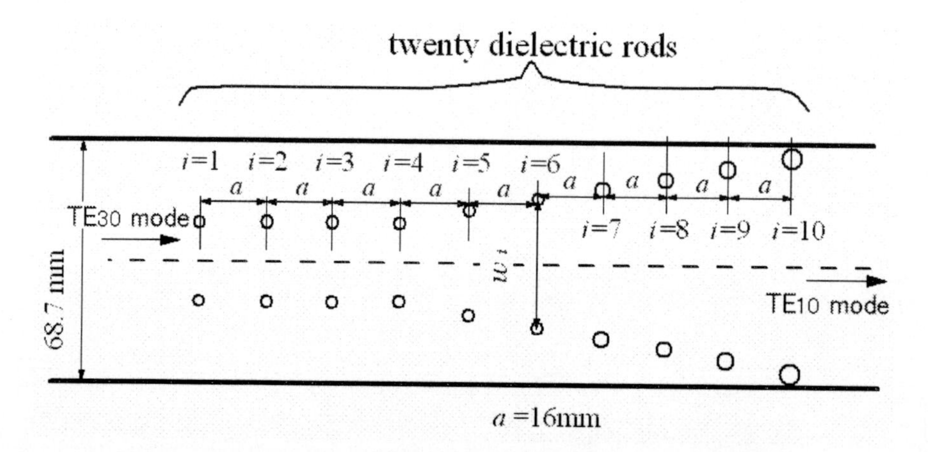

Figure 29. The proposed structure of the TE_{30} to TE_{10} mode converter.

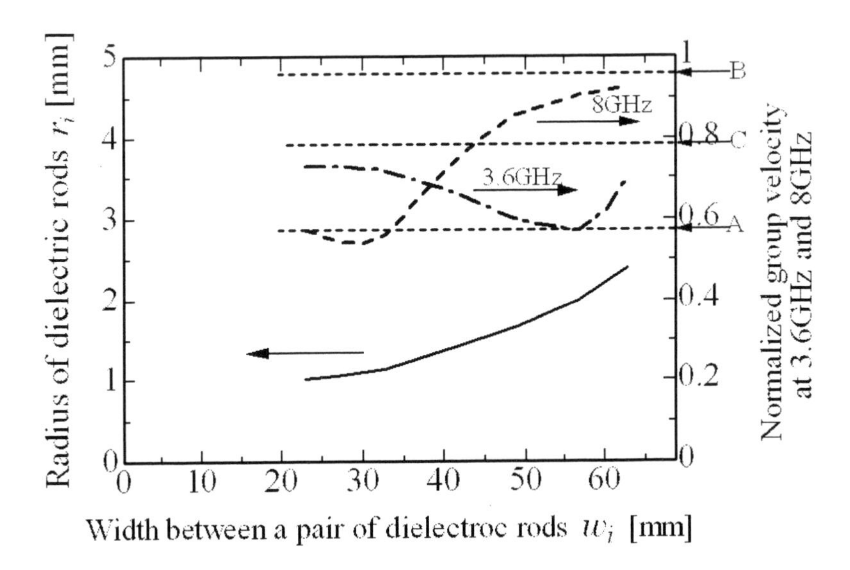

Figure 30. The group velocity (doted lines) in a metallic waveguide having a periodic array of dielectric rods with various distances w_i between the rods and corresponding radii r_i (solid line) of the rods, at 3.6 GHz and 8 GHz.

If the group velocity is normalized using light velocity in a vacuum, v_g/c_0 is the same as the gradient of the characteristic curve. Therefore, when d and r are fixed to certain values, v_g/c_0 is calculated for the periodic structure of the dielectric rods at a specific frequency. If group velocity A is gradually changed to B for $1.5<\omega w/2\pi c<2$ when d is varied, and group velocity C is not changed for $0.5<\omega w/2\pi c<1$, then one unit of each pair of d and r connects to its respective pair to form a structure shown in Figure 29.

The width of the metallic waveguide is assumed to be 68.7 mm, which is three times wider than the WR-90 waveguide (22.9 mm×10.2 mm, $f_c \approx 6.55$ GHz), and period a is fixed at 16 mm. Figure 30 shows a sample of calculated results of normalized velocity along the axis of the waveguide at 3.6 GHz and 8 GHz for dielectric rods ($LaAlO_3$: $\varepsilon_r= 24$, radius r [mm]) aligned at a distance w [mm], between two arrays. It is desirable that the normalized velocity A (TE_{30}: v_g/c_0 =0.574) monotonically decreases to B (TE_{10}: v_g/c_0 =0.962) at 15 GHz and normalized velocity C (TE_{10}: v_g/c_0 =0.795) is not changed at 3.6 GHz. However, at 8 GHz, such a condition is not found around w =30 mm, because the placement of the dielectric rod at the center of the waveguide is the same as that where the electric field becomes a minimum. On the other hand, placement of the dielectric rod near the sidewall of the waveguide is the same as that where the electric field becomes a maximum. At the transition region, around w =30 mm, the characteristics are complex. The design takes priority, in order to not vary the group velocity at 3.6 GHz. Since the group velocity must become slow with dielectric material at 3.6 GHz and becomes slowest at w =55 mm, the design takes priority at 8 GHz for mode conversion, because both 3.6 GHz and 8 GHz conditions cannot be satisfied at the same time. After calculating v_g with pairs of w_i and r_i, if group velocity A is gradually changed to B, then each pair of w_i and r_i are combined, as shown in Figure 29.

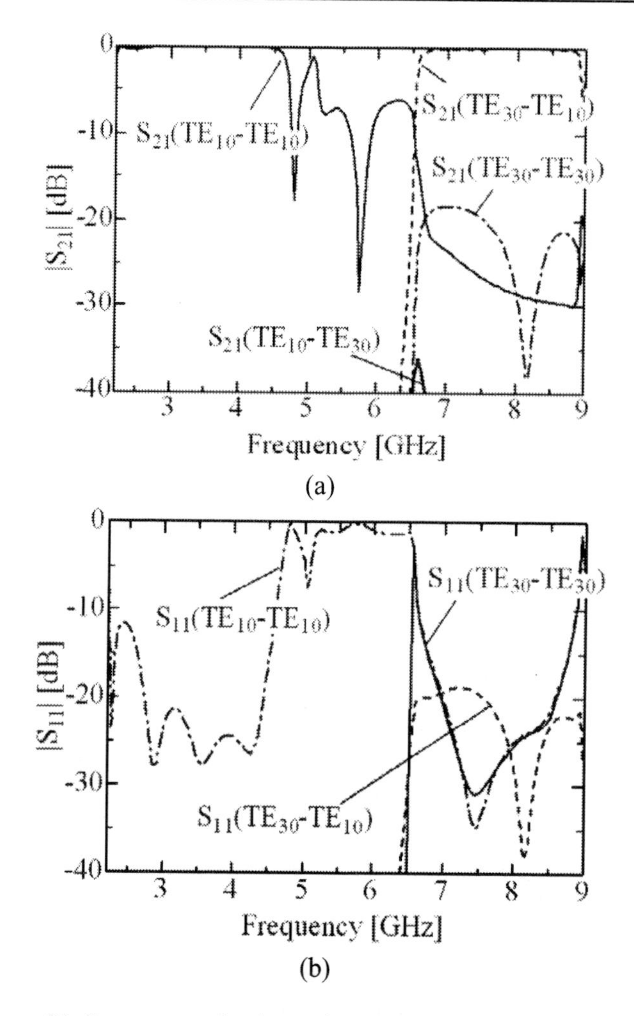

Figure 31. S parameters for the mode converter; (a) $|S_{21}|$ and (b) $|S_{11}|$.

Table 1. Location and radius of the dielectric rods.

Rod Number i	Distance w_i between a pair of the rods [mm]	Radius r_i of the dielectric rod [mm]
1	22.9	0.728
2	22.9	1.03
3	22.9	1.03
4	22.9	1.03
5	26.7	1.06
6	32.7	1.15
7	40.7	1.39
8	48.7	1.65
9	56.7	2.00
10	62.7	2.40

The first pair of rods has $r = 0.728$ mm and the remainder have a half cross-section radius of 1.03 mm in order to decrease electromagnetic reflection. Twenty dielectric rods are placed from one third the width of the waveguide to near the sidewall with increasing radius of the rods r and with constant $a = 16$ mm. Table 1 shows the relation between the distance w and radius r. The S parameters between the input port (Port 1) and output. port (Port 2) are calculated using the HFSS software by Ansoft and the results are shown as in Figures. 31(a) and (b). The electromagnetic waves pass through as the TE_{10} mode for 2.5-4.5 GHz and TE_{30} mode is converted to the TE_{10} mode for 6.8-8.8 GHz under a condition of over 95% efficiency.

7. Conclusion

The PC waveguide we developed will have low dielectric losses compared with NRD waveguides and it has an acceptable bending loss at millimeter-wave frequencies.

Arrayed metallic rods do not usually possess bandgaps for H-polarization, but do have bandgaps for E-polarization. Photonic crystals using metamaterials have bandgaps for H-polarization. Therefore, bandgaps for both E- and H-polarization can be designed independently.

Attenuation constants when the E-planes in a metallic waveguide have been replaced with MPC were calculated for a fixed cutoff frequency. The propagation losses decrease when the normalized radius of the rods r/a increases.

If the wall of the waveguide is made of a metal array with one row removed, the propagation bandwidth is narrowed when the normalized radius r/a is reduced because the upper edge of the bandgap is decreased. The bandwidth, however, is extended by widening the normalized width of the waveguide w/a.

A new waveguide with in-line dielectric rods has been investigated, in which an electromagnetic wave is propagated without higher modes above $2f_c$. This type of waveguide uses $LaAlO_3$ as a low loss material in order to reduce the attenuation constant, because many rods are placed at the center of the waveguide. The attenuation constants when using $LaAlO_3$ are not so high compared with conventional waveguides, and are much smaller than ridge waveguides. The normalized bandwidth of this waveguide, $W_B = (\Delta\omega_1+\Delta\omega_2+\Delta\omega_3)/\omega_0 \approx 0.95$, is wider than that reported in references [19] and [20].

We have investigated a new configuration of waveguide with dual in-line dielectric rods for which electromagnetic waves propagate without higher modes above $2f_c$. An economical method of fabrication consists in inserting dielectric rods only at the bent portions because mode conversion does not occur along straight segments. As an example, holes with diameters of 0.2 mm were drilled through the top of a waveguide and dielectric rods were inserted. The S parameters were calculated using HFSS software and the results for different structural conditions were found to be similar.

We have previously reported that single-mode propagation is available for a metallic waveguide with dielectric rods arrayed at the center of a waveguide using the TE_{10} mode, and the TE_{30} mode. In this investigation, a mode converter is proposed which passes through the TE_{10} mode for the low frequency range and converts TE_{30} to the TE_{10} mode for the high frequency range by small variation of the group velocity. It was shown that the electromagnetic waves pass through as the TE_{10} mode for 2.5-4.5 GHz and are converted to the TE_{30} mode for 6.8-8.8 GHz under a condition of over 95% efficiency.

References

[1] Yoneyama, T. & Nishida, S. (1981). Nonradiative dielectric waveguide for millimeter-wave integrated circuits. *IEEE Transactions on Microwave Theory and Techniques*, vol. MTT-29, 1188-1192.

[2] Cohn, S. B. (1947). "Properties of Ridge Wave Guide. *Proceedings of the I.R.E.*, Vol. *35*, 783-788, Aug.

[3] Kokubo, Y. (2008). Wide Band Metallic Waveguide with In-Line Dielectric Rods. *IEEE Microwave and Wireless Components Letters*, Vol. 18 No.2, 79-81.

[4] Kokubo, Y. (2010). Frequency Range Dependent TE_{10} to TE_{20} Mode Converter. *Microwave and Optical Technology Letters*, Vol. 52, No. 1, 169-171.

[5] Mekis, A., Chen, J. C., Kurland, I., Fan, S., Villeneuve, P. R. & Joannopoulos, J. D. (1996). High Transmission though Sharp Bends in Photonic Crystal Waveguides. *Physical Review Letters*, Vol.77, 18, 3787-3790.

[6] Kokubo, Y., Kawai, T. & Ohta, I. (2003). A New Type of Photonic Crystal Waveguide at Millimeter and Submillimeter Wave. *IEICE Transactions on Electronics*, Vol. J86-C, No.3, 314-315. (Japanese edition)

[7] Taflove, A. & Brodwin, M. E. (1975). Numerical Solution of Steady-State Electromagnetic Scattering Problems Using the Time-Dependent Maxwell's Equations. *IEEE Transactions on Microwave Theory and Techniques*, Vol. MTT-23, 623-630.

[8] Nakashima, Y., Kokubo, Y., Ohta, I. & Kawai, T. (2005). Characteristics of the 2D Photonic Crystals with the Metallic Meta Materials. *IEICE Transactions on Electronics*, Vol. J88-C, NO.12, 1180-1183.(Japanese edition)

[9] Pendry, J. B., Holden, A. J., Robbins, D. J. & Stewart, W. J. (1999). Magnetism from Conductors and Enhanced Nonlinear Phenomena. *IEEE Transactions on Microwave Theory and Techniques*, Vol. 47, No.11, 2075-2084.

[10] Kokubo, Y. (2008). Analysis of Propagation Loss of Metallic Photonic Crystal Waveguides. *Microwave and Optical Technology Letters*, Vol.*50*, No.11, 2942-2945.

[11] Sievenpiper, D. F., Sickmiller, M. E. & Yablonovitch, E. (1996). 3D Wire Mesh Photonic Crystals. *Physical Review Letters*, Vol. 76, No.14, 2480-2483.

[12] Pendry, J. B. (1996). Extremely Low Frequency Plasmons in Metallic Mesostructures. *Physical Review Letters*, Vol.76, No.25, 4773-4776.

[13] Qiu, M. & He, S. (2000). Guided modes in a two-dimensional metallic photonic crystal waveguide. *Physics Letters A*, 266, 425-429.

[14] Hirokawa, J. & Ando, M. (1999). Model Antenna of 76GHz Post-wall Waveguide- Fed Parallel Plate Slot Arrays. *IEEE AP-S International Symposium and USNC/URSI National Radio Science Meeting*, Vol.1, 146-149.

[15] Deslandes, D. & Wu, K. (2001). Integrated microstrip and rectangular in planar form. *IEEE Microwave and Wireless Components Letters*, Vol. 11, No. 2, 68-70.

[16] Benisty, H. (1996). Modal analysis of optical guides with two-dimensional photonic band-gap boundaries. *Journal of Applied Physics*, 79(10), 15, 7483-7492.

[17] Boroditsky, M., Coccioli, R. & Yablonovitch, E. (1998). Analysis of photonic crystals for light emitting diodes using the finite difference time domain technique. *Proceedings of SPIE*, Vol. 3283, 184-190.

[18] Kokubo, Y. & Kawai, T. (2009). 90-Degree H-Plane Bent Waveguide Using Dielectric Rods. *Microwave and Optical Technology Letters*, Vol. 51, No. 9, 2015-2017.

[19] Shibano, T., Maki, D. & Kokubo, Y. (2006). Dual Band Metallic Waveguide with Dual in-line Dielectric Rods. *IEICE Transactions on Electronics*, Vol.J89-C No.10, 743-744. (Japanese edition); Correction and supplement. (2007). Ibid. Vol. J90-C No.3, 298. (Japanese edition).

[20] Kokubo, Y., Maki, D. & Kawai, T. (2007). Dual-Band Metallic Waveguide with Low Dielectric Constant Material. *37th European Microwave Conference (EuMC2007),* 890-892.

[21] Collin, R. E. (1990). 9.5 Capacitively Loaded Rectangular Waveguide. *Field Theory of Guided Waves,* 2nd edition, IEEE Press, 621-624.

[22] Plihal, M., Shambrook, A. & Maradudin, A.A. (1991). Two-dimensional photonic band structure. *Optics Communications*, 80(3.4), 199-204.

[23] Kokubo, Y. (2007). Wide Band Metallic Waveguide with In-Line Dielectric Rods. *IEICE Transactions on Electronics.,* Vol. J90-C, no. 9, 642-643.

[24] Ansoft Corporation, (2005). Introduction to the Ansoft Macro Language. *HFSS v10.*

[25] Kokubo, Y. (2009). Rectangular TE_{30} to TE_{10} Mode Converter. *IEICE Transactions on Electronics*, Vol. E92-C, No.8, 1087-1090.

In: Photonic Crystals
Editor: Venla E. Laine, pp. 95-114

ISBN: 978-1-61668-953-7

Chapter 4

PHYSICS OF PHOTONIC CRYSTAL COUPLERS AND THEIR APPLICATIONS

Chih-Hsien Huang,[1,2] Forest Shih-Sen Chien[3], Wen-Feng Hsieh[1,2,*] and Szu-Cheng Cheng[4,*]

[1] Institute of Electro-Optical Science & Engineering, National Cheng Kung University, No.1, University Rd., Tainan City 701, Taiwan, R. O. C.

[2] Department of Photonics & Institute of Electro-Optical Engineering, National Chiao Tung University, 1001 Tahsueh Rd., Hsinchu 30050, Taiwan, R. O. C.

[3]Department of Physics, Tunghai University, Taichung 407, Taiwan.

[4]Department of Physics, Chinese Culture University, Yang-Ming Shan, Taipei 111, Taiwan, R. O. C.

Abstract

Properties of the directional coupler made by photonic crystals (PCs) are studied by the tight-binding theory (TBT), which considers the coupling between defects of PCs. Based on this theory, the amplitude of the electric field in a photonic crystal waveguide (PCW) can be expressed as an analytic evolution equation. As an identical PCW is inserted with one or several partition rods away, the PC coupler is created. The nearest-neighbor coupling coefficient α between defects of the coupler causes the splitting of dispersion curves, whereas the next-nearest-neighbor coefficient β causes a sinusoidal modulation to dispersion curves. The sign of α determines the parity of fundamental guided modes, which can be either even or odd, and the inequality $|\alpha| < 2|\beta|$ is the criterion for occurring crossed dispersion curves. There is no energy transferred between the PCWs at the frequency of the crossing point, named as the decoupling point, of dispersion curves.

By translating the defect rods along the propagation axis of the coupler, blue shift (red shift) in the frequency of the decoupling point occurs to the square (triangular) lattice. Therefore, the frequency of the decoupling point and coupling length of the coupler can be adjusted by moving defect rods. By applying propagating fields having a frequency at the decoupling point and another frequency where the coupler has an ultra short work region, we designed a dual-wavelength demultiplexer with a coupling length of only two wavelengths and power distinguish ratio as high as 15 dB.

An analytic formula can also be derived by the TBT for asymmetric couplers made of two different coupled PCWs. The asymmetric coupler possesses the following properties: (1)

Its dispersion curves will not cross at the "decoupling point" and the electric field would only localize in one PCW of the coupler; (2) The eigenfield at the high (low) dispersion curve always mainly localizes in the PCW that possesses high (low) eigenfrequency, even though the symmetry of eigenmodes has changed. As the field with a given frequency is incident into one of the PCWs of the coupler, both the energy transfer between two waveguides and the corresponding coupling length can be given. If an optical Kerr medium is introduced into a symmetric coupler, the coupler becomes an optical switch through the modification of the refractive index of the coupler by sending a high-intensity field.

1. Introduction

Photonic crystals (PCs) arise from the cooperation of periodic structures with light.[1] PCs are interesting owing to whose band structure possesses a complete photonic band gap (PBG).[2-4] Light with frequency set within the range of a PBG cannot propagate inside PCs. A point defect is created as only a single rod inside the PBG structure is modified by changing the dielectric constant, radii of the rod or removing the rod. In general, a bound state is always existed in one-dimensional (1D) PC defect but not necessary existed in two-dimensional (2D) or three-dimensional (3D) PC point defects.[5] At the frequency of the bound state, due to the electromagnetic (EM) wave being strongly localized around a defect, the defect can own a high Q value[1] and become a filter, resonance cavity, or modulator of light.[6-10] When the refractive index of the defect structure increases, the defect mode will shift to low frequency and there could have multimode oscillating in the cavity. As defects in a 2D or 3D PC structure are arranged periodically and distant away, a waveguide named as the coupled resonance optical waveguide (CROW)[11-13] is created. The EM wave in the CROWs can be expressed as the superposition of bound states from point defects. This approach is named as the tight-binding approximation that is commonly used in condensed matter physics. Similarly, a 2D photonic crystal waveguide (PCW)[14,15] can be created by changing the radius, dielectric constant of the dielectric rods or the radius of periodic air holes in a dielectric slab sequentially.[16,17] The PCWs are the most promising elements of PCs for building large-scale photonic integrated circuits (PICs),[18,19] because the EM waves can be transmitted,[15] sharply bent,[20] split,[21] and dispersion-compensated[22] by means of specific PCW designs. The PCWs can also be expressed as the superposition of defect modes as more coupling terms corresponding to the vicinal defects are considered. Therefore, the tight-binding approximation is also suitable to treat the wave propagation in the PCWs.

Waveguide coupler made of two vicinal single waveguides is a fundamental device for designing optical filters and switches for PICs. The PC directional coupler,[23-27] in which two PCWs are separated with a (or several) participation rods, has received much attention, since numerous photonic devices based on this mechanism have been proposed and demonstrated, e.g., filters,[28] switches[29] and multiplexers/demultiplexers.[30-32] Typically, the waveguide coupling is handled by the coupled-mode theory (CMT),[33] which is a straightforward and simple method to give approximate solutions in coupled waveguide systems. According to the CMT, the fundamental guided modes are specified by even-parity and odd-parity modes; the dispersion curves of two modes show splitting and no crossing point. However, the dispersion curves derived from the plane wave expansion method (PWEM)[34,35] reveals that the properties of fundamental guided modes are quite different from the prediction of the CMT and the dispersion curves do cross, i.e., the eigenmodes are degenerate and the waveguides are decoupling.[30] Therefore, another theory is needed to describe the EW wave propagation of the PC coupler.

In this chapter, we used the tight binding theory (TBT)[36-39] to derive the electric field evolution equation in a single PCW and the coupling equations of symmetric and asymmetric couplers. From the derived equations, we will firstly discuss the properties of symmetric couplers in which the defect rods of waveguides are identical. Secondly, in order to design a device made by the PC coupler with proper coupling length, we tuned the defect rods transversely or longitudinally along the waveguide direction to modify the coupling length and to tune the decoupling point in the symmetric coupler. Finally, the physical properties of an asymmetric coupler, whose defects have different radii on each waveguide, are studied.

2. Tight Binding Theory

We first consider an optical waveguide, which consists of a periodic sequence of identical defects in the PC with lattice constant a as shown in Figure 1(a). Field distributions of a point defect from triangular and square lattices are shown in Figure 2. An isolated point defect, formed by reducing the radii of a dielectric rod, has a localized electric-field distribution with a bound mode of frequency, ω_0. We can express the electric and magnetic fields of this bound mode as $\mathbf{E}(\mathbf{r},t) = \mathbf{E}_0(\mathbf{r})exp(-i\omega_0 t)$ and $\mathbf{H}(\mathbf{r},t) = \mathbf{H}_0(\mathbf{r})exp(-i\omega_0 t)$. The presence of other defects near this point defect perturbs permittivity from $\varepsilon(\mathbf{r})$ to $\varepsilon'(\mathbf{r})$ and the perturbed fields of the waveguide $\mathbf{E}'(\mathbf{r},t) = \mathbf{E}_0'(\mathbf{r},t)e^{-i\omega_0 t}$ and $\mathbf{H}'(\mathbf{r},t) = \mathbf{H}_0'(\mathbf{r},t)e^{-i\omega_0 t}$ are expressed as a superposition of bound modes, i.e., $\mathbf{E}_0'(\mathbf{r},t) = \sum_m u_m'(t)\mathbf{E}_0(\mathbf{r}-ma\hat{\mathbf{x}})$ and $\mathbf{H}_0'(\mathbf{r},t) = \sum_m u_m'(t)\mathbf{H}_0(\mathbf{r}-ma\hat{\mathbf{x}})$, where m is an integral number to specify the position of defects. Since the waveguide fields should obey the Maxwell's equations, by using the reciprocal (divergence) theory:[12]

$$\int ds(\mathbf{E}_0^{*}\times\mathbf{H}_0' + \mathbf{E}_0'\times\mathbf{H}_0^{*}) = \int dv(i\omega_0\Delta\varepsilon\mathbf{E}_0^{*}\cdot\mathbf{E}_0' - \varepsilon'\mathbf{E}_0^{*}\cdot\partial\mathbf{E}_0'/\partial t - \mu_0\mathbf{H}_0^{*}\cdot\partial\mathbf{H}_0'/\partial t) \qquad (1)$$

with $\Delta\varepsilon=\varepsilon'-\varepsilon$ and $u_m'(t) = u_m(t)e^{-i\omega_0 t}$, we obtain

$$i\frac{du_n}{dt} + (-\omega_0 + P_0)u_n + \sum_{m=1} P_m(u_{n+m} + u_{n-m}) = 0. \qquad (2)$$

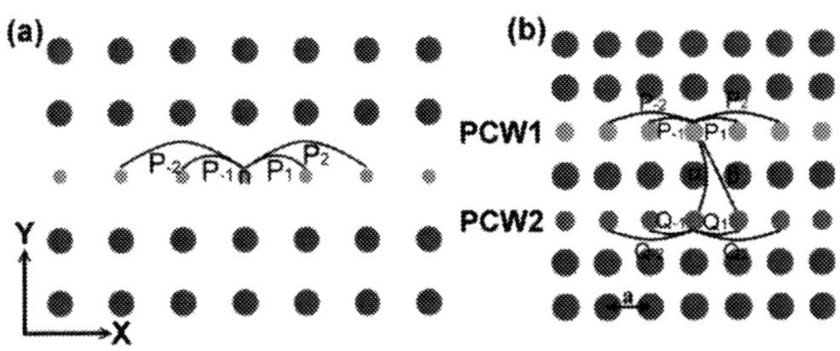

Figure 1. The structures of (a) a PCW, and (b) a coupler with one separation rods. P_m's and Q_m's are the coupling coefficients between defects within single waveguides, α and β are the coupling coefficients between waveguides, respectively. (After Ref. [40];©2009 IOP.)

Here the coupling coefficient P_m is defined as

$$P_m = \frac{\omega_0 \iiint dv \Delta\varepsilon \mathbf{E}_{0n} \cdot \mathbf{E}_{0n+m}}{\iiint dv(\mu_0 |\mathbf{H}_{0n}|^2 + \varepsilon |\mathbf{E}_{0n}|^2)}. \tag{3}$$

P_0 is a frequency shift arising from the perturbed polarization induced by the self-field from neighbor defects. If a plane wave with wave vector k and frequency ω is incident into this waveguide, the dispersion of the propagating wave is:

$$\omega(k) = \omega_0 - P_0 - 2\sum_{m=1} P_m \cos(mka). \tag{4}$$

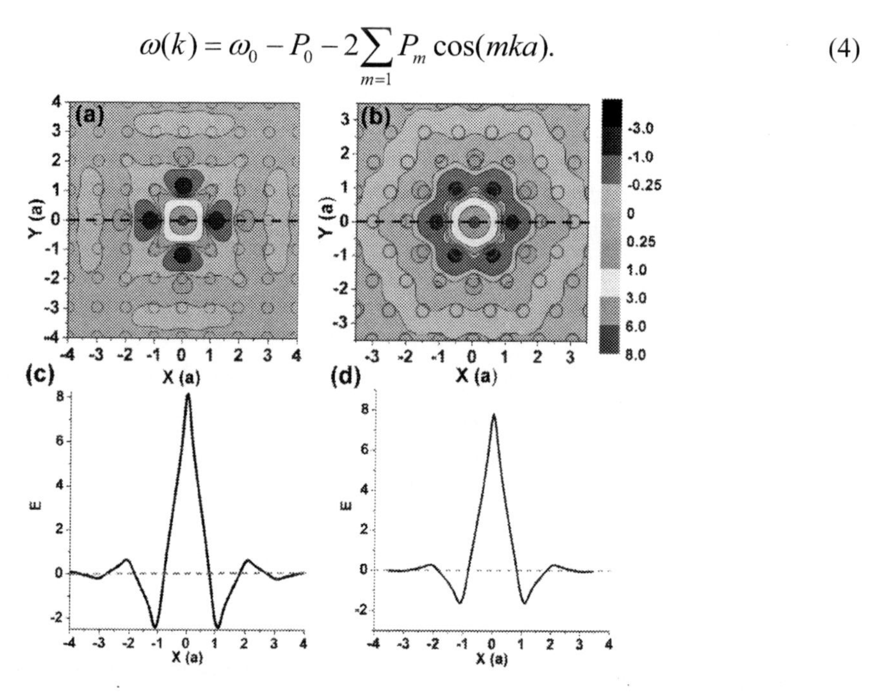

Figure 2. The electric field distribution (E_z) of a point defect with reduce rod (r_d = 0.1a) (a) in square lattice for ω_0 = 0.325 c/a and (b) in triangular lattice for ω_0 = 0.333 c/a. Graphs (c) and (d) are field distributions of square and triangular lattices along the x-axis, respectively.

The electric field distribution (E_z) of a single point defect, simulated by the PWEM, in the square and triangular lattices with the dielectric constant and radii of dielectric rods being 12, 0.2a are shown in Fig. 2. The radius (r_d) of the defect rods and eigenfrequency ω_0 in a square lattice are 0.1a and 0.325 c/a, respectively, where c is the speed of light in vacuum; those in a triangular lattice are 0.1a and 0.333 c/a. The field profiles along the x-axis in Fig. 2(c) and (d) have the opposite sign when the field extends to the odd nearest-neighbor rod(s) ($\mathbf{E}_0(0,0)^*\mathbf{E}_0(xa,0) < 0$, x = 1,3,5,…) and have the same sign when the field extends to the even nearest-neighbor rods[22,41]. To maintain a single mode propagating in the waveguides, the radius or the refraction index of the rods in the waveguides is reduced and $\Delta\varepsilon$ becomes negative in the following discussion. Since the electric field mainly localizes around the dielectric rods of the waveguides, we can use the maximal values to replace the integral values of Eq. (3) for a simple estimation. Coupling coefficients P_1 and P_3 are positive, also P_2 and P_4 are negative. Coefficients P_3 and P_4 are two orders of magnitude smaller than P_1, and thus only P_1 and P_2 are taken into account when calculating the dispersion relation. From the dispersion relation in Eq. (4), the frequency increases as k increases in PCWs.

For an asymmetric coupler partitioned by a perfect row of rods in a PC, there are two rows of periodic defect rods, shown as PCW1 and PCW2 in Figure 1(b). The field distribution of the eigenmode of the isolated point defect in each PCW can be written as the product of time-varying and spatial-varying functions, i.e., $\mathbf{E}_{10}(\mathbf{r},t)=\mathbf{E}_{10}(\mathbf{r})\exp(-i\omega_1 t)$ in PCW1 and $\mathbf{E}_{20}(\mathbf{r},t)=\mathbf{E}_{20}(\mathbf{r})\exp(-i\omega_2 t)$ in PCW2, where ω_1 and ω_2 are the frequencies of localized modes of a single point defect in each PCW.

The evolution equation of an isolated PCW1 can be written as

$$i\frac{\partial}{\partial t}u_n = (\omega_1 - P_0)u_n - \sum_{m=1} P_m(u_{n+m} + u_{n-m}), \tag{5}$$

and P_m is given by Eq. (3). Note that P_m is also equal to C_m^{11}, where C_m^{ij} is the coupling coefficient between the site n of the ith PCW and the site $n+m$ of the jth PCW, and is defined as

$$C_m^{ij} = \frac{\omega_i \int_{-\infty}^{\infty} dv \Delta\varepsilon(\mathbf{r})\mathbf{E}_{in} \cdot \mathbf{E}_{jn+m}}{\int_{-\infty}^{\infty} dv[\mu_0 |\mathbf{H}_{in}|^2 + \varepsilon |\mathbf{E}_{in}|^2]}. \tag{6}$$

Let k and $\bar{\omega}_1$ be the wavevector and its corresponding eigenfrequency of PCW1, respectively, we obtain the dispersion relation of PCW1:

$$\bar{\omega}_1(k) = \omega_1 - P_0 - \sum_{m=1} 2P_m \cos(mka). \tag{7}$$

Similarly, the evolution equation and dispersion relation of an isolated PCW2 are shown below:

$$i\frac{\partial}{\partial t}v_n = (\omega_2 - Q_0)v_n - \sum_{m=1} Q_m(v_{n+m} + v_{n-m}), \tag{8}$$

$$\bar{\omega}_2(k) = \omega_2 - Q_0 - \sum_{m=1} 2Q_m \cos(mka), \tag{9}$$

where $Q_m = C_m^{22}$. $v_n(\mathrm{t})$ and $\bar{\omega}_2$ are the time-varying function and the eigenfrequency of the isolated PCW2, respectively.

Due to the field distributions of defect modes being not strongly localized around defects, we shall consider the coupling effect of two asymmetric PCWs up to the second nearest-neighboring defects, with coupling coefficient $\alpha = C_0^{12} = C_0^{21}$ and $\beta = C_{\pm1}^{12} = C_{\pm1}^{21}$ shown in Figure 1(b). The coupled equations of an asymmetric coupler are given by:[40,42]

$$i\frac{\partial}{\partial t}u_n = (\omega_1 - P_0)u_n - \sum_{m=1} P_m(u_{n+m} + u_{n-m}) - \alpha v_n - \beta(v_{n+1} + v_{n-1}), \tag{10}$$

$$i\frac{\partial}{\partial t}v_n = (\omega_2 - Q_0)v_n - \sum_{m=1} Q_m(v_{n+m} + v_{n-m}) - \alpha u_n - \beta(u_{n+1} + u_{n-1}). \tag{11}$$

When the stationary solutions of coupled Eqs. (10) and (11) are taken as $u_n = U_0 \exp(ikna - i\omega t)$ and $v_n = V_0 \exp(ikna - i\omega t)$, we obtain the characteristic equations of the coupler:

$$(\omega - \bar{\omega}_1)\, U_0 + g(ka)\, V_0 = 0, \tag{12}$$

$$(\omega - \bar{\omega}_2)\, V_0 + g(ka)\, U_0 = 0, \tag{13}$$

where $g(ka) = \alpha + 2\beta Cos(ka)$. The eigenfrequencies (dispersion relations) and eigenvectors (field amplitudes) of Eqs. (12) and (13) are

$$\omega^{\pm}(k) = \frac{(\bar{\omega}_1 + \bar{\omega}_2)}{2} \pm \sqrt{\Delta^2 + (g(ka))^2}, \tag{14}$$

$$\chi^{\pm} = (V_0 / U_0)^{\pm} = -\frac{\Delta \pm \sqrt{\Delta^2 + (g(ka))^2}}{g(ka)}, \tag{15}$$

where $\Delta = (\bar{\omega}_2 - \bar{\omega}_1)/2$ and $\chi^{\pm}$ are the amplitude ratios corresponding to frequencies $\omega^{\pm}(k)$. Note that $\chi^+\chi^- = -1$ is due to the orthogonality of these two eigenmodes at a given wave vector k.

3. Physical Properties of a Symmetric Coupler

In symmetric waveguides, $\bar{\omega}_2 = \bar{\omega}_1$, Eqs. (14) and (15) become[43]

$$\omega^{\pm}(k) = \bar{\omega}_1 \pm g(ka) = \bar{\omega}_1 \pm (\alpha + 2\beta cos(ka)), \tag{16}$$

$$\chi^{\pm} = (V_0 / U_0)^{\pm} = \mp 1. \tag{17}$$

Obviously, the nearest coupling between two waveguides only leads to a relative shift of eigenfrequency by $\pm\alpha$, and the next-nearest neighbor coupling leads to the sinusoidal modulation $\pm 2\beta cos(ka)$ of the dispersion curves. If $|\alpha| < |2\beta|$, it is possible that $\alpha + 2\beta cos(ka) = 0$, and the intersection of dispersion curves occurs at wave number

$ka = \cos^{-1}(-\alpha/2\beta)$. Accordingly, three dispersion curves of $\overline{\omega}_1$, ω^+ and ω^- are equal and degenerate at the crossing point, where the coupling length becomes infinite and the coupler is decoupled. The inequality $|\alpha| < |2\beta|$ is a necessity for the dispersion crossing that originates from the cross coupling of the next-nearest neighbor defects due to the non-uniform nature along the propagation direction of the mode function of individual PCWs. Such a cross coupling does not exist in conventional waveguides, since the wave functions of conventional waveguides are uniform along the propagation direction. Therefore, the dispersion curves of two identical conventional waveguides can never cross.

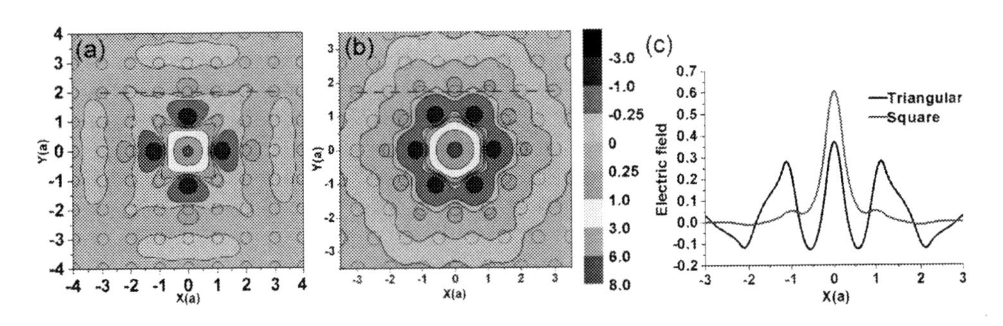

Figure 3. The electric field distribution (E_z) of a point defect mode in the square lattice for (a) ω_0 $=0.325c/a$ with a reduced-rod (r_d=0.1a) defect; that in the triangular lattice for (b) eigenfrequency ω_0 = 0.365 c/a with a defect rod ε_r= 2.56; and (c) The electric field distributions along dashed lines of graphs (a) and (b). (After Ref. [40];©2009 IOP.)

From the electric field distributions of defects in the square and triangular lattices, shown in Figure 3, we find that the electric field at the site (x = 0, y = 0) of the square lattice has the same polarity as its nearest-neighbor site (x = 0, y = 2a) and the next-nearest-neighbor site (x = a, y = 2a). Because $\Delta\varepsilon < 0$ for the air-defect PCWs in both square and triangular lattices, the coupling coefficients α and β are both negative.

Now that $|g(ka)|$ has a maximal value at k = 0 from Eq. (16), one should expect that the dispersion curves have the largest splitting there. As α and β are negative values discussed before, $g(ka)$ always is a negative value for all k if $|2\beta/\alpha| < 1$, and its value can change from the negative to the positive sign as k increases from 0 to π when $|2\beta/\alpha|>1$. Under the condition of $|2\beta/\alpha| > 1$, the dispersion relation of the coupler can cross when $g(k_d a) = 0$ at the crossing point k_d with eigenfrequencies $\omega^- = \omega^+ = \overline{\omega}_1$;, where is that is, the field launched in PCW1 will always be confined in PCW1 without being transfered to PCW2, and vice versa. We can simply use the ratio of the maximal field values instead of integrals as Eq. (3) to estimate coefficients α, β and $|2\beta/\alpha|$ by assuming the field distribution is strongly localized near the dielectric rods. Thus, $\zeta = |2\beta/\alpha| \sim 2E(\pm a, 2a)/E(0, 2a)$ in the square lattice and $\sim 2E(\pm a, \sqrt{3}\,a)$/E(0, $\sqrt{3}\,a$) in the triangular lattice. From the field distribution of the point defect at the dash red line of the point defect in square (Figure 3(a)) and triangular lattices (Figure 3(b)), as shown in Figure 3(c), the value of ζ in the triangular lattice would be larger than that in the square lattice, such meaning there would be easier to have a cross point in triangular than in square lattice.

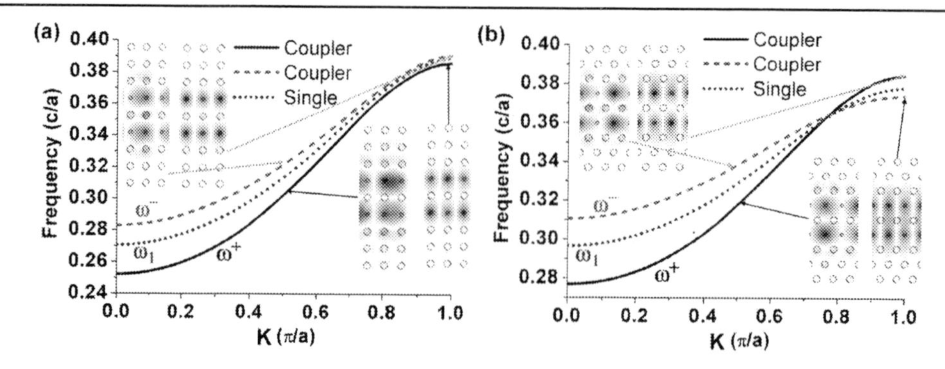

Figure 4. Dispersion relations of the couplers (ω^+ and ω^-) and singe PCWs (ω_1) simulated by PWEM in square (a) and triangular (b) lattices. The radii and dielectric constant of the dielectric rods in the perfect crystals are 0.2a and 12. The radii of the defect rods are 0.1a. The insets are the electric field distribution of eigenmodes

From the simulation results of PWEM, shown in Figure 4, there is a crossing point in a triangular lattice but no crossing point in a square lattice with the radii of dielectric rods being 0.1a. Due to α, β and $g(ka)$ all being negative in a square lattice, frequency ω^+ is less than frequency ω^-. From Eq. (17), the parities of low-frequency and high-frequency modes must be odd and even, respectively. However, for the coupler in a triangular lattice, $g(ka)$ becomes positive for $k > k_d$. Therefore, the mode parities could interchange beyond the crossing point.

There are two dispersion curves with one even mode and one odd mode in a symmetric directional coupler. The amplitude of the EM wave propagating along the coupler can be expressed as the linear combination of these eigenmodes and the coupling length of a coupler is defined as $\pi/\Delta k$, where Δk is the wave vector mismatch of these eigenmodes.[43] When the frequency of the incident wave is chosen at the crossing point, i.e., Δk=0, the coupling length is infinite and there is no coupling between waveguides. On the other frequency of the incident wave, Δk is not zero and waveguides would couple each other. This property was used in designing a multiplexer/demultiplexer. [30]

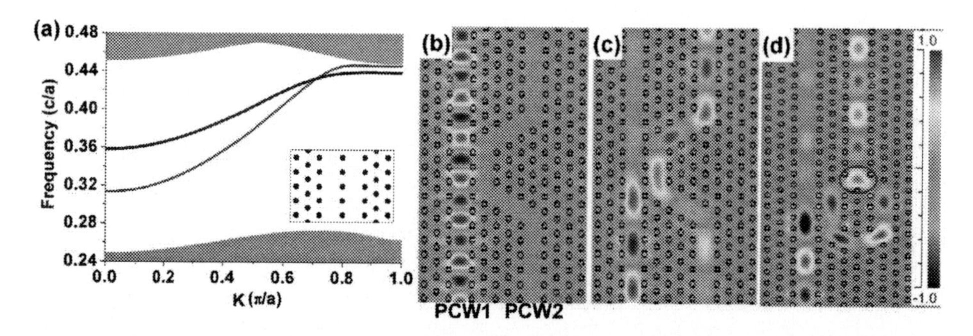

Figure 5(a). Dispersion relations of a coupler simulated by the PWEM. The radii and dielectric constant of the dielectric rods in the perfect crystals are 0.2a and 12. The FDTD simulation with (b) f_A =0.432 c/a, (c) f_B = 0.361 c/a, and (d) the optimum design with f_B = 0.361 c/a. (After Ref. [30];©2004 OSA.)

Considering a directional coupler made by removing rods along the waveguides, the dispersion relations and PC structure are shown in Figure 5(a). Let f_A =0.432 c/a corresponds to λ_A = 1300 nm in which the coupling length is infinite, and f_B = 0.361 c/a for λ_B = 1550 nm thus $\Delta k = 0.27\ \pi/a$ in which the coupling length is $3.6a$. The FDTD simulated optical wave at f_A shows no coupling from one PCW into the other PCW in Figure 5(b), but the wave at f_B shows coupling in Figure 5(c). However, in this coupler, there exists a backward propagating wave. To reduce the backward wave amplitude, a hexagonal loop can be used to merge the backward wave with the forward wave after traveling the loop. In phase or constructive interference of the forward and backward waves at the merging point is crucial for maximum output power. If there is a phase mismatch, part of the coupled wave will transfer back to the PCW1. The interference can be tuned by changing the shape or size of the loop. The optimum design in Figure 5(d) is to make the hexagonal loop three rows in width and five rows in length, so that the interference is constructive. The output power ratio of f_B is increased to 16.2 dB as the backward wave is eliminated and the output power ratio of f_A is 15.6 dB.

4. Tuning the Decoupling Point of a Photonic-Crystal Directional Coupler

As discussed before, in a triangular lattice PCs made of dielectric rods in the air, the dispersion curves of a coupler will cross at a particular wave vector. At this crossing point, named as the decoupling point, the coupling length becomes infinite so that the EM wave incident from one PCW will never be coupled into the other PCW. In a square lattice of PCs, however, only the coupler made of removed rods or reduced rods with a smaller radii will make dispersion curves crossed at the wave vector near the Brilluoin-zone boundary.[40,43] Using the infinite coupling length for one wavelength and a finite coupling length for another, we have designed a miniature bidirectional coupler[30,44], in which knowing the crossing point and coupling length in advance is an important advantage in designing a coupler. Here, we propose using the TBT to control these two parameters and derive the design rules for a coupler.[45]

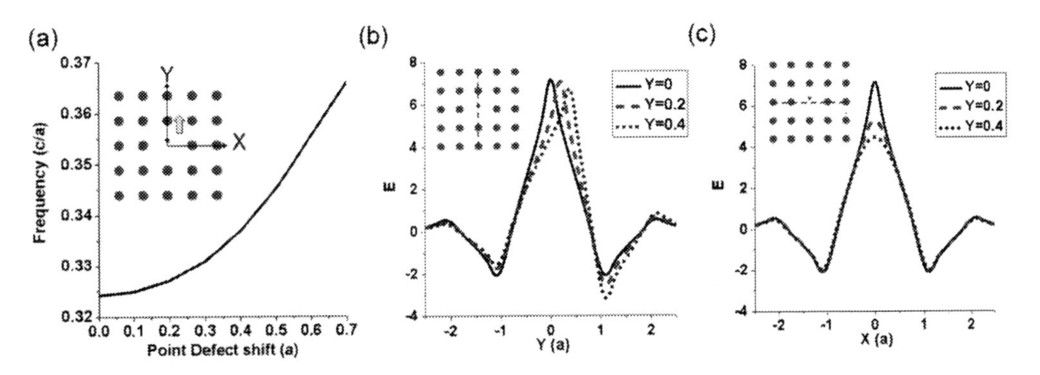

Figure 6 (a). The eigenfrequencies, (b) the electric field along the y-axis, and (c) the electric field along the x-axis of the point-defect modes with a defect rod located at different positions along the y-axis, where c is the speed of light in vacuum. (After Ref. [45];©2009 OSA.)

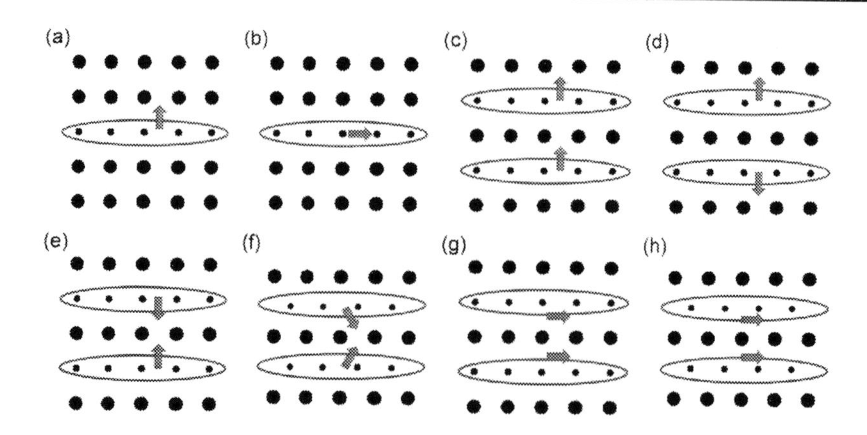

Figure 7. The ways of moving the defect rods in single or coupled PCWs. (After Ref. [45];©2009 OSA.)

The dispersion relation of the single line defect derived by the TBT is described as $\bar{\omega}_1(k) = \omega_1 - P_0 - \sum_{m=1} 2P_m \cos(mka)$. Moving the point defect along ±*y*-direction (or ±*x*-direction) in the square lattice will shift the eigenfrequency (ω_1) of a point defect toward the higher frequency as shown in Figure 6(a). Such a blue shift in frequency is caused by less concentration of electric fields in the dielectric defect rod[5] when the defect rod moves away from the center as shown in Figure 6(b) for field distribution along the *y*-direction and Figure 6(c) for that along the *x*-direction. From Figure 6(c) the field distribution is almost unchanged along the propagation axis of PCWs so that it would remain unaffected. Therefore, the dispersion curve should just show a blue shift after moving all the defect rods along *y*-direction in the PCW (see Figure 7(a)). However, if moving all the defect rods of the PCW along the *x*-direction, shown in Figure 7(b), one would expect increase in both ω_1 and P_0 but only slightly change P_m's. Therefore, we would expect that the dispersion curve is almost unchanged at small wave vector *k* and slightly increases at the larger wave vector *k* by translating all the defects along the *x*-direction.

When a second identical waveguide is created to make a symmetric directional coupler as shown in Figure 1(b), dispersion relations of the coupler are $\omega(k) = \bar{\omega}_1(k) \pm (\alpha + 2\beta\cos(ka))$, where the plus sign stands for the odd mode and the minus sign for the even mode. The dispersion relations of the coupler split from $\bar{\omega}_1(k)$ with their frequency difference being determined by coefficients α and β. Whether the dispersion curves cross at a point or not is determined by the ratio $\zeta = |2\beta/\alpha|$.[42] As the PCWs are formed by moving all the defect rods along the *y*-direction, shown in Figure 7(c), we would expect that there exists a larger ω_1 and barely changes in α and β because the distance of the defect rods between two waveguides is unchanged. Therefore, the dispersion curves of the coupler would show only a blue shift. As enlarging the distance between two line defects in the coupler by oppositely moving all the defect rods along the *y*-direction as shown in Figure 7(d), we would expect that coefficients α and β become smaller and there is a reduced coupling of PCWs. Thus it would reduce the frequency separation between two dispersion curves, which shift together toward the higher frequency. On the other hand, as reducing the separation of the line defects, shown in Figure 7(e), we can increase the separation of the dispersion curves, which both shift to the higher frequency. In addition, similar effects showing in a square lattice are reproduced by reducing the separation of the line defects of the coupler in a triangular lattice, shown in Figure 7(f).

Moving all the defect rods along the x-axis, shown in Figures. 7(g) and (h), will change the intensity ratios of electric fields at $(0,2a)$ and $(\pm 1a, 2a)$ in the square lattice so that coefficients α and β would also change and the decoupling point move to a different wave vector k. The larger ζ makes the decoupling point moved to the smaller wave vector k. Let the ratio ζ and the electric field before (after) moving all the defect rods along the x-axis as ζ_1 (ζ_2) and E_1 (E_2), respectively. Assuming the field distribution is strongly localized at the dielectric rods, we can simply use the ratio of the maximal field values instead of integrals as Eq. (2.9) to estimate coefficients α, β and ζ. Thus,

$$\zeta_2/\zeta_1 \approx (E_2(\text{-}a,2a)/E_2(0,2a))/(E_1(\text{-}a,2a)/E_1(0,2a)). \quad (18)$$

From the field distribution before and after shifting the defect rods shown as in Figure 8, we obtained $\zeta_2/\zeta_1 > 1$ in square lattice, but $\zeta_2/\zeta_1 < 1$ in triangular lattice. Therefore, moving all defects along the x-direction in the square lattice will create crossed dispersions and further shift toward the smaller wave vector k with increasing the moving distance in the x-direction. However, in the triangular lattice, the decoupling point shifted toward the larger wave vector k. These phenomena are due to the lattice structure of the former case being getting close to a triangular one, which possesses crossed dispersions on translating the line defects along the x-direction; while that of latter case tends to become a rectangular one.

In the previous section, we have used the TBT to analyze the variation of dispersion curves by moving defects in directional couplers that causes the change of eigenfrequency of a single PCW and coupling coefficients P_m, α and β. Here, we will use the PWEM and the finite difference time domain (FDTD) method[46] to examine the proposed design rule for a PC with the radius and dielectric constant of the dielectric rods being $0.2a$ and 12, respectively. The radii of defect rods in square and triangular lattices are $0.1a$ and $0.09a$, respectively.

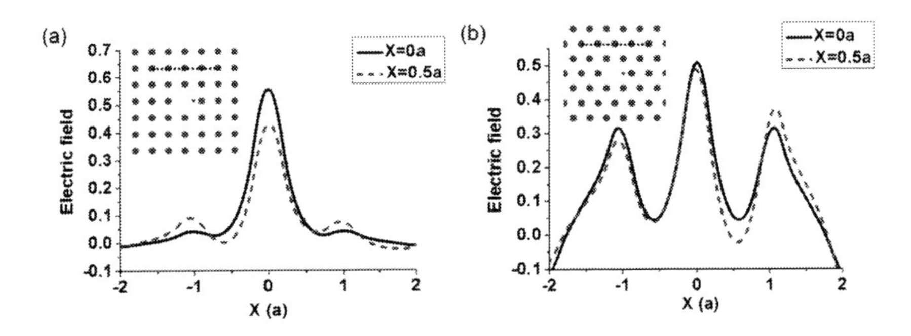

Figure 8. The electric field distribution of the point defect mode before (x = $0a$) and after moving the defect rods by $0.5a$ along the x-axis (x=$0.5a$). (a) The electric field located at y = $2a$ in the square lattice and (b) the electric field located y = $\sqrt{3}a$ in the triangular lattice. (After Ref. [45];©2009 OSA.)

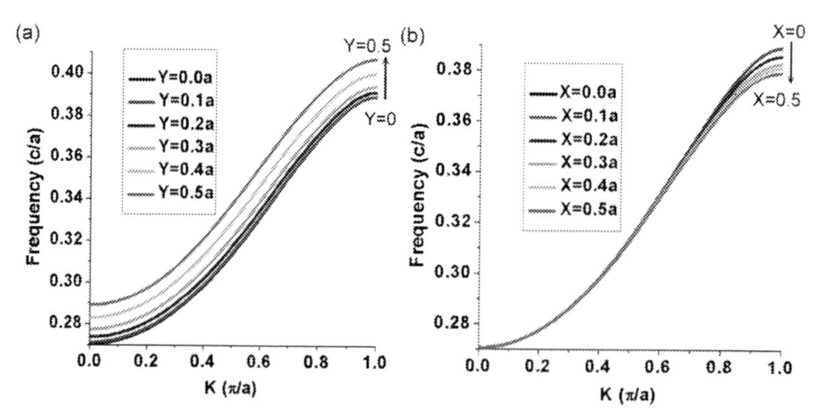

Figure 9. Dispersion curves of a single PCW with all the defect rods moving along (a) the y-direction, and (b) along the x-direction. (After Ref. [45];©2009 OSA.)

Firstly, we examined a single PCW with all line defect rods being transversely moved (along the y-axis) in the square lattice using the PWEM. We can see that the dispersion curve in Figure 9(a) shifts completely to the higher frequency, which means that the coupling coefficients P_m between defect rods are unaffected during transversely moving all defect rods. The frequency shift is mainly dominated by the variation of the eigenfrequency (ω_1) of the point defect as moving the defect rod. However, as the defect rods moved along the x-axis shown in Figure 9(b), both ω_1 and P_0 would increase and cancel out the effect of changing eigenfrequency $\bar{\omega}_1$ in the regime of small wave vectors, whereas the dispersion curve bends down in the regime of large wave vectors, due to the high-order terms of $P_m\cos(mka)$.

Secondly, by transversely moving defect rods separately, figure 10 shows how the dispersion curve varies as changing the structure of the coupler. Here, we used the square lattice as a demonstration because there are similar effects in the triangular lattice. As simultaneously moving two line defects off the center of the original PCWs and keeping the separation between PCWs fixed, we find the dispersion curves shift toward the higher frequency as shown in Figure 10(a). On the other hand, the reduction of the separation of the line defects to decrease the coupling between PCWs pushes the dispersion curves apart (see Figure 10(b)); whereas, symmetrically enlarging the separation of the line defects not only shifts the dispersion curves toward high frequencies but also makes two dispersion curves closer (see Figure 10(c)).

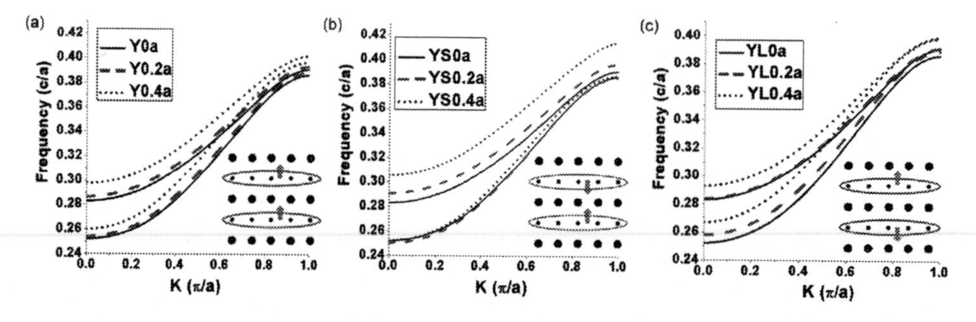

Figure 10. The dispersion curves of the shifted DCs, whose structures are indicated in the insets. (After Ref. [45];©2009 OSA.).

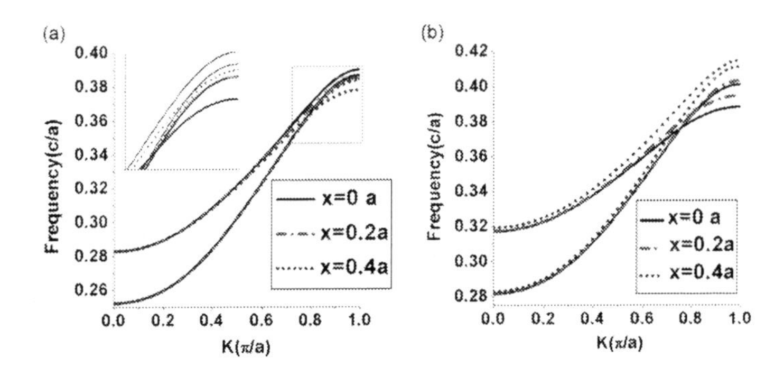

Figure 11. The dispersion curves as moving the defect rods along the x-direction in the square lattice (a) and triangular lattice (b). (After Ref. [45];©2009 OSA.).

Thirdly, we examined the effects of moving all the defect rods along the x-direction in Figure 11. We found that the dispersion relations, originally showing no decoupling point, in a square lattice have a crossing point at a high frequency as the symmetry is broken. And the decoupling point moves toward the lower frequency or the smaller wave vector k as the rods are moved further. Contrarily, the decoupling point moves toward the higher frequency or the larger wave vector k in the triangular structure and eventually without crossing.

Finally, we used the FDTD method to simulate the electric field transferring between these two shifted coupled PCWs. When an EM wave with the given frequency (0.362 c/a) is incident upon one (PCW1) of these two channels, the coupling length, for which the energy completely couples to another channel (PCW2), is defined as $\pi/\Delta k$, where Δk is the wave vector mismatch between two modes of the coupler at the incident frequency. Using the square lattices as examples, we have shown the dispersion curves of a coupler in the square lattice (see Figure 12(a)). Due to rather smaller $\Delta k \sim 0.0346\ \pi/a$ for the original coupler without moving defects, as shown in Figure 12(b) the coupling length is 29a, which is quite long but finite. As the incident EM wave whose frequency is at the decoupling point formed by longitudinally shifting all defects a 0.5a distance, the electric field will propagate in the incident channel without leaking into another channel (see Figure 12(c)). In addition, moving the line defects close to each other by 0.5a makes $\Delta k \sim 0.23\ \pi/a$, which is a larger vector mismatch, so that the coupling length becomes shorter (~ 4a see Figure 12(d)). These FDTD simulation results verify the coupling length can be tuned properly by moving positions of defect rods to shift the decoupling point of dispersion curves and agree with those calculated by the PWEM.

The proposed TBT can also be applied to other structures, e.g., TM polarization in a PC with air holes in a dielectric slab or transversely moving a perfect rod. In the case of a PC with air holes in dielectric slab, the radii of the holes must be increased to insure single mode existing in the PCW and there would be also a decoupling point in this structure. On the other hand, the TBT can also well predict the propagation of an EM wave in a single line defect or a directional coupler created by transversely moving a row or two rows of perfect dielectric rods without changing the radii or dielectric constant of the dielectric rods.

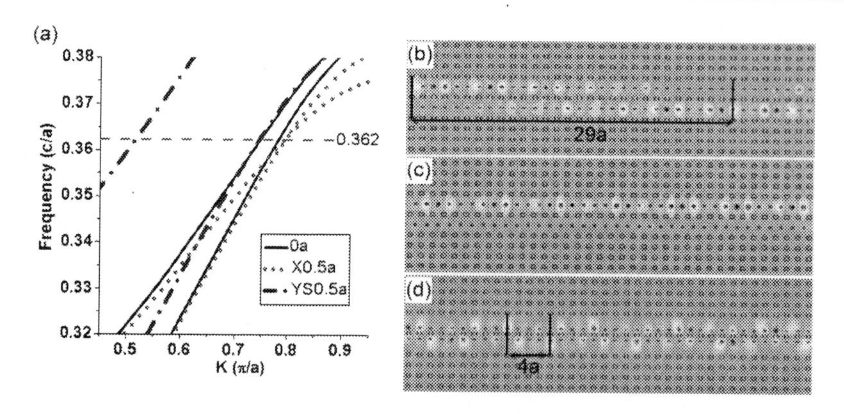

Figure 12. The dispersion curves (a) and the FDTD simulation results of the directional coupler without moving defects in (b), longitudinally moving the defect rods by $0.5a$ in (c), and transversely moving the defects closer by $0.5a$ in (d). (After Ref. [45];©2009 OSA.).

5. Physical Properties of an Asymmetric Coupler

In section 2, we derived the dispersion relation and the eigenmode ratios of the coupler as

$$\omega^{\pm}(k) = \frac{(\bar{\omega}_1 + \bar{\omega}_2)}{2} \pm \sqrt{\Delta^2 + (g(ka))^2}, \tag{14}$$

$$\chi^{\pm} = (V_0/U_0)^{\pm} = -\frac{\Delta \pm \sqrt{\Delta^2 + (g(ka))^2}}{g(ka)}. \tag{15}$$

As we know, coupling coefficients α and β are negative. Here, we assume $\bar{\omega}_2 > \bar{\omega}_1$ in the following discussion; therefore, we shall call the waveguide 2 (waveguide 1) as the high-frequency PCW2 (low-frequency PCW1).

Because $g(ka) < 0$ for $0 \le k < k_d$ under $|2\beta/\alpha| > 1$ (or for all k under $|2\beta/\alpha| < 1$), the lower frequency mode (ω^-) has $-1 < \chi^- < 0$; namely, the PCW1 and PCW2 electric fields not only are out-of-phase but also concentrate on the low-frequency PCW1. This odd-like fundamental (low-frequency) mode is called the "anti-bonding" mode, borrowed from the molecular physics of two atoms. On the other hand, the high-frequency and even-like mode called the "bonding" mode has $\chi^+ > 1$; thus, it is superimposed by the in-phase electric fields from both PCWs, where the field strength is concentrated on the high-frequency PCW2.

However, as $k > k_d$ under $|2\beta/\alpha| > 1$, $g(ka)$ becomes positive and $0 < \chi^- < 1$. The fundamental mode is a bonding mode, which is superimposed by the in-phase electric fields from both PCWs, where the field strength is concentrated on the low-frequency PCW1. And the high frequency antibonding mode with $\chi^+ < -1$ has the field strength concentrated on the high-frequency PCW2. We find that the fundamental modes of the asymmetric coupler contain no degenerate state (anti-crossing dispersion relations) and can switch from the antibonding to bonding mode as k varies crossing the decoupling point k_d . As the previous study on the symmetric coupler, we simply can set $\Delta = 0$ to obtain $\chi^{\pm} = 1$ or -1 at all k, i.e., the fundamental mode is either odd or even depending upon the sign of $g(ka)$. The dispersion curves of the symmetric coupler can cross at the decoupling point if $|2\beta/\alpha| > 1$. Furthermore, upon increasing the separation of PCWs to two rows apart,[47] from Eq. (3), coupling coefficients α and β become positive values and are smaller than coupling coefficients of the one-row-separation PCWs. The fundamental mode becomes a bonding mode, and whether or not mode switching would happen is still determined by the criterion: $|2\beta/\alpha| > 1$.

In order to prove that the derived formula by the TBT can explain phenomena shown by the PWEM, we consider for example a 2D triangular lattice PC made by dielectric rods with dielectric constant $\varepsilon_r = 12$ and radius = $0.2a$ in the air. The line defects forming the PCW1 and PCW2 are created by setting the dielectric constants of defect rods at 2.56 and 2.25, respectively. The eigenfrequencies of a point defect with transverse electric field (TE), whose electric field is parallel to the dielectric rods, are $\omega_1 = 0.365\ (2\pi c/a)$ and $\omega_2 = 0.371\ (2\pi c/a)$, respectively. The decoupling point is located at $k_d = 0.73\pi/a$ where the eigenfrequencies of the PC couplers decouple in the eigenfrequency in single line-defect PCWs, shown in Figure 13(a). Note that the dispersion curves do not cross in the asymmetric coupler. As shown in Figure 13(b), the eigenmode of the high (low) frequency band at the wave vectors $k < k_d$ are the bonding (anti-bonding) modes, but these modes switch when $k > k_d$, namely, the eigenmode of high (low) frequency band being anti-bonding (bonding). And the electric field is concentrated on the PCW2 for the high-frequency ($\omega^+(k_d)$) mode and on the PCW1 for the low frequency ($\omega^-(k_d)$) mode at the decoupling point k_d. The mode switching phenomenon at k_d is shown easily by plotting the ratios of the eigenmodes ($\chi = V_0/U_0$) obtained either by the PWEM. We observe that χ's change sign at the decoupling point k_d (see Figure 13(c)).

6. Electric Field Distribution and Energy Transfer

After obtaining the eigenfrequencies (dispersion relations) and eigenvectors (field amplitudes) of the coupler, we shall calculate the energy transfer between the coupled PCWs. If an EM wave with a given frequency propagates in the coupler, the wave function or field distribution at site n in each of the coupled PCWs can be expressed as the superposition of the eigenmodes of the coupler,

$$U_n(na) = Ae^{ik_a na} + Be^{ik_b na}, \tag{19}$$

$$V_n(na) = A\chi^a e^{ik_a na} + B\chi^b e^{ik_b na}, \tag{20}$$

where the propagation constants of the anti-bonding mode k_a and bonding mode k_b and their corresponding amplitude ratios of χ^a and χ^b can be obtained from Eqs. (14) and (15). Note that $\chi^a\chi^b$ is not necessarily equal to -1 for a given frequency because the mode patterns of the coupler at a given frequency are not the eigenmodes of the same system. Let $x = na$, one can rewrite Eqs. (19) and (20) as the following continuous equations,

$$U(x) = Ae^{ik_a x} + Be^{ik_b x}, \tag{21}$$

$$V(x) = A\chi^a e^{ik_a x} + B\chi^b e^{ik_b x}. \tag{22}$$

Taking derivatives of $U(x)$ and $V(x)$ with respect to x, we have coupled PCW equations.

$$\frac{dU(x)}{dx} = iM_1U(x) + i\kappa_{12}V(x), \tag{23}$$

$$\frac{dV(x)}{dx} = iM_2V(x) + i\kappa_{21}U(x), \tag{24}$$

where $M_1 = \left(k_a\chi^b - k_b\chi^a\right)/\left(\chi^b - \chi^a\right)$ and $M_2 = \left(k_a\chi^a - k_b\chi^b\right)/\left(\chi^a - \chi^b\right)$ are the effective propagation constants of PCW1 and PCW2 of the directional coupler, $\kappa_{12} = (k_a - k_b)/(\chi^a - \chi^b)$ and $\kappa_{21} = -\kappa_{12}\chi^b\chi^a$ are the effective coupling coefficients between PCWs. The solutions of the coupled PCW equations are.

$$\begin{bmatrix} U(x) \\ V(x) \end{bmatrix} = \begin{bmatrix} e^{iM_1x} & 0 \\ 0 & e^{iM_2x} \end{bmatrix} \begin{bmatrix} D_{11}\eta & iD_{12}\eta \\ iD_{21}\eta^* & D_{22}\eta^* \end{bmatrix} \begin{bmatrix} U(0) \\ V(0) \end{bmatrix}. \tag{25}$$

Here $U(0)$ and $V(0)$ are the electric field amplitudes at $x = 0$, $D_{12} = \left(\kappa_{12}\ Sin(fx)\right)/f$, $D_{21} = \left(\kappa_{21}\ Sin(fx)\right)/f$, $D_{11} = D_{22}^* = Cos(fx) - i\delta Sin(fx)/f$, and $\eta = \exp(i\delta x)$, with $f = (k_a - k_b)/2$ and $\delta = f\left(\chi^a + \chi^b\right)/\left(\chi^a - \chi^b\right)$. The maximum energy transferred from PCW1 to PCW2 is proportional to $|\kappa_{21}/f|^2 = 4(\chi^b\chi^a)^2/(\chi^b - \chi^a)^2$ and that from PCW2 to PCW1 is proportional to $|\kappa_{12}/f|^2 = 4/(\chi^b - \chi^a)^2$. There are maximum energy transfers into the other waveguides at $fL=\pi/2$, so coupling length L is equal to $\pi/|k_a - k_b|$. There are no crossing points in asymmetric PCWs to make the coupling length becoming infinite as shown in Figure 14(a), but the smallest energy transfer still occurs around the decoupling points. The energy will transfer completely into the other waveguide only in symmetric ones because it happens only at $\delta = 0$.

For an incident wave frequency ω, the wave vector k_a for the anti-bonding mode should be larger than k_b for the bonding mode for $k < k_d$. Since $|\chi^a|$ is smaller for large wave vector, if we denote the mode ratio of the lower frequency band at k_b as $\chi^a(k_b)$, we have $|\chi^b\chi^a| \le |\chi^b\chi^a(k_b)| = 1$ and $4(\chi^b\chi^a)^2/(\chi^b - \chi^a)^2 \le 4/(\chi^b - \chi^a)^2 \le 1$, shown in Figures. 14(b) and (c), which are obtained by the PWEM. Therefore, the maximum energy transfer from PCW1 to PCW2 should be smaller than that transfer from PCW2 to PCW1. However $|\kappa_{12}/f|^2$ can be larger than 1, meaning the output peak energy can be larger than the input peak energy. It results from the different strength of electric field localization in each waveguide.

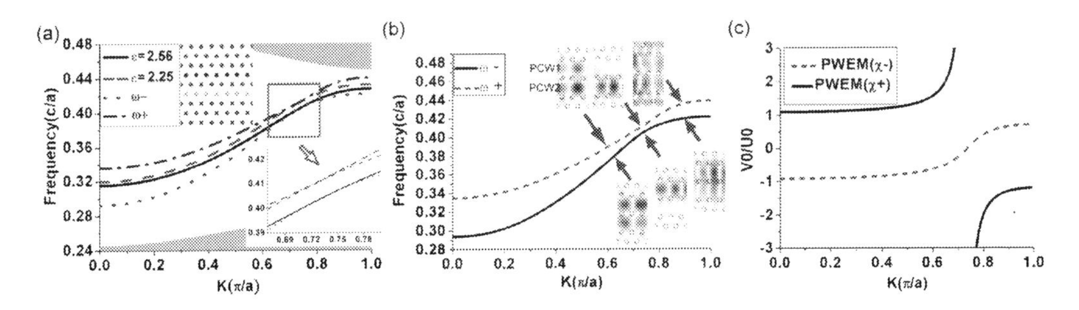

Figure 13. Simulation results of PWEM. (a) Dispersion relations of two isolated PCWs (ε =2.56 and ε=2.25) and the directional coupler in the triangular lattice (shown as the inset). (b) The dispersion curves of the directional coupler and its eigenmode profiles below, above and at the decoupling point. (c) The mode amplitude ratios of the coupler. (After Ref. [40];©2009 IOP.).

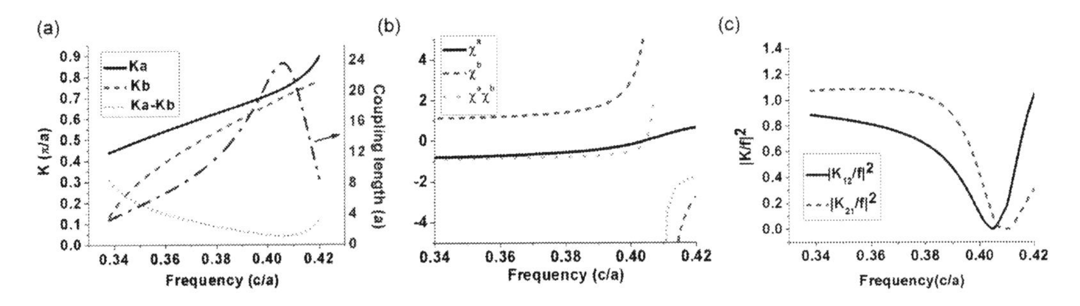

Figure 14. (a) The wave vectors of the bonding (antibonding) mode and coupling length of the PC couplers for different frequencies. (b) The mode amplitude ratio (χ=V_0/U_0) of the bonding and antibonding mode. (c) The ratios of the maximum energy transferred from the PCW2 to PCW1 ($|\kappa_{12}/f|^2$) and from the PCW1 to PCW2 ($|\kappa_{21}/f|^2$). (After Ref. [40];©2009 IOP.)

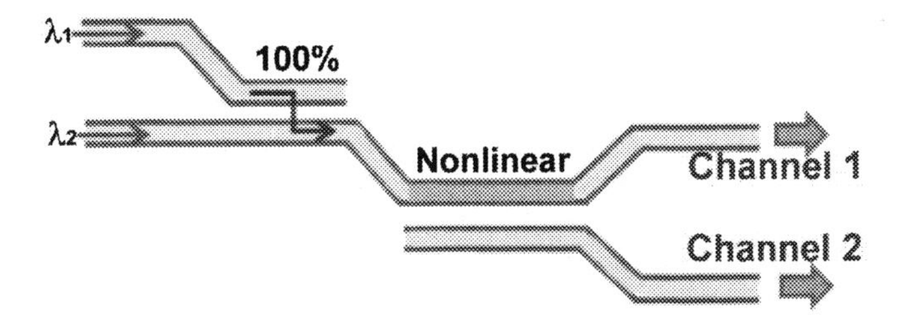

Figure 15. The structure of a switcher made of the directional coupler.

For application of asymmetric coupler, it can be used to design as a switcher, shown as Figure 15. When nonlinear material such as the quantum dots is added in one branch of the symmetric coupler, the weak single beam with wavelength λ_2 would couple into the other waveguide at the operation frequency near but not at the decoupling point at the coupling length. The single beam will output at channel two. However, as the strong-intensity control beam with wavelength λ_1 couples into the coupler, the coupler will become asymmetric. When the operating frequency is around the decoupling point, the coupling efficient is very low. Fields with wavelength λ_2 will output from the channel one. Therefore, the signal beam is switched by the control beam in a directional coupler.

7. Conclusion

In this chapter, the TBT is used to study the wave propagation of PCWs. The basic advantage of this method is that an analytic solution can be derived to describe wave evolution in the single waveguides or couplers. When the other identical waveguide is carved into the single PCW with one or several partition rods, we found the dispersive behavior of couplers can be interpreted by the inter-waveguide coupling coefficients α and β from the nearest neighbor and the next-nearest neighbor defects of PCWs, respectively. The former leads to the splitting of dispersion curves of couplers and determines the parity of the fundamental modes. The dispersion curves are further modulated by the coefficient β. The inequality $|\alpha| < |2\beta|$ is the criterion for occurring crossed dispersions at $\cos(ka)= -\alpha/(2\beta)$, where the PCWs are decoupled.

The crossing point of the dispersion relations and coupling length is tuned by moving the defect rods. When the defect rods are moved along the propagation direction, the crossing point shifts toward the high frequency in triangular lattice PCs and toward the low frequency in square lattice PCs. As the two rows of the defect rods in the waveguides are moved close to each other, the coupling length descends and the coupling length ascends as moving two rows of the defect rods away from each other. These phenomena can be well explained by the TBT and are consistent with simulation results from the FDTD method.

As waveguides are not identical, the symmetry breaks and these two dispersion curves will never cross, even with the criterion $|\alpha| < |2\beta|$. But fundamental modes still switch their parities at this criterion. At the higher frequency of the dispersion curves, the electric field distribution of the eigenmodes would localize mainly at the PCW with a higher eigenfrequency and vice versa, which corresponds to an incomplete EM field transfer between waveguides. For a given frequency, the incident electric-field distributions and energy transfer of the coupler can be expressed analytically by the wavevector and derived amplitude ratios of bonding and anti-bonding modes. The coupling length at the decoupling point becomes finite, but there still exists the smallest energy transfer around there. Although a complete energy transfer into the other waveguides is impossible in asymmetric waveguides, the peak power in the output dielectric rods can be larger than that in the input ones due to the electric fields having different strength of electric field localization in each waveguide.

References

[1] Johnson, SG. *Photonic crystals : the road from theory to practice* (Kluwer Academic, Boston, 2002).
[2] Yablonovitch, E. *J. Phys. Condens. Matter*, 1993, 5, 2443-2460.
[3] Yablonovitch, E. *J. Opt. Soc. Am. B*, 1993, 10, 283-295.
[4] Soukoulis, CM. *Phtonic crystals and light localization in the 21st century* (Kluwer Academic, Dordrecht, The Netherlands, 2000).
[5] Joannopoulos, JD; Johnson, SG; Winn, JN; Meade, RD. *Photonic crystals* (Princeton University Press, New Jersey, 2008).

[6] Asano, T; Song, BS; Akahane, Y; Noda, S. *IEEE J. Sel. Top.* Quantum Electron, 2006, 12, 1123-1134.
[7] Ren, C; Tian, J; Feng, S; Tao, HH; Liu, YZ; Ren, K; Li, ZY; Cheng, BY; Zhang, DZ; Yang, HF. *Opt. Express*, 2006, 14, 10014-10020.
[8] Djavid, M; Abrishamian, MS. *Opt. and Quantum Electron*, 2007, 39, 1183-1190.
[9] Song, BS; Asano, T; Noda, S. *Nano*, 2007, 2, 1-13.
[10] Yang, XD; Wong, CW. *Opt. Express*, 2007, 15, 4763-4780.
[11] Yariv, A; Xu, Y; Lee, RK; Scherer, A. *Opt. Lett*, 1999, 24, 711-713.
[12] Christodoulides, DN; Efremidis, NK. *Opt. Lett*, 2002, 27, 568-570.
[13] Jagerska, J; Le Thomas, N; Zabelin, V; Houdre, R; Bogaerts, W; Dumon, P; Baets, R. *Optics Letters*, 2009, 34, 359-361.
[14] Meade, RD; Devenyi, A; Joannopoulos, JD; Alerhand, OL; Smith, DA; Kash, K. *J. Appl. Phys.*, 1994, 75, 4753-4755.
[15] Mekis, A; Chen, JC; Kurland, I; Fan, SH; Villeneuve, PR; Joannopoulos, JD. *Phys. Rev. Lett*, 1996, 77, 3787-3790.
[16] Johnson, SG; Manolatou, C; Fan, SH; Villeneuve, PR; Joannopoulos, JD; Haus, HA. *Opt. Lett*, 1998, 23, 1855-1857.
[17] Mekis, A; Fan, SH; Joannopoulos, JD. *Phys. Rev. B*, 58, 4809-4817.
[18] Almeida, VR; Barrios, CA; Panepucci, RR; Lipson, M. *Nature*, 2004, 431, 1081-1084.
[19] Bogaerts, W; Taillaert, D; Luyssaert, B; Dumon, P; Van Campenhout, J; Bienstman, P; Van Thourhout, D; Baets, R; Wiaux, V; Beckx, S. *Opt. Express*, 2004, 12, 1583-1591.
[20] Tokushima, M; Kosaka, H; Tomita, A; Yamada, H. *Appl. Phys. Lett*, 2000, 76, 952-954.
[21] Lin, SY; Chow, E; Bur, J; Johnson, SG; Joannopoulos, JD. *Opt. Lett*, 2002, 27, 1400-1402.
[22] Hosomi, K; Katsuyama, T. *IEEE J. Quantum Electron*, 2002, 38, 825-829.
[23] Kuchinsky, S; Golyatin, VY; Kutikov, AY; Pearsall, TP; Nedeljkovic, D. *IEEE J. Quantum Electron*, 2002, 38, 1349-1352.
[24] Yamamoto, N; Ogawa, T; Komori, K. Opt. *Express*, 2006, 14, 1223-1229.
[25] Chung, LW; Lee, SL. Opt. *and Quantum Electron*, 2007, 39, 677-686.
[26] Yu, TB; Wang, MH; Jiang, XQ; Liao, QH; Yang, JY. Opt. J. A, *Pure Appl. Opt*, 2007, 9, 37-42.
[27] Ha, SW; Sukhorukov, AA; Dossou, KB; Botten, LC; Lavrinenko, AV; Chigrin, DN; Kivshar, YS. Opt. *Express*, 2008, 16, 1104-1114.
[28] Qiu, M; Jaskorzynska, B. Appl. *Phys. Lett*, 2003, 83, 1074-1076.
[29] Sharkawy, A; Shi, S; Prather, DW. *Appl. Opt*, 2002, 41, 7245-7253.
[30] Chien, FSS; Hsu, YJ; Hsieh, WF; Cheng, SC. Opt. *Express*, 2004, 12, 1119-1125.
[31] Miller, DAB. *J. Opt. Soc. Am. B*, 2007, 24, A1-A18.
[32] Momeni, B; Yegnanarayanan, S; Soltani, M; Eftekhar, AA; Hosseini, ES; Adibi, A. J. *Nanophotonics*, 2009, 3.
[33] Nishihara, H; Haruna, M; Suhara, T. *Optical Integrated Circuits* (McGraw-Hill, New York, 1989).
[34] Leung, KM; Liu, YF. *Phys. Rev. Lett*, 1990, 65, 2646-2649.
[35] Johnson, SG; Joannopoulos, JD. *Opt. Express*, 2001, 8, 173-190.
[36] Bayindir, M; Temelkuran, B; Ozbay, E. *Phys. Rev. Lett*, 2000, 84, 2140-2143.
[37] Kamalakis, T; Sphicopoulos, T. *IEEE J. Quantum Electron*, 2005, 41, 1419-1425.
[38] Mookherjea, S. Opt. *Lett*, 2005, 30, 2406-2408.

[39] Neokosmidis, I; Kamalakis, T; Sphicopoulos, T. *IEEE J. Quantum Electron*, 2007, 43, 560-567.

[40] Huang, CH; Hsieh, WF; Cheng, SC. *J. Opt. A, Pure Appl. Opt*, 2009, 11, 015103.

[41] Huang, CH; Lai, YH; Cheng, SC; Hsieh, WF. *Opt. Express*, 2009, 17, 1299-1307.

[42] Huang, CH; Hsieh, WF; Cheng, SC. *J. Korean Phys. Soc.*, 2008, 53, 1246-1250.

[43] Chien, FSS; Tu, JB; Hsieh, WF; Cheng, SC. *Phys. Rev. B*, 2007, 75, 125113.

[44] Chien, FSS; Cheng, SC; Hsu, YJ; Hsieh, WF. *Opt. Commun*, 2006, 266, 592-597.

[45] Huang, CH; Hsieh, WF; Cheng, SC. *J. Opt. Soc. Am. B*, 2009, 26, 203-209.

[46] Taflove, A; Hagness, SC. *Computational electrodynamics: the finite-difference time-domain method* (Artech House, Norwood, MA, 2000).

[47] Koponen, T; Huttunen, A; Torma, P. *J. Appl. Phys.*, 2004, 96, 4039-4041.

In: Photonic Crystals
Editor: Venla E. Laine, pp. 115-133

ISBN: 978-1-61668-953-7

Chapter 5

A PHOTONIC BAND GAP IN QUASICRYSTAL-RELATED STRUCTURES

***P.N. Dyachenko*[1] *and Yu.V. Miklyaev*[2]**
[1]Image Processing Systems Institute RAS, Samara, Russia
Samara State Aerospace University, Samara, Russia.
[2]South Ural State University, Chelyabinsk, Russia

Abstract

The most interesting phenomena in photonic crystals stipulate the presence of omnidirectional band gap, i.e. overlapping of stop bands in all directions. A higher rotational symmetry and an isotropy of quasicrystals in comparison with ordinary crystals give a hope to achieve a gap opening at lower dielectric contrasts. But nonperiodic nature of quasicrystals makes the size of stop bands lower than in the case of an ordinary periodic crystal. Another possibility to get a higher lattice isotropy (symmetry) is to use the so called quasicrystal approximants and periodic structures with a large unit cell. Within such a unit cell quasicrystal approximants are nearly identical to the actual quasicrystal. Outside of that unit cell, they qualitatively resemble the quasicrystal, but in a periodic manner. This chapter is devoted to investigation of a photonic band structure of such lattices.

The transition from periodic structure to nonperiodic one is studied for 2D case by considering quasicrystal approximants of growing period. It is done in order to weigh advantages and disadvantages of quasicrystals. It is shown, that higher isotropy in the 2D case allows one to reduce the refractive index threshold necessary for the gap opening

For the 3D case we consider different approximants of icosahedral quasicrystals with 8 and 32 "atoms" per unit cell and Si-34 structure with 34 "atoms" per unit cell. All considered structures demonstrate a photonic gap opening and a high isotropy of the band structure. At the same time there are no structures with refractive index threshold for gap opening lower than the best cases of periodic lattices, in particular, with diamond symmetry.

Introduction

Quasicrystal structures found in metal alloys in the early 80s possess point symmetry groups incompatible with a periodicity [1]. In a comparison with crystals, they possess a higher rotational symmetry, such as icosahedral, decagonal, and the like. This discovery has significantly changed the understanding of the role of aperiodic ordering in a condensed state physics [2], stimulating the search of physical properties peculiar to aperiodic structures. Electron and phonon properties experience drastic changes, as Bloch's theorem is inapplicable in this case. As a result, the electron band structures and quasicrystal lattice oscillations may become rather exotic. It remained a hot topic of discussion for many years [3 – 5].

Similar problems arise when studying the interaction of photons with aperiodic dielectric structures. Similar to crystals, photons experience Bragg's diffraction. It leads to the formation of photonic band gaps (PBGs). In 1998 it was shown that two-dimensional photonic quasicrystals could possess a PBG [6], in which propagation of electromagnetic radiation of a certain polarization is forbidden in any direction. The photonic crystal concept was proposed in papers by Y. Yablonovitch [7] and S. John [8] and in the early work of V.P. Bykov [9]. In these papers it was pointed out, that with periodically structured dielectric there is a possibility to manipulate the properties of the spontaneous emission. In the subsequent papers, numerous prospective practical applications of photonic crystals were discussed [10].

A complete PBG is formed when Bragg's band gaps (stop-bands) are overlapped in all directions. When the dielectric distribution is three-dimensionally periodic, we have a different periodicity in different directions and, hence, different frequencies of the band gap centers. The overlapping band gaps can be obtained, on the one hand, with a large-size band gaps, i.e. with high-amplitude spatial harmonics of the permittivity distribution. This is accomplished by synthesizing dielectric-air lattices in materials with a high refractive index. On the other hand, the overlapping band gaps are obtained more easily when the periodicity is more isotropic, i.e. when the Brillouin zone is near-spherical in shape [11]. Among the familiar photonic crystal structures, the lattices with the diamond symmetry [12] have the least permittivity modulation contrast necessary for the band gap formation.

The quasicrystals have a high rotational symmetry and, hence, their band structure can be near-isotropic. Thus, it may be suggested that such structures are preferable in terms of a formation of the complete PBGs. In Ref. [6] the two-dimensional photonic quasicrystals of the eighth order were shown to have large PBGs for TM-polarization (the magnetic field parallel to structure plane) and TE-polarization (the electric field parallel to structure plane). Besides, it was noted that the defect states in photonic quasicrystals were more complex and promising in terms of possibilities for flexible state parameters adjustment. The same authors pointed out the necessity for studying three-dimensional photonic crystals. The first two-dimensional photonic crystal featuring a two-dimensional complete PBG (for both TE and TM polarization) was proposed in Ref. [13]. Such a structure was theoretically and experimentally shown [13] to have a low dielectric contrast threshold for the gap opening.

For the 3D case investigations of photonic quasicrystals started from experimental works. In Ref. [14] a cubic icosahedral quasicrystal with large band gaps in certain directions in microwave range was fabricated using a stereolithography technique. Its transmission coefficients in the microwave range were measured, but the theoretical analysis was not conducted. In the recent publications, fabrication of three-dimensional photonic quasicrystals for infrared [15] and visible [16] range was reported, suggesting that the topic of photonic quasicrystals gains momentum [17].

A theoretical study of 3D photonic quasicrystals was postponed in comparison with their successful synthesis. Photonic quasicrystals have no translational symmetry consequently, robust methods for calculating their optical properties have not been offered yet. Note that these methods will require considerable computational efforts. In particular, it will be challenging to use the method of the plane waves decomposition. The problem can be tackled by studying approximants of photonic quasicrystals. In Ref. [18] it was shown that even the lowest-order two-dimensional approximants could have a near isotropic PBG, with the PBG's location and size being practically independent of the approximant order. The authors of Ref. [19] showed that higher-order quasicrystal approximants had the band gap threshold equal to that in quasicrystals.

This chapter is devoted to band structure investigations of 2D and 3D approximants of photonic quasicrystals and crystals with a large number of "atoms" per the unit cell. There are several methods to calculate lattice sites positions in the quasiperiodic case [20]. Two most known are multigrid and "cut-and-project" methods. The "cut and project" construction, is that a quasicrystal consists of a slice (an intersection with one or more hyperplanes) of a higher-dimensional periodic pattern. In order that the quasicrystal itself be aperiodic, this slice must not be a lattice plane of the higher-dimensional lattice. (For example, Penrose tilings can be viewed as two-dimensional slices of five-dimensional hypercubic structures.). From the position of synthesis the multigrid method have a close analogy in a holographic (interference) lithography [16, 21,22].

Part 1) is devoted to investigation of a photonic band structure of approximants of 2D photonic quasicrystals with different size and symmetry those that can be obtained by an interference lithography.

For 3D quasicrystals the experimental realization of interference lithography method has not such success as in 2D case till now. Most of the work was concentrated on icosahedral symmetry and two-photon absorption lithography as a method of synthesis where position of lattice size is calculated on the base of cut-and-project method [20]. In the part 2) we consider the band structure of different approximants of icosahedral quasicrystal that can be synthesized by this method of lithography.

In part 3) we consider the band structure of photonic crystal with a Si-clatrate symmetry. In this case we get higher symmetry of crystal properties due to the large number of "atoms" per unit cell.

1. 2D Approximants of Photonic Quasicrystals Obtainable by Holographic Lithography

1.1. Approximants Configuration

Following the method, described in [22], we define the spatial dielectric distribution in our structure as threshold function of the absorbed light energy. Through multiple exposures of the dual beam interference we get the absorbed energy distribution that has the form:

$$I(\vec{r}) = \sum_{i=1}^{N} \cos^2(\vec{k}_i \cdot \vec{r}), \qquad (1)$$

where $\vec{k_i}$ is the 2D grating vector with the magnitude $\vec{k_i} = (kx_i, ky_i)$. If we have

$$\vec{k}_i = \left|\vec{k}\right|\left(\cos(2\pi \cdot i / N), \sin(2\pi \cdot i / N)\right), \qquad (2)$$

we get energy distribution corresponding to the quasicrystal with a 2N-fold rotational symmetry. It means that for 8-fold quasicrystal we need 4 exposures, and for 12-fold symmetry 6 exposures are required. In this case grating vectors have the same length. To get a periodic distribution all vectors $\vec{k}_i$ must belong to the 2D periodic lattice – a square or a hexagonal one. The degree of quasicrystal approach will be defined by the deviation of grating vectors from the distribution given by Eq.(2). To obtain a better approximation of the 8-fold quasicrystal for grating vectors of approximants it is reasonable to choose a square lattice, and for the 12-fold one the hexagonal lattice is more suitable. For 8-fold quasicrystal we have studied three approximants of different size. The four corresponding grating vectors have the following coordinates in the reciprocal space:

our first approximant - $\vec{k}_1 = (1,0)$, $\vec{k}_2 = (3/4, 3/4)$, $\vec{k}_3 = (-3/4, 3/4)$, $\vec{k}_4 = (0,1)$;

the second approximant - $\vec{k}_1 = (1,0)$, $\vec{k}_2 = (5/7, 5/7)$, $\vec{k}_3 = (-5/7, 5/7)$, $\vec{k}_4 = (0,1)$;

the third approximant - $\vec{k}_1 = (1,0)$, $\vec{k}_3 = (-7/10, 7/10)$, $\vec{k}_4 = (0,1)$.

The points corresponding to the grating vectors of 8-fold approximants in the reciprocal space are shown in Figure 1a).

For the 12-fold quasicrystal we have studied two approximants. Corresponding grating vectors have the form:

the first approximant: $\vec{k}_1 = (1,0)$, $\vec{k}_2 = (9/10, 3\sqrt{3}/10)$, $\vec{k}_3 = (1/2, \sqrt{3}/2)$, $\vec{k}_4 = (0, 3\sqrt{3}/5)$, $\vec{k}_5 = (-1/2, \sqrt{3}/2)$, $\vec{k}_6 = (-9/10, 3\sqrt{3}/10)$;

the second approximant: $\vec{k}_1 = (1,0)$, $\vec{k}_2 = (6/7, 4\sqrt{3}/7)$, $\vec{k}_3 = (1/2, \sqrt{3}/2)$, $\vec{k}_4 = (0, 4\sqrt{3}/7)$, $\vec{k}_5 = (-1/2, \sqrt{3}/2)$, $\vec{k}_6 = (-6/7, 4\sqrt{3}/7)$. The corresponding points of the hexagonal lattice in the reciprocal space are shown in Figure 1b).

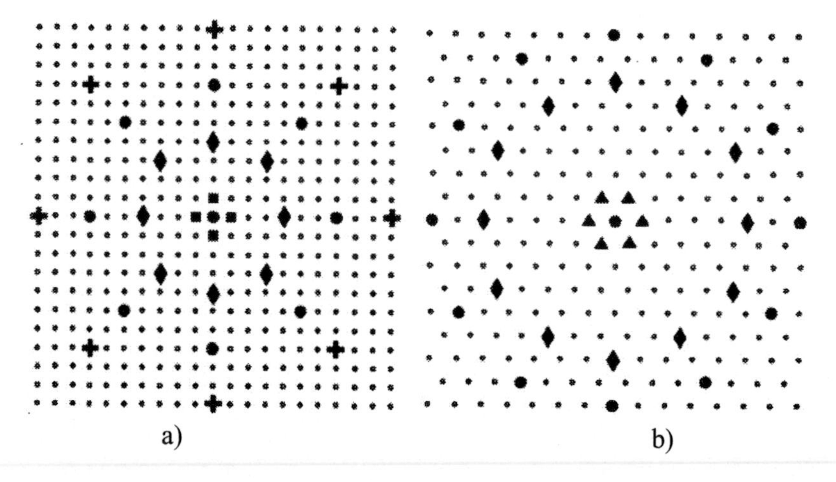

Figure 1. a). Grating vectors for creating of 8-fold quasicrystal :♦ - the first approximant, ● - the second, ✚ - the third approximants, points marked by ■ correspond to square lattice, b) grating vectors for creating of 12-fold approximants: ♦ - the first approximant, ● - the second one, ▲ - hexagonal lattice.

Figure 2. The dielectric distribution for a) the first approximant of 8-fold quasicrystal, b) second approximant of 8-fold quasicrystal, c) third approximant of 8-fold quasicrystal, d) the first approximant of 12-fold quasicrystal, f) the second approximant of 12-fold quasicrystal.

The superposition of such four and six gratings (two-wave interference patterns) gives us the required deposited energy distribution. We assume that regions with the energy level higher than the threshold value will be converted into dielectric material and the rest will be filled with an air. Examples of the dielectric distribution for different approximants are shown on Figures 2. White regions correspond to the dielectric and the black ones to the air.

The band structure calculations were based on a standard plane-wave expansion method with MIT Photonic-Bands Package [23], following the directions in the irreducible Brillouin zone, indicated on Figure 3. We restrict ourselves by consideration of TM modes. This is well known that periodic crystals constructed from dielectric rods readily open gaps in TM modes, and air rods in dielectric background is more favorable for the gaps in TE modes. As one can see from Figures 2, photonic crystal "atoms" in our case do not have the form of rods, but one can say that in our case we have disconnected structure that is not favorable for the existence of the gap in TE modes. In correspondence with this statement we have not found sizeable gaps for TE polarization for these structures.

1.2. Results

On Figures 4-6 band diagrams of TM modes are shown. All results here correspond to the optimal filling factor. This factor is defined by the choice of the deposited energy threshold. The dielectric permittivity in calculations was taken equal to that of a silicon (ε=12). For this permittivity we get the following optimal filling ratio dielectric/air: 14.3% for the first, 17% for the second and the third approximants of the 8-fold quasicrystal, 18.8% for the first and 19.1% for the second approximants of the 12-fold quasicrystal. The optimal filling factors for the square and the hexagonal lattices are 14% and 13%, correspondingly.

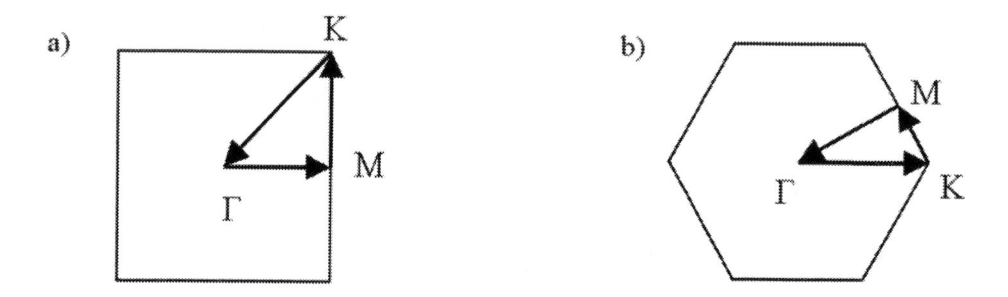

Figure 3. First Brillouin zones for a) the square and 8-fold quasicrystal approximants and b) the hexagonal and the 12-fold quasicrystal approximants. Thick lines correspond to the crystal directions probed in the band diagram calculations.

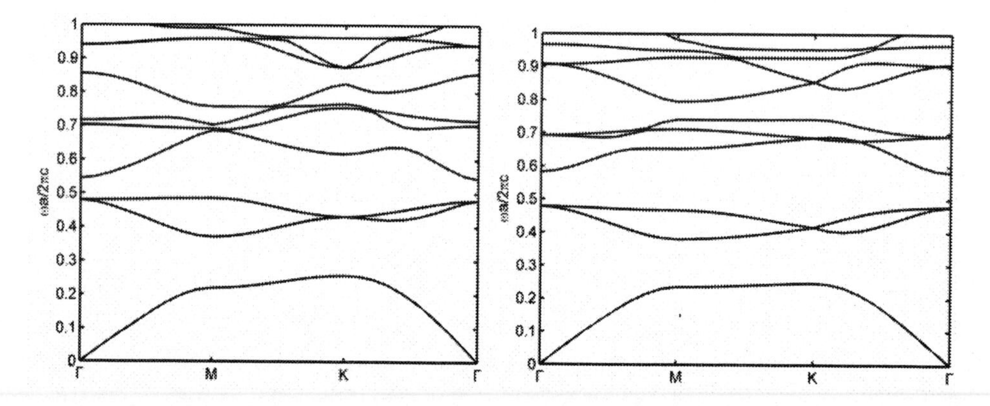

Figure 4. a) The Photonic band structure of the holographic square lattice, b) The Photonic band structure of the holographic hexagonal lattice.

As one can see, for even the smallest approximants of quasicrystals the location of the band gap and its width are distributed much more isotropic than that of the square and the hexagonal lattices. Interesting to note, that the midgap frequency for lattices with 8-fold symmetry is almost coincident with that of 4-fold symmetry, and the midgap frequency of hexagonal lattice coincides with that of 12-fold quasicrystal.

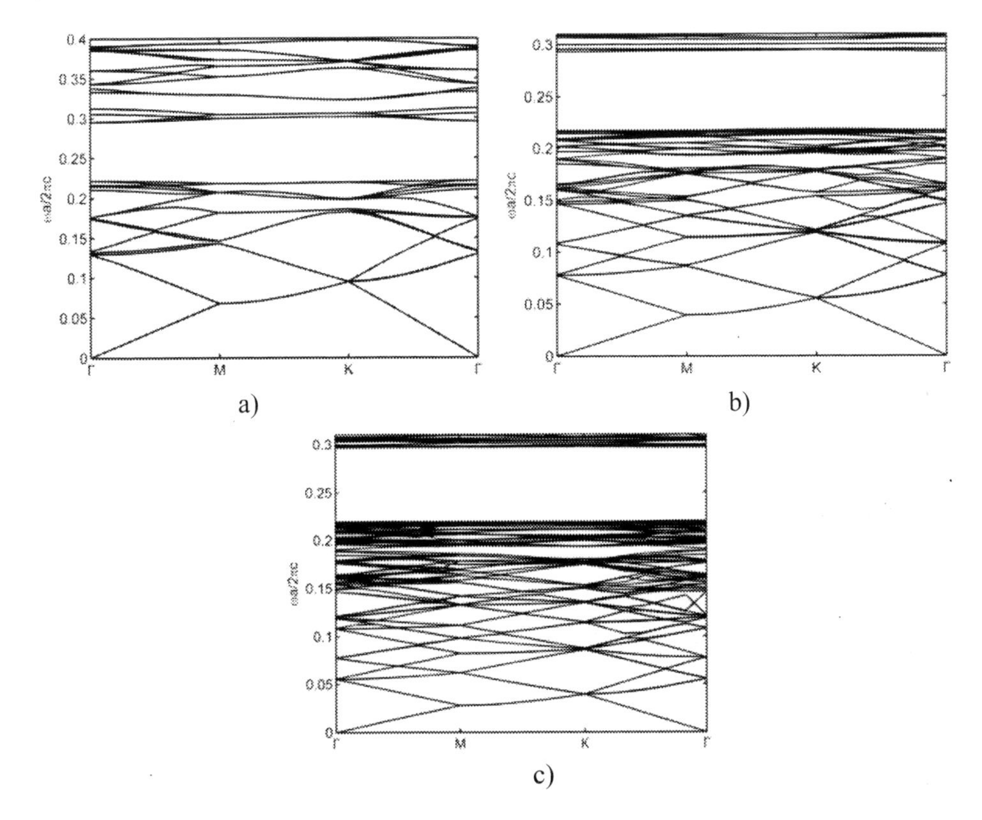

Figure 5. a) The Photonic band structure of the first 8-fold approximant, b): The Photonic band structure of the second 8-fold approximant, c): The Photonic band structure of the third 8-fold approximant.

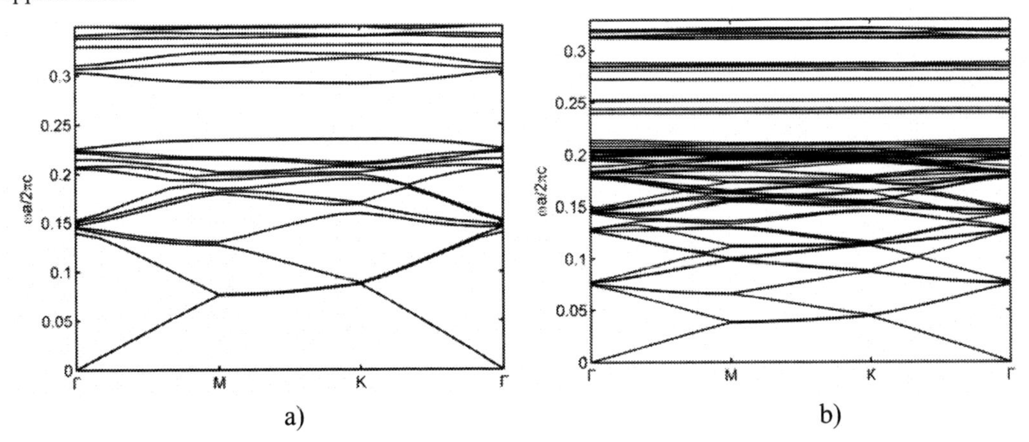

Figure 6. a) The Photonic band structure of the first 12-fold approximant, b) The Photonic band structure of the second 12-fold approximant.

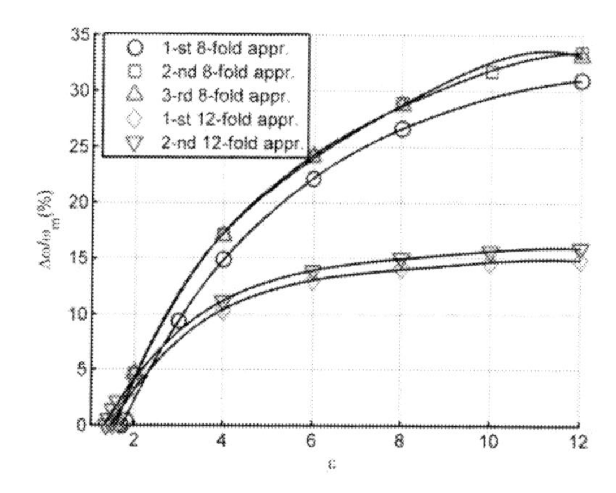

Figure. 7. The Size of the complete band gap $\Delta\omega / \omega_m$ vs. the relative permittivity ε.

On Figures 7 the dependencies of the gap size on the dielectric contrast for the 8-fold and the 12-fold approximants are shown. As previously, the results correspond to the optimal filling factors. We have also found the threshold dielectric contrasts for the gap opening in TM modes for the case of holographic square lattice ($\varepsilon_{th} = 3.0$) and the hexagonal one ($\varepsilon_{th} = 1.75$).

These results of the band structure calculations can be summarized into the table depicted on Figure 8.

As one can see from comparison of Figures 7, a higher degree of isotropy of a 12-fold quasicrystal than that of an 8-fold one leads to lower thresholds of dielectric contrast for the gap opening. But for the high dielectric contrast, the band gap of 12-fold quasicrystal is smaller than that of the 8-fold quasicrystal. It can be explained as follows. More isotropic structure makes stop band overlapping easier for all directions. At the same time lower periodicity in that case makes each stop band smaller. For a large dielectric contrast each stop band is much larger than the difference between lowest and highest midgap frequencies, so the isotropy is not governing factor in this case. To our knowledge, the threshold dielectric contrast ε_{th}=1.35 for the gap opening in TM modes (the second approximant of the 12-fold quasicrystal) is the lowest of ever reported. For comparison, the threshold dielectric constant required for a band gap in TM polarization for the square lattice of rods is ε_{th}=3.8 [10]. For the octagonal quasicrystal of rods the threshold value is ε_{th}=1.6.

Rotation symmetry	4	6	8 (1-st appr.)	8 (2-nd appr.)	8 (3-rd appr.)	12 (1-st appr.)	12 (2-nd appr.)
Gap opening dielectric contrast	3.0	1.75	1.7	1.55	1.55	1.5	1.35
Gap/Midgap ratio for $\varepsilon = 12$ (%)	36	48	31	33	33	15	16

Figure 8. The Comparison of the gap size and the band gap opening dielectric threshold for quasicrystals with different rotational symmetries and periodicities

There is a discontinuity in the data sequence given on the Figure 8. With growing size of the unit cell for the 8-fold and the 12-fold symmetry we obtain a decrease of the dielectric contrast threshold, but do not observe the bandgap size decay. On the contrary, we obtain an increase of the gap size with an increase of an approximant size. This fact can be explained by the following. A different order of rotational symmetry of the unit cell and different point group symmetry makes it impossible to support all the Fourier components of the unit cell by a translational periodicity. In spite of a larger amplitude of Fourier components in a part of an angular spectrum we get no increase in the size of the omnidirectional band gap.

2. A Complete Photonic Band Gap in Icosahedral Quasicrystals

In Refs. [24, 25] it was shown that the really existing BC8 phase in silicon and germanium and the hypothetical BC32 structure were, respectively, the 1/0 and 1/1 approximants of an icosahedral quasicrystal with the six-dimensional body-centered cubic (bcc) lattice, so that all atom coordinates could be derived by projecting a six-dimensional lattice on 3D space. The approximant 1/0 contains 8 atoms in a primitive rhombohedral cell of the bcc lattice and the approximant 1/1 32 atoms - hence the structures' names. Consequently, the cubic unit cells 1/0 and 1/1 contain 16 and 64 atoms, respectively. In the structure of the approximant 1/0, all atoms are found in equivalent crystallographic positions 16 *(c)* with the coordinates *xxx*, where $x = x_{ic} = \tau^{-2} / 4$ and τ is the golden mean. Each atom of the 1/0 approximant has a quadruple coordination, which makes such a structure locally similar to a diamond structure, and this is precisely the structure for which the lowest threshold of the complete PBG formation has been obtained.

Based on the 1/0 structure, it is possible to form a 1/1 approximant whose unit cell is τ times larger. Its 64 atoms occupy 16 *(c)* positions with $x = x_{ic}$ and 48 *(c)* positions with $x = (1 - 2x_{ic}) / 2$, $y = (2\tau - 1)x_{ic}$ $z = x_{ic}$. Atoms in the 16 *(c)* positions have a triple coordination. As shown in Refs. [24, 25], it is possible to form a 1/1 approximant in which all the atoms have the coordination number of 4. Hereafter, such an approximant will be denoted as 1/1F.

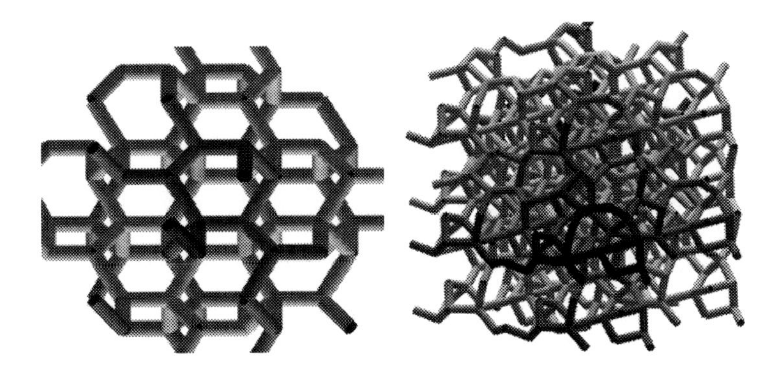

Figure 9. a). Pictorial representation of 1/0DR, b)Pictorial representation of 1/1FDR.

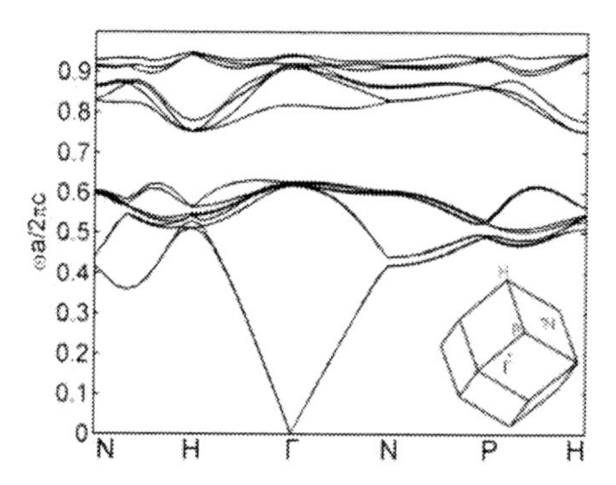

Figure 10. Band structure of the approximant 1/0DR (the first 15 bands). The dielectric filling factor: f=22.7%, the permittivity: $\varepsilon=12$. Size of the complete band gap: $\Delta\omega/\omega_m=17.6\%$.

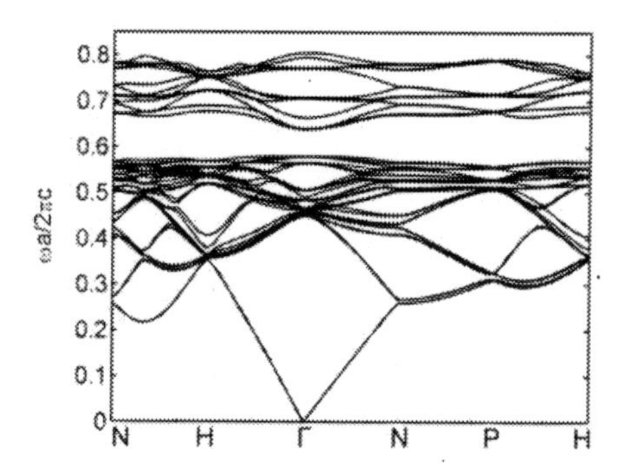

Figure 11. Band structure of the approximant 1/1DR (the first 40 bands). The dielectric filling factor: f=26%, the permittivity: $\varepsilon=12$. The size of the complete band gap: $\Delta\omega/\omega_m=9.8\%$.

For each approximant, two lattice site types were explored, a DR-lattice with dielectric rods and an AS-lattice with air spheres. In Type-I structure, dielectric rods of a definite radius connect neighboring atomic positions in the lattice. In the following, such structures are termed as 1/0DR (see Figure 9a)), 1/1FDR (see Figure 9b)), and 1/1DR. Type-II represents air spheres found in dielectric, with the corresponding approximants denoted as 1/0AS, 1/1FAS, and 1/1AS. For the sake of the simplicity, the lattice material was assumed to be non-absorbing, nonmagnetic, and isotropic. The eigenmodes of Maxwell's equations were found using a method of the field decomposition into the plane waves [23].

2.1. Results

Figure 10 shows a zone structure of the approximant 1/0DR for the permittivity of $\varepsilon=12$ (silicon in the near IR range of wavelengths) and the estimated optimal dielectric filling factor of f=22.7% (the percentage of the dielectric volume in the entire primitive cell volume). The frequency is given in dimensionless units of $\omega a/2\pi c$, where ω is the cyclic frequency, a is the size of cubic cell of 1/0DR, and c is the speed of light in the vacuum. Marked on the x-axis are the high-symmetry points of Brillouin's zones of the bcc lattice (see the inset in Figure 10). A complete PBG of the size $\Delta\omega/\omega_m=17.6\%$ takes place between the zones 8 and

9 (between the frequencies 0.6293($\omega a/2\pi c$) and 0.7509($\omega a/2\pi c$), the dielectric filling factor being *f*=22.7%. At this value of the dielectric filling factor, the ratio of the dielectric rod radius to the cubic cell size is *r/a*=0.09. The size of the PBG $(i)-(i+1)$ is defined by the following relationship:

$$\Delta\omega_{i,i+1}/\omega_m = 2\frac{\min(\omega_{i+1})-\max(\omega_i)}{\min(\omega_{i+1})+\max(\omega_i)}100\%, \quad (3)$$

where $\max(\omega_i)$ and $\min(\omega_{i+1})$ are the maximal and the minimal frequencies for the zones (i) and $(i+1)$, respectively. For 1/0AS, the PBG size was found to be equal to $\Delta\omega/\omega_m$=11% between zones 12 and 13 (between the frequencies 0.8014 and 0.8948), at ε=12 and the optimal dielectric filling factor of *f*=18.9%. This dielectric filling factor is obtained when the ratio of the air sphere radius to the primitive cell size is *r/a*=0.25. The band structure 1/0AS is shown on Figure 12. It is noteworthy that the PBG for 1/0AS lies in higher frequencies. It is also interesting that the PBG band numbers are different for 1/0AS and 1/0DR.

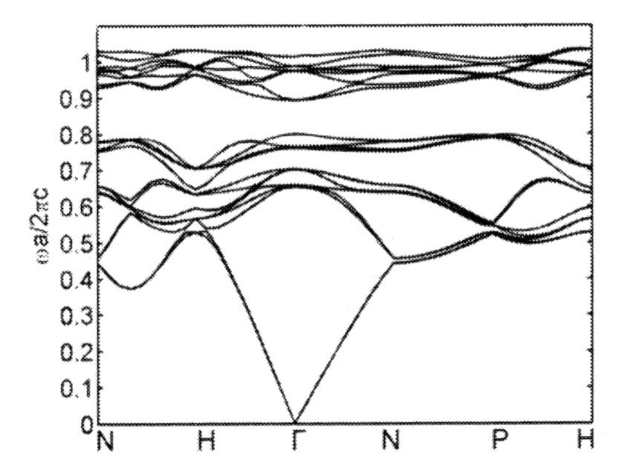

Figure 12. Band structure of the approximant 1/0AS (the first 20 bands). Dielectric filling factor: *f*=18.9%, permittivity: ε=12. Size of the complete band gap: $\Delta\omega/\omega_m$=11%.

Now we continue with higher-order approximants of photonic quasicrystals. All atoms of the approximant 1/1FAS have the coordination number of 4 and, therefore, its structure is locally similar to that of diamond. When the dielectric filling factor is f~22%, the structure of 1/1FAS has an unbound dielectric lattice, which is impossible to fabricate experimentally. Besides, the approximant 1/1FAS does not have the complete PBGs. Hence, we restrict our further analysis to the 1/1FDR structure. Such lattices are easier to generate experimentally, as was demonstrated in Refs. [14,15]. The band structure of 1/1FDR at ε=12 and the optimal dielectric filling factor of *f*=23.8% is shown in Figure 14. Similarly to the case with 1/0DR, the frequency is shown in Figure 10 in dimensionless units, with *a* being the cubic cell size of the 1/0DR in both cases. The dielectric filling factor equals *f*=23.8% when *r/a*=0.095. The size of the complete PBG is $\Delta\omega/\omega_m$=10.3%. It is located between the bands 32 and 33 (between frequencies 0.6342($\omega a/2\pi c$) and 0.7029($\omega a/2\pi c$)). It is interesting that the optimal filling factor is nearly the same in both approximants, whereas the size of the complete PBG for 1/1FDR appears to be nearly half of that for the 1/0DR. This is due to the fact that the Fourier harmonics of the permittivity spatial distribution for 1/1DR are of smaller amplitude than for 1/0DR, thus requiring a further structure optimization.

The atoms of the approximant 1/1 have different coordination numbers and, thus, comparing the properties of 1/1DR and 1/1FDR we can find in which way the atom coordination affects the band gap size and the threshold. As distinct from 1/1FAS, 1/1AS has a bound dielectric lattice. Figure 13 shows the band structure of 1/1AS at $\varepsilon=12$ and f=20.5%. The complete PBG is found between the zones 56 and 57 (between the frequencies 0.8019 and 0.8666), the PBG size being $\Delta\omega/\omega_m$=7.76%. The ratio of the air sphere radius to the size of 1/0DR's primitive cell is r/a= 0.252.

The band structure of 1/1DR is shown in Figure 11 for $\varepsilon=12$ and f=26%. The PBG is found between the zones 28 and 29 (between the frequencies 0.5783 and 0.6379), with the PBG size being $\Delta\omega/\omega_m$=9.8%. At the above value of the dielectric filling factor, the ratio of dielectric rod radius to the cubic cell size is r/a=0.1.

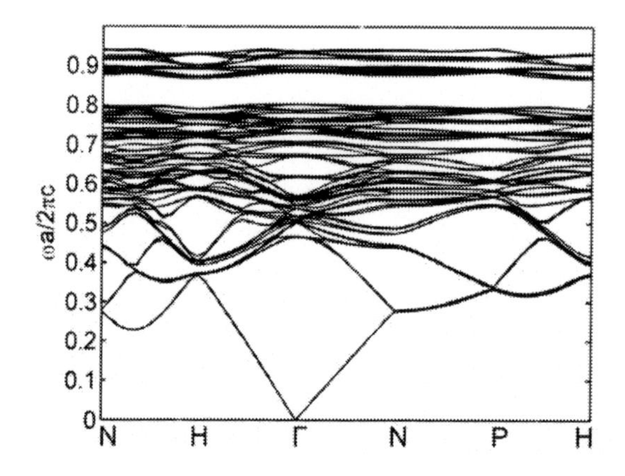

Figure 13. Band structure of the approximant 1/1AS (the first 70 bands). The dielectric filling factor: f=20.5%, the permittivity: $\varepsilon=12$. Size of the complete band gap: $\Delta\omega/\omega_m$=7.76%.

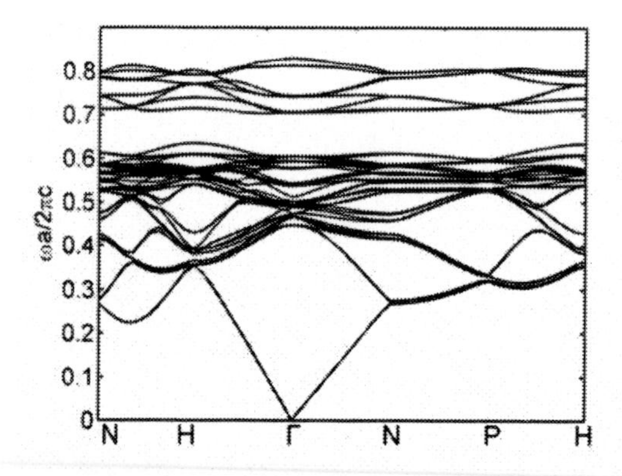

Figure 14. Band structure of the approximant 1/1FDR (the first 40 bands). The dielectric filling factor: f=23.8%, the permittivity: $\varepsilon=12$. Size of the complete band gap: $\Delta\omega/\omega_m$=10.3%.

The size of the complete PBGs was found to be nearly the same for 1/1DR and 1/1FDR, meaning that in this case the atom coordination does not have an essential effect. This fact may prove useful when designing photonic quasicrystals with a six-dimensional simple cubic lattice [24,25], in which atoms have different coordination. Also, it is interesting to note that in 1/1AS the PBG is found in higher bands and frequencies, compared to 1/1DR. This is favorable for reducing the value of the PBG threshold.

Let us now find indices of those reflections *hkl* that produce the complete PBG in 1/0DR. In view of Bragg's condition for backscattering and the mean permittivity, we have $K^2 = 4[1+f(\varepsilon-1)](a\omega/2\pi c)^2$, where $K^2 = h^2+k^2+l^2$. For the band center in Figure 10, we have $a\omega/2\pi c$=0.7, whence, K^2=6.85. Whence it follows that the complete PBG is mainly formed by the type-211 (K^2=6) and type-220 (K^2=8) reflections. However, at point *P* (third-order axis) the band gap is very wide, so that the type-222 reflections (K^2=12) are also likely to give their contribution. Calculation suggests that just above the threshold of complete PBG formation (ε=6, *f*=0.296) we have $a\omega/2\pi c$=0.78; hence, K^2=6.04, which implies that the PBG is formed owing to reflections 211 and 200, which in a quasicrystal correspond to reflections directed along the icosahedron's second-order axes (there are thirty reflections of this type on the sphere). For the 1/0AS, the results are as follows. For the band center in Figure 12, we have $a\omega/2\pi c$=0.85, whence K^2= 8.9. Consequently, the complete PBG is mainly formed by the type-221 reflections (K^2=9). When the permittivity is near-threshold, ε=6, for the 1/0AS we obtain $a\omega/2\pi c$=1 at *f*=0.238, and, hence, K^2= 8.76. Thus, the complete PBG is formed by the reflections 220 (K^2=8) and 221 (K^2=9). As Figures 10 and 12 suggest, the PBG is more isotropic in the 1/0AS than in the 1/0DR. It is due to the fact that the former, reflections responsible for the PBG have closer lengths of the inverse lattice vectors. The reason is that in the 1/0AS the PBG lies in higher frequencies, compared to the 1/0DR.

Let us analyze the 1/1FDR, for which the Bragg's condition is given by $K^2 = 4[1+f(\varepsilon-1)](\tau a\omega/2\pi c)^2$, where $\tau = (1+\sqrt{5})/2$ is the golden mean. If ε=12, *f*=0.238 and at the band center, we have $a\omega/2\pi c$=0.67, then, K^2=17.05. Concerning this approximant, the theory suggests that a strong contribution should come from the reflections 400 (K^2=16) and 321 (K^2=14), however, the reflections 411 (K^2=18) and 420 (K^2=20) also seem to give their contribution, although they do not correspond to any strong reflections in quasicrystals. At ε=6, we have *f*=0.3123 and the band center is 0.793, so that K^2=16.87. In percentage, when changing from ε=6 to ε=12, the characteristic parameter K^2 does not change as much as it does in the 1/0DR. It should be noted that in the 1/1DR the PBG is more isotropic than in the 1/0DR, which is because in the former, reflections responsible to the PBG have closer lengths of the inverse lattice vectors. This results from the fact that with increasing approximant order the structure approaches a quasicrystal geometry. As can be seen in Figures 10 and 11, the frequency bands in which the PBGs lie do overlap. Consequently, with increasing approximant order, the complete PBG is retained and, hence, it will also be peculiar to the corresponding quasicrystal [18,19].

For 1/1DR and 1/1AS, Bragg's condition looks similarly to the case of 1/1FDR. At *f*=0.26 and ε=12, the band center of 1/1DR is $a\omega/2\pi c$=0.608, then K^2=14.94. Thus, the major contribution comes from the type 321 (K^2=14) and type 400 (K^2=16) reflections. When ε=6, the approximant 1/1DR does not have a complete PBG. This can be explained by the fact that in 1/1DR the PBG is less isotropic than in 1/1FDR, which is owing to a lesser value of the modulus of the inverse lattice vector. For 1/1AS, at ε=12 and *f*=0.205, the band center is $a\omega/2\pi c$=0.8342, then we have K^2=23.72. Consequently, the PBG is mainly formed by type 422 (K^2=24) reflections. At ε=6 and *f*=0.297, the band center is $a\omega/2\pi c$=0.9538, hence, K^2=23.67. In percentage, when changing from ε=6 to ε=12, the value of K^2 does not change as much as in 1/0AS. Also, Figures 12 and 13 suggest that the PBG frequency ranges are overlapped for 1/0AS and 1/1AS.

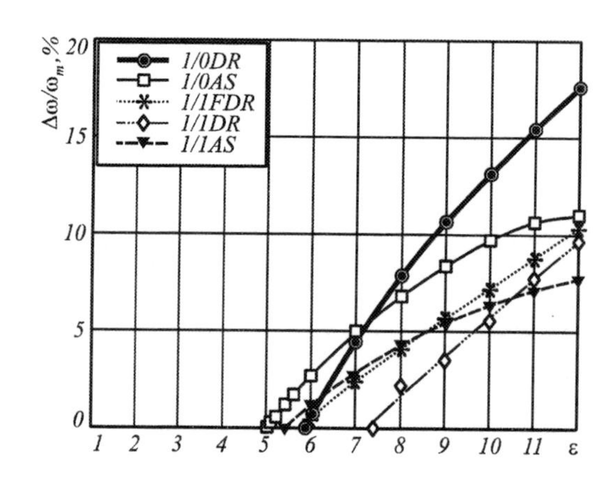

Figure 15. The Size of the complete band gap $\Delta\omega/\omega_m$ vs. the relative permittivity ε .

Figure 15 shows the band gap size as a function of permittivity. For each value of the permittivity, we deduced an optimal dielectric filling factor at which the complete PBG has a maximal size. For the approximants 1/0AS and 1/0DR, the permittivity threshold of the PBG formation was found to be ε_{th}=5 and ε_{th}=5.8, respectively. For the higher-order approximants 1/1AS, 1/1DR, and 1/1FDR, the threshold values were found to be ε_{th}=5.3, ε_{th}=7.4, and ε_{th}=5.8, respectively. For 1/0DR and 1/1FDR, the PBG threshold was found to be practically the same, which means that further increase of the approximant order brings us to the same, or a bit smaller, threshold value [19]. Similar conclusions can be made with respect to the approximants 1/0AS and 1/1AS. It can be noted that for 1/1DR, the PBG threshold turned out to be considerably larger than for 1/1FDR. This is due to the influence of the atom coordination number. Thus, we can infer that for the quasicrystals composed of dielectric rods the preferable atomic arrangement is that in which atoms have the same coordination number. Besides, we can infer that when realizing photonic crystals with low PBG threshold, the preferable configuration is air spheres in dielectric. In principle, it is also possible to construct approximants with different sets of strong reflections that will have lower thresholds of the complete PBG formation [26,27].

3. Band Structure of A Photonic Crystal with the Clathrate Si-34 Lattice

As an alternative to photonic quasicrystals and their approximants one can consider crystals with many atoms per-unit-cell as a lattice of higher isotropy. In 2D case it was shown that such ~~a~~ crystals can have wide and isotropic band gap [28,29]. In 3D case a pyrochlore crystal lattice with 4 "atoms" in the unit cell was theoretically studied [30]. The PhC was shown to have PBGs comparable in size with the diamond lattice's PBG.

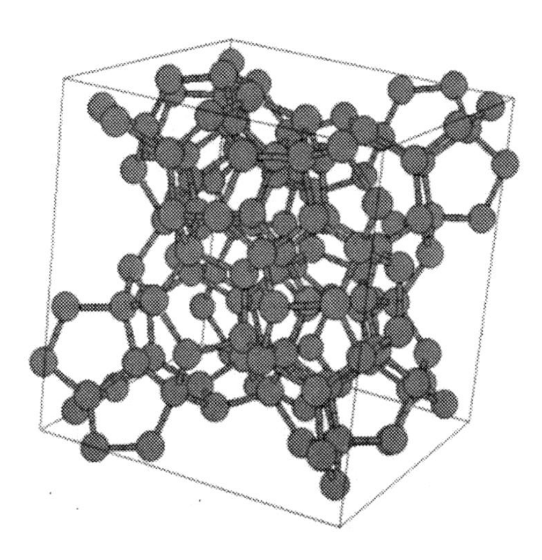

Figure 16. Four unit cells of the Si-34 lattice.

Below, a PhC with a large number of per-unit-cell sites with isotropic properties is exemplified by a crystal with the clathrate Si-34 crystal lattice [31]. This lattice belongs to a class of face-centered cubic (fcc) lattices, containing 34 sites per cell (see Figure 16). The lattice is obtained by packing a pentagonal dodecahedron and tetrakisdecahedron. In Si-34, each "atom" is bonded with four neighbors by distorted tetrahedral bonds. The diamond crystal lattice, known to have tetrahedral atom bonds, allows obtaining the largest PBG among all photonic crystals. In the Si-34 clathrate lattice, atoms are arranged in the very isotropic manner and the Brillouin zone is closest in shape to a sphere. This is favoring for the opening of the complete PBG. All of the preceding suggests that studies into the clathrate crystal lattices as candidates for photonic crystal applications show great promise. The objective of this work is to look into the possibility of creating a complete PBG in Si-34-based photonic crystals.

In our studies, we used the following model. The dielectric material is supposed to be non-absorbing, non-magnetic, and isotropic, i.e. the refractive index of the medium is $n=\sqrt{\varepsilon}$, where ε is the permittivity of the dielectric in the optical range. The eigenmodes of Maxwell's equations with periodic boundary conditions were computed by preconditioned conjugate-gradient minimization of the block Rayleigh quotient in a plane-wave basis, using the freely available software package MIT Photonic-Bands (MPB) [23].

Three cases were analyzed: (1) the lattice sites are presented by dielectric spheres in air, (2) the lattice sites are presented by hollow spheres in dielectric, and (3) neighboring lattice sites are connected with dielectric rods. As a result of calculations, we have found that the Si-34-based photonic crystal composed of dielectric spheres in air has no large band gaps (on the order of 5% of the permittivity $\varepsilon=12$), where as the photonic crystal composed of air spheres in the dielectric has no band gaps at all. At the same time for the lattice composed by dielectric rods and with the same value of ε the size of the PBG is about 16%. This can be pointed out as a distinguishing feature of the studied structure, because usually air spheres in dielectric matrix is much more favorable for the gap opening than dielectric spheres and comparable with dielectric rod structure. It can be explained as follows. In comparison with, for example, fcc or diamond grating, the studied structure has rather nonuniform distribution of atoms in spite of its high isotropy. For reasonable value of the ratio between dielectric and air volumes the structure of air spheres in dielectric makes dielectric mesh disconnected. From the same reason the dielectric spheres at the lattice sites also do not allow to get a connected structure. That leads to decrease of the dielectric mode frequency and closes the PBG. It is well known, that a connectivity of two interpenetrating dielectric and air lattices is favourable for the gap opening [10]. Such connectivity we obtain only for the case of dielectric rods, connecting lattice sites. Therefore, our further research was focused on studying the photonic crystal composed of the dielectric rods in the air, connecting the neighboring lattice sites and thus forming tetrahedral bonds.

Such a structure can be synthesized from a photopolymer the by two-photon absorption lithography with the length of bonds, corresponding to the PBG in the near infrared region [15]. To get a sufficiently high dielectric contrast this connected polymeric lattice should be replicated in silicon, by the means of a double inversion procedure, as it was demonstrated for the case of photonic quasicrystals [15].

Figure 17 shows the band structure of a photonic crystal at $\varepsilon=12$ and an optimal filling factor of f=22% (the ratio of the dielectric volume to the cell's total volume). Such a filling factor is obtained with the radius of rods equal to r/a=0.045, where a is the size of the cubic cell. The length of the dielectric rods is equal l/a=0.16. The frequency is shown in the plots in the dimensionless units of $\omega a / 2\pi c$, where ω is the cyclic frequency, and c is the velocity of light in the vacuum. On the abscissa, the high-symmetry points of the fcc lattice's Brillouin zone are plotted. The complete PBG is found between the 34-th and 35-th bands. The position of the band gap corresponds to the length of the tetrahedral atom bonds. It can be seen from the comparison of the ratio of mid-gap frequencies of diamond and Si-34 structures and the ratio of distances between the neighbor sites of these two lattices. In this sense the gap in Si-34 corresponds to the fundamental gap of a diamond grating. The band gap has the size of $\Delta\omega / \omega_m$=15.6%, the filling factor being f=22%. Thus, it is a photonic crystal with a clathrate lattice symmetry with 34 "atoms" per unit cell that has a complete PBG.

With the aim of determining a minimal value of the permittivity ε_{th} that leads to the emergence of a complete PBG, i.e. determining the complete PBG threshold, the relation between the band gap width and the permittivity was derived. To obtain this relation, at each permittivity value ε we derived the dielectric filling factor f corresponding to the maximal complete PBG, as shown in Figure 18. Figure 18 shows that the complete PBG threshold is found at ε_{th}=5.0. This threshold value is higher than the threshold of ε_{th}=4.0 for a photonic crystal with the diamond lattice symmetry, but lower than that of the photonic crystal with the inverted opal lattice [32]. The reason is that the Si-34-based PBG structures are more isotropic, so that the frequency of their boundary bands (bands bounding the PBG) weakly depends on the direction of the electromagnetic wave propagation [33].

We have studied the band gap isotropic properties. For the photonic band *(i)*, the isotropy parameter F was defined as follows [34] $F = 2(\max(\omega_i) - \min(\omega_i)) / (\max(\omega_i) + \min(\omega_i))$. Of the earlier known PhC structures, the most isotropic PBG we found in the crystal with the fcc lattice, composed of air spheres in dielectric. At f=27% and ε=12, such a lattice has F=0.08 for the "lower" band and F=0.066 for the "upper" band. At the same conditions (r/a=0.051), a PhC with the Si-34 crystal lattice has the isotropy parameter of F=0.022 for the "lower" band and F=0.056 for the "upper" band. Thus, the PBG of the clathrate Si-34 lattice demonstrates higher isotropy. We also found out, that the parameter F is monotonically decreases with increasing permittivity.

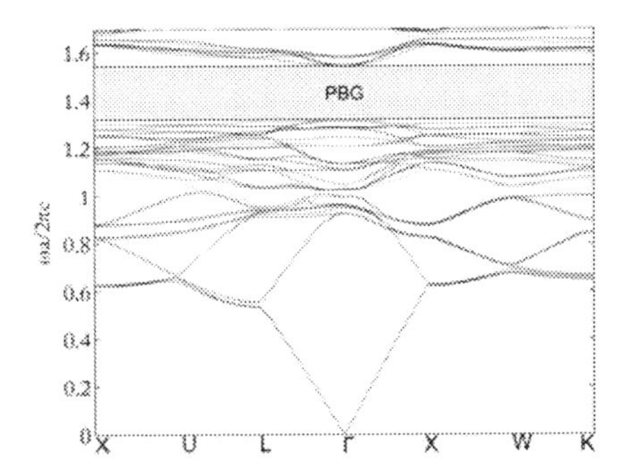

Figure 17. Band structure of Si-34. Permittivity: ε=12. Filling factor: f=22%. Band gap size: $\Delta\omega / \omega_m$=15.6%.

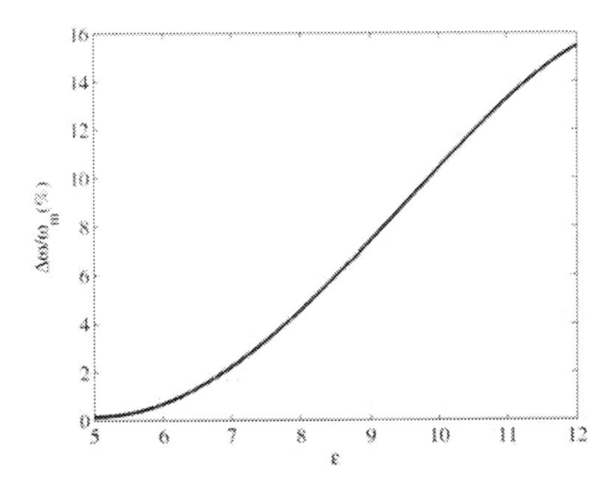

Figure 18. The band gap size $\Delta\omega / \omega_m$ *vs* the permittivity ε.

It has been known that the group velocity of the electromagnetic waves can become zero at the PBG boundary. Physically, this means that waves scattered at the PBG boundary form a standing wave, thus making possible the emission of coherent light [35]. In the conventional PhCs this can be achieved only in certain directions, as the location of the PBG boundary is a function of the direction of the electromagnetic wave propagation. For the radiation to be coherent irrespective of the propagation direction, photonic crystals with an isotropic PBG need to be used. Isotropic photonic bandgaps can also be used for obtaining the omnidirectional negative refraction. This phenomenon was studied for 2D photonic crystals [36] and quasicrystals [37].

Conclusion

Summing up, quasiperiodic structures and periodic structures with many atoms per unit cell demonstrate an existence of a complete photonic band gap. In both cases obtained gaps are very isotropic. In 2D photonic structures their higher isotropy allows gap opening at lower dielectric contrasts in comparison with ordinary periodic gratings. For the 3D case we have a number of structures with many atoms per unit cell, that demonstrate PBG. But till now there are no such a structure with lower a threshold of dielectric contrast for the gap opening, as in the best ordinary 3D periodic gratings. In the 3D case in comparison with 2D we have a large variety of choices (possibilities) for quasiperiodic lattice symmetries, their approximant configurations, etc. So, we can say, that a question of a competition of quasiperiodisity versus periodicity in photonic band gap materials is still open.

References

[1] Shechtman, D; Blech, I; Gratias, D; et al., *Phys. Rev. Lett*, 1984, 53, 1951-1953.
[2] Macia, E. Rep. *Prog. Phys.*, 2006, 69, 397-441.
[3] Quilichini, M; Janssen, T. *Rev. Mod. Phys.*, 1997, 69, 277-314 .
[4] Vekilov, Yu. Kh; Isaev, EI; Arslanov, SF. *Phys. Rev. B*, 2000, 62, 14040-140048.
[5] Krajci, M; Hafner, J. *Phys. Rev. B*, 2007, 75, 024116.
[6] Chan, YS; Chan, CT; Liu, ZY. *Phys. Rev. Lett*, 1998, 80, 956-959.
[7] Yablonovitch, Y. *Phys. Rev. Lett*, 1987, 58, 2059-2062.
[8] John, S. *Phys. Rev. Lett*, 1987, 58, 2486-2489.
[9] Bykov, VP. Sov. *Phys. JETP*, 1972, 35, 269-273.
[10] Johnson, SJ; Joannopoulos, JD. Photonic Crystals: *The Road from Theory to Practice*, Kluwer Academic Publishers, London, 2003.
[11] Joannopoulos, JD; Meade, RD; Winn JN. Photonic Crystals: *Molding the Flow of Light*, Princeton University Press, Singapore, 1999.
[12] Ho, KM; Chan, CT; Soukoulis, CM. *Phys. Rev. Lett.*, 1990, 65, 3152.
[13] Zoorob, ME; Charlton, MDB; Parker, GJ. et al., *Nature*, 2000, 404, 740-743.
[14] Man, W; Megens, M; Steinhardt, PJ. et al., *Nature*, 2005, 436, 993-996.
[15] Lidermann, A; Cademartiri, L; Hermatschweiler, M; et al., *Nature Mater*, 2006, 5, 942-945.
[16] Xu, J; Ma, R; Wang, X. et al, Opt. *Express*, 2007, 15, 4287-4295.
[17] Peach, M. *Materials Today*, 2006, 9, n 7-8, 44-47.
[18] Wang, K; David, S; Chelnokov, A; et al., *J. Mod. Opt*, 2003, 50, 2095-2105.
[19] Dyachenko, PN; Miklyaev, Yu.V. *Proc. of SPIE*, 2006, 6182, 61822I.
[20] Janot, C. Quasicrystals: *A Primer*, Oxford Univ. Press, New York, 1992.
[21] Campbell, M; Sharp, DN; Harrison, MT; et al., *Nature*, 2000, 404, 53-56.
[22] Gauthier, R; Ivanov, A. *Opt. Express*, 2004, 12, 990-1003.

[23] Johnson, SG; Joannopoulos, JD. *Opt. Express*, 2001, 8, 173-190.

[24] Dmitrienko, VE; Kleman, M. Philos. Mag. *Lett*, 1999, 79, 359-367.

[25] Dmitrienko, VE; Kleman, M; Mauri, F. *Phys. Rev. B*, 1999, 60, 9383-9389.

[26] Dyachenko, PN; Miklyaev, Yu. V; Dmitrienko, VE. *JETP Letters*, 2007, 86, 240-243.

[27] Dyachenko, PN; Miklyaev, Yu. V; Dmitrienko, VE; Pavelyev, VS. *Proceedings of SPIE*, 2008, 6989, 69891T.

[28] David, S; Chelnokov, A; Lourtioz, JM. *Opt. Lett*, 2000, 25, 1001-1003.

[29] Lee, TDM; Parker, GJ; Zoorob, ME; Cox, SJ; Charlton, MDB. *Nanotechnology*, 2005, 16, 2703-2706.

[30] Garcia-Adeva, AJ. *Phys. Rev. B*, 2006, 73, 073107.

[31] Adams, GB; O'Keeffe, M; Demkov, AA; Sankey, OF; Huang, YM. *Phys. Rev. B*, 1994, 49, 8048.

[32] Bush, K; John, S; *Phys. Rev. E*, 1998, 58, 3896.

[33] Dyachenko, PN; Kundikova, ND; Miklyaev, Yu. V. *Phys. Rev. B*, 2009, 79, 233102.

[34] Takeda, H; Takashima, T; Yoshino, K. *J. Phys*., Condens. Matter, 2004, 16, 6317.

[35] Meier, M; Mekis, A; Dodabalapur, A; Timko, A; Slusher, RE; Joannopoulos, JD. *Appl. Phys. Lett*, 1999, 74, 7.

[36] Gajic, R; Meisels, R; Kuchar, F; Hingerl, K. *Phys. Rev. B*, 2006, 73, 165310.

[37] Feng, Z; Zhang, X; Wang, Y; Li, ZY; Cheng, B; Zhang, DZ. *Phys. Rev. Lett*, 2005, 94, 247402.

In: Photonic Crystals
Editor: Venla E. Laine, pp. 135-147

ISBN: 978-1-61668-953-7

Chapter 6

SPECTRUM ENGINEERING WITH ONE-DIMENSIONAL PHOTONIC CRYSTAL

Anirudh Banerjee[*]
Department of Electronics and Communication,
Amity School of Engineering and Technology,
Lucknow-226010, Amity University, Uttar Pradesh, India.

Abstract

In this chapter, the effect of different controlling parameters such as number of periods of the structure, refractive index contrast, filling fraction and angle of incidence of light on the structure on the output spectrum of one-dimensional (1D) photonic crystal has been shown and discussed. By knowing the effect of these parameters on the output spectrum, one can engineer the output spectrum of the 1D photonic crystal to utilize them in various optical applications such as filters, reflectors etc., in a desired and optimum way.

1. Introduction

Periodic media like photonic crystals are well acknowledged for their capability to control the propagation and emission of electromagnetic waves. In these photonic crystals, generally, two different optical materials are periodically arranged. It is common to distinguish photonic crystals as one-dimensional (1D), two-dimensional (2D) and three-dimensional (3D) by the number of dimensions within which the periodicity has been introduced into the structure. Out of these 1D, 2D and 3D photonic crystals, the 1D photonic crystals are most popular and easiest to fabricate. A periodic 1D photonic crystal can be easily fabricated by sequential deposition of two materials. Though, various structures and fabrication methods have been proposed for photonic crystals of two dimensions and three dimensions, but these are considerably more difficult to fabricate.

[*] E-mail address: anirudhelectronics@yahoo.com. (Corresponding author)

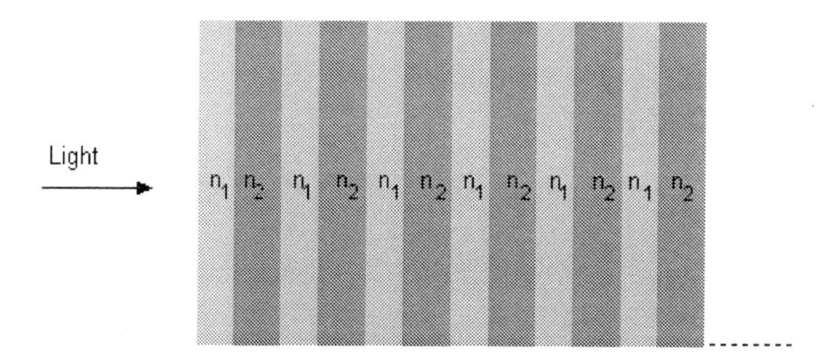

Figure. 1. Schematic of layer-by-layer 1D photonic crystal.

Although photonic crystals were studied in one form or another since 1887, the term "photonic crystal" was first used over hundred years later, after Eli Yablonovitch and Sajeev John published two milestone papers on photonic crystals in 1987 [1-2]. Before 1987, 1D photonic crystals were studied in the form of periodic multi-layer dielectric stacks (such as the Bragg mirror). In 1887, Lord Rayleigh in his study [3] showed that such systems have a spectral range of large reflectivity, known as a stop-band. Later on, a detailed theoretical study of 1D optical structures was performed by Bykov [4], who was the first to investigate the effect of a photonic band-gap on the spontaneous emission from atoms and molecules embedded within the photonic structure. Bykov also speculated as to what could happen if 2D or 3D periodic optical structures were used [5]. However, these pioneering works and some other related works by Ohtaka [6] did not gain the attention they deserved at the time. It was only after the appearance of the papers of Yablonovitch [1] and John [2] these periodic structures became popular.

These photonic crystals are characterized by three parameters: the lattice topology, the spatial period and the dielectric constants of the constituent materials. By suitable selection of these parameters, a gap in the electromagnetic dispersion relation can be created, within which the propagation of electromagnetic waves of certain wavelengths is forbidden. The forbidden wavelength range is called the photonic band gap. When the wavelength of an incident wave does not lie within the band gap it passes through the structure. The structure acts as an optical band pass filter for wavelength ranges falling outside the stop band, while at the same time it acts as a reflector for the photonic band wavelengths. Actually, when an incident wave enters a periodic array of dielectric films, the wave is partially reflected at the boundaries of the dielectric layers. If the partially reflected waves are in phase and superimposed, they form a total reflected wave, and the incident wave is unable to enter the medium. On the other hand, when the wavelength of an incident wave does not lie within the band gap, destructive interferences occur and partially reflected waves cancel one other. Consequently, reflection from the periodic structure does not happen, and the light passes through the structure. However, the ranges of transmission, transmission intensity and transmission band width depend upon a number of parameters such as number of periods in the structure, refractive index contrast, filling fraction and angle of incidence of light on the structure. It is possible to get the desired ranges of electromagnetic spectrum transmitted or reflected through the structure by changing these parameters. Hence, these structures find potential applications as optical filters and reflectors also [7]. In this chapter, the effect of different controlling parameters such as number of periods of the structure, refractive index contrast, filling fraction and angle of incidence of light on the structure on transmission spectra of the 1D photonic crystals has been shown and discussed. By knowing the effect of these parameters on output spectra, one can engineer the output spectra of these 1D photonic band gap structures to utilize them as optical filters or reflectors etc., in a desired and optimum way.

2. Mathematical Approach to Spectrum Computation of 1d Photonic Crystal

The simplest 1D photonic crystal is a periodic structure consisting of alternate layers of materials with different refractive indices, n_1 and n_2 with thicknesses d_1 and d_2 respectively, as depicted in figure 1. Here, the subscripts 1 and 2 are used for low and high index layers, respectively. $d = d_1 + d_2$ is the period of the lattice. It is assumed that the incident and exit media are both air $(n_0 = 1.0)$. Light is incident on the multilayer at an angle θ_0.

For the *s* wave, the characteristic matrix [8] $M[d]$ of one period is given by

$$M[d] = \prod_{i=1}^{l} \begin{bmatrix} \cos\beta_i & \dfrac{-i\sin\beta_i}{p_i} \\ -ip_i\sin\beta_i & \cos\beta_i \end{bmatrix} = \begin{bmatrix} M_{11} & M_{12} \\ M_{21} & M_{22} \end{bmatrix}, \tag{1}$$

where $l = 2$ (1 and 2 signify the layers of refractive indices n_1 and n_2 respectively) $\beta_i = \frac{2\pi n_i d_i \cos\theta_i}{\lambda_0}$, λ_0 is the free space wavelength, $p_i = n_i \cos\theta_i$, θ_i is the ray angle inside the layer of refractive index n_i and thickness d_i, and is related to the angle of incidence θ_0 by

$$\cos\theta_i = \left[1 - \frac{n_0^2 \sin^2\theta_0}{n_i^2}\right]^{\frac{1}{2}}. \tag{2}$$

The matrix $M[d]$ in equation (1) is unimodular as $|M[d]| = 1$.

For an *N* period structure, the characteristic matrix of the medium is given by [8]

$$[M(d)]^N = \begin{bmatrix} M_{11}U_{N-1}(a) - U_{N-2}(a) & M_{12}U_{N-1}(a) \\ M_{21}U_{N-1}(a) & M_{22}U_{N-1}(a) - U_{N-2}(a) \end{bmatrix} \equiv \begin{bmatrix} m_{11} & m_{12} \\ m_{21} & m_{22} \end{bmatrix},$$

where

$$M_{11} = \left(\cos\beta_1 \cos\beta_2 - \frac{p_2}{p_1} \sin\beta_1 \sin\beta_2 \right),$$

$$M_{12} = -i\left(\frac{\cos\beta_1 \sin\beta_2}{p_2} + \frac{\sin\beta_1 \cos\beta_2}{p_1} \right),$$

$$M_{21} = -i(p_1 \sin\beta_1 \cos\beta_2 + p_2 \cos\beta_1 \sin\beta_2),$$

$$M_{22} = \left(\cos\beta_1 \cos\beta_2 - \frac{p_1 \sin\beta_1 \sin\beta_2}{p_2} \right),$$

U_N are the Chebyshev polynomials of the second kind

$$U_N(a) = \frac{\sin[(N+1)\cos^{-1} a]}{\left[1 - a^2\right]^{\frac{1}{2}}}, \quad (3)$$

where

$$a = \frac{1}{2}[M_{11} + M_{22}]. \quad (4)$$

The transmission coefficient of the multilayer is given by

$$t = \frac{2p_0}{(m_{11} + m_{12}p_0)p_0 + (m_{21} + m_{22}p_0)}, \quad (5)$$

and the transmissivity for this structure can be written in terms of transmission coefficient as

$$T = |t|^2, \quad (6)$$

where $p_0 = n_0 \cos\theta_0 = \cos\theta_0$. The transmissivity of the structure, for p wave can be obtained by using expressions (1)-(6) with the following values of p

$$p_i = \frac{\cos\theta_i}{n_i} \text{ and } p_0 = \frac{\cos\theta_0}{n_0}.$$

3. Effect of Structural Parameters on Transmission Spectrum

3.1. Effect of Number of Periods

Considering a stack of quarter-wave dielectric layers of alternate low and high index. The low index layers are those of Na_3AlF_6 (n_1=1.34) and the high index layers are those of SiO_2 (n_2=1.46). The films are quarter wave at $\lambda_0 = 0.650\mu m$. Thus, $n_1 d_1 = n_2 d_2 = \lambda_0/4 = 0.650\mu m/4$

from which d_1 and d_2 are calculated to be equal to 0.121 μm and 0.111 μm respectively. Using the transfer matrix method described in Section 2 the transmission spectra for different number of periods have been calculated for normal incidence of light on the structures.

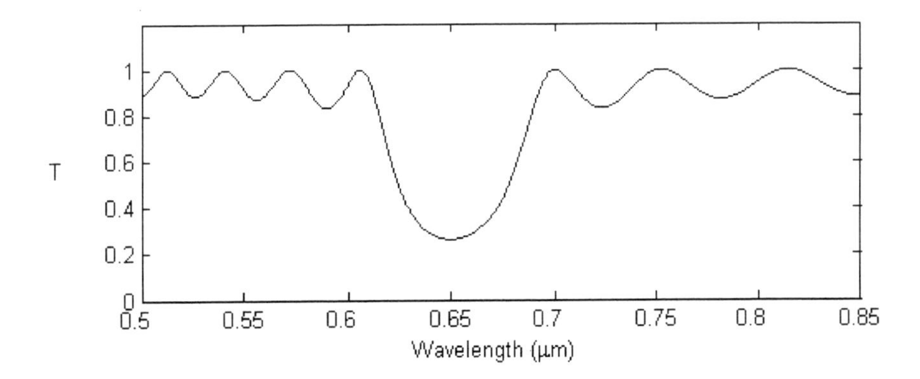

Figure 2(a). Transmission spectrum of Na_3AlF_6 (n_1=1.34 and d_1=0.121μm) / SiO_2 (n_2=1.46 and d_2=0.111μm) quarter wave stack with N=15.

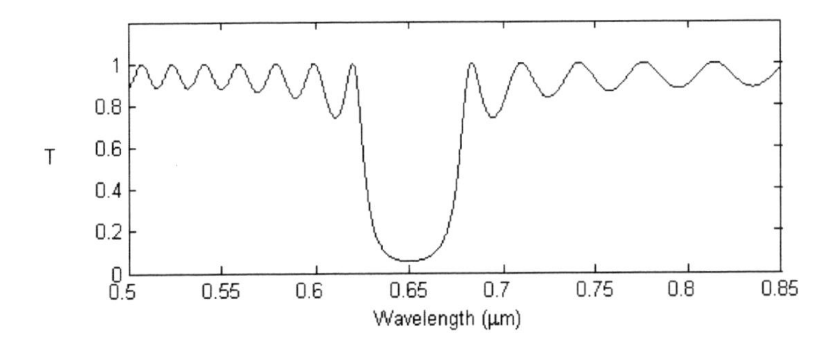

Figure 2(b). Transmission spectrum of Na_3AlF_6 (n_1=1.34 and d_1=0.121μm) / SiO_2 (n_2=1.46 and d_2=0.111μm) quarter wave stack with N=25.

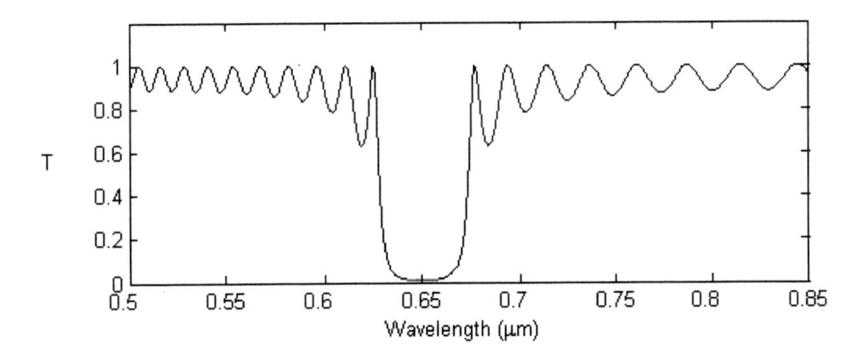

Figure 2(c). Transmission spectrum of Na_3AlF_6 (n_1=1.34 and d_1=0.121μm) / SiO_2 (n_2=1.46 and d_2=0.111μm) quarter wave stack with N=35.

Figures 2 (a), (b) and (c) are plots of transmittance versus wavelength for N=15, 25 and 35 respectively. Comparison of these figures shows that the reflectance of the stop band increases with increase in the number of periods.

3.2. Effect of Refractive Index Contrast

The transmission spectra for three quarter wave 35 period structures made up of alternate layers of Na_3AlF_6/SiO_2, SiO_2/Tin (IV) Sulfide and Na_3AlF_6/Tin (IV) Sulfide with refractive index contrasts (i.e. ratio of high and low refractive indices n_2/n_1) 1.08, 1.78 and 1.94 respectively are shown in figures 3 (a), 3 (b) and 3 (c) respectively. Light is incident on these structures normally. Comparison of these figures shows that the wavelength range of 100% reflectance (known as the photonic band gap) increases with increasing refractive index contrast. Therefore, due to wide stop bands, photonic band gap structures with high refractive index contrasts are specially useful as reflectors.

Each photonic band gap structure has a threshold value of refractive index contrast to exhibit a photonic band gap. The higher the refractive index contrast, the fewer are the number of layers required to exhibit sufficient photonic bandgap effect. The reason is that each layer of the photonic band gap structure partially reflects the propagating wave. As a consequence, if each layer reflects more waves due to a higher refractive index contrast, sufficient net reflections can be achieved by fewer layers than a structure with the same configuration but with a lower refractive index contrast.

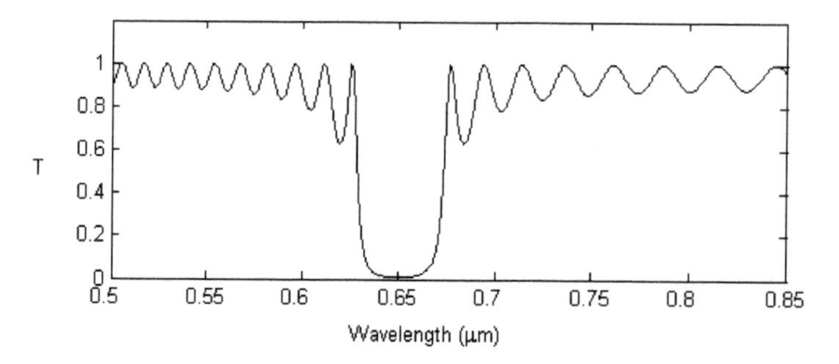

Figure 3(a). Transmission spectrum of Na_3AlF_6 (n_1=1.34 and d_1=0.121μm) / SiO_2 (n_2=1.46 and d_2=0.111μm) quarter wave stack with N=35 and n_2/n_1=1.08.

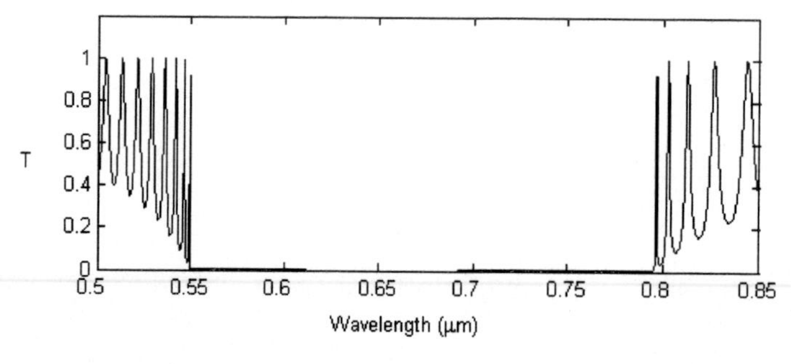

Figure 3(b). Transmission spectrum of SiO_2 (n_1=1.46 and d_1=0.111μm) / Tin (IV) Sulfide (n_2=2.6 and d_2=0.062μm) quarter wave stack with N=35 and n_2/n_1=1.78.

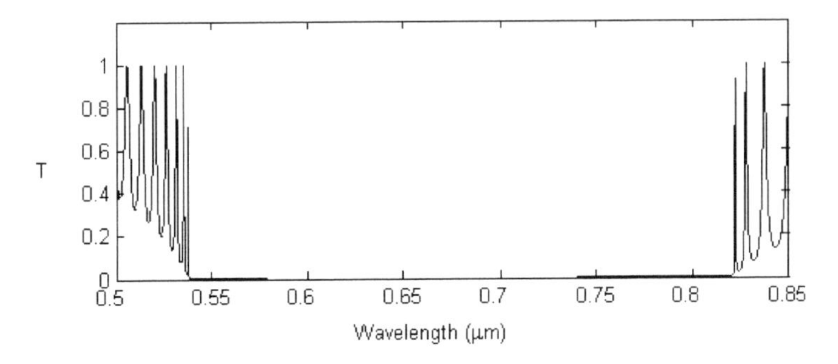

Figure 3(c). Transmission spectrum of Na_3AlF_6 (n_1=1.34 and d_1=0.121μm) / Tin (IV) Sulfide (n_2=2.6 and d_2=0.062μm) quarter wave stack with N=35 and n_2/n_1=1.94.

Table 1.Transmission bands obtained at different values of filling fraction with $Na_3Al\ F_6$ / Tin Sulfide structure.

Filling fraction (η)	Allowed bands (μm)	Width of transmission bands (μm)
0.1	0.500-0.520 0.555-0.627 0.667-0.775 0.840-0.950	0.020 0.072 0.108 0.110
0.2	0.521-0.586 0.634-0.728 0.804-0.950	0.065 0.094 0.146
0.3	0.500-0.570 0.583-0.691 0.747-0.900	0.070 0.108 0.153
0.4	0.500-0.530 0.565-0.865	0.030 0.300
0.5	0.525-0.617 0.674-0.950	0.092 0.276
0.6	0.500-0.576 0.622-0.758 0.846-0.950	0.076 0.136 0.104

3.3. Effect of Filling Fraction

In Sections 3.1 and 3.2, quarter wave films were considered. The analysis is now extended to films with a lattice period $d=d_1+d_2$ equal to 0.650 μm, the mid wavelength of the range was investigated. Normal incidence of light was assumed. Using equations (1)-(6), the transmission spectra for 25 layered Na_3AlF_6/Tin Sulfide and Silica (SiO_2) / Tin Sulfide structures are calculated for different values of filling fraction: $\eta = d_1/(d_1+d_2)$. The transmission bands corresponding to each value of the filling fraction are thus determined for the two structures. These bands are listed in tables 1 and 2, from which it is inferred that different transmission bands of different band widths are obtained for different values of filling fraction. Therefore, by changing the filling fraction, the width of the transmission band of the structure can be changed.

Table 2. Transmission bands obtained at different values of filling fraction with SiO_2 / Tin Sulfide structure.

Filling fraction (η)	Allowed bands (μm)	Width of transmission bands (μm)
0.1	0.413-0.491 0.530-0.645 0.725-0.950	0.078 0.115 0.225
0.2	0.400-0.435 0.455-0.535 0.565-0.730 0.735-0.950	0.035 0.080 0.165 0.215
0.3	0.400-0.462 0.475-0.568 0.605-0.748 0.820-0.950	0.062 0.093 0.143 0.130
0.4	0.400-0.407 0.423-0.485 0.512-0.610 0.635-0.790 0.880-0.950	0.007 0.062 0.098 0.155 0.070
0.5	0.400-0.435 0.446-0.516 0.538-0.632 0.691-0.880 0.888-0.950	0.035 0.070 0.094 0.189 0.062
0.6	0.400-0.452 0.476-0.541 0.576-0.695 0.701-0.900	0.052 0.065 0.119 0.199

3.4. Effect of Angle of Incidence

Effect of angle of incidence of light on Na_3AlF_6/Tin Sulfide and Silica (SiO_2) / Tin Sulfide structures is then investigated. The transmittance curves are plotted for these structures at different angles of incidence. Figures 4(a)-(h) show the transmission spectra of 35 layered Na_3AlF_6/Tin Sulfide $(\eta = 0.6)$ and Silica (SiO_2) / Tin Sulfide $(\eta = 0.4)$ structures at different angles of incidence. The solid curves representing *s* polarization and the dotted curves representing *p* polarization, coincide at normal incidence. Transmission bands obtained are listed in tables 3-4. The transmission bands obtained at normal incidence become narrower and shift towards the lower wavelength side when the angle of incidence is increased. Since the ranges of transmission depend on the values of control parameters N, n_2/n_1, filling fraction and angle of incidence, therefore it is possible to get the desired ranges of the electromagnetic spectrum transmitted through the structure by changing the values of these parameters and/or by varying the angle of incidence of light on the structure.

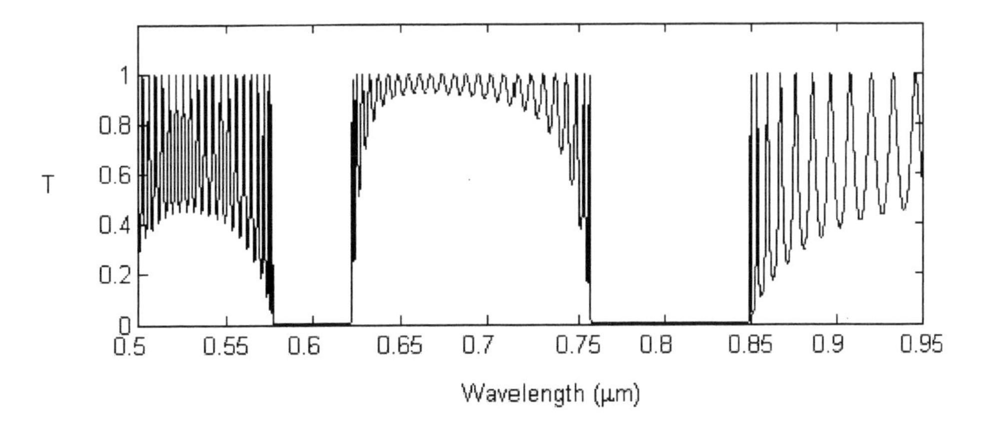

Figure 4(a). Transmission curve for Na_3AlF_6 / Tin Sulfide structure for η=0.6 at 0 degree.

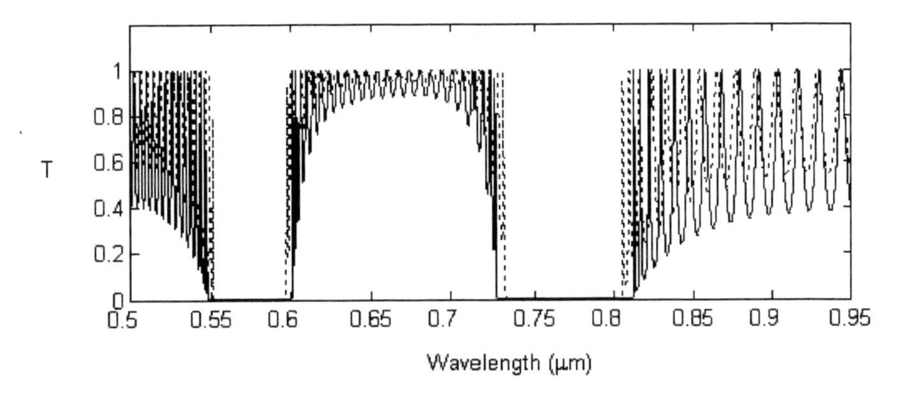

Figure 4(b). Transmission curve for Na_3AlF_6 / Tin Sulfide structure for η=0.6 at 30 degree.

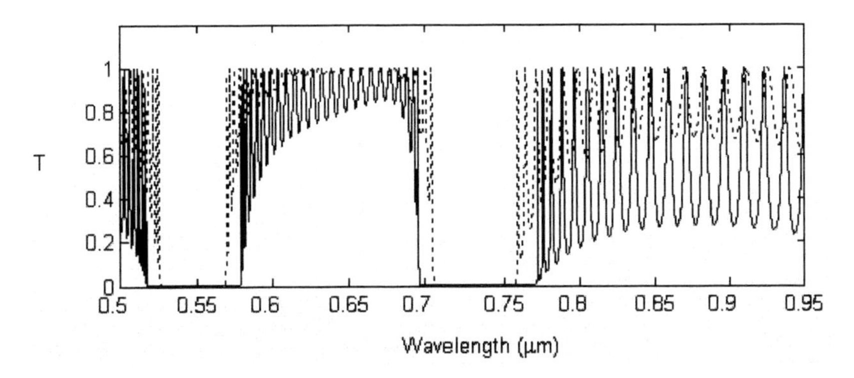

Figure 4(c). Transmission curve for Na_3AlF_6 / Tin Sulfide structure for η=0.6 at 45 degree.

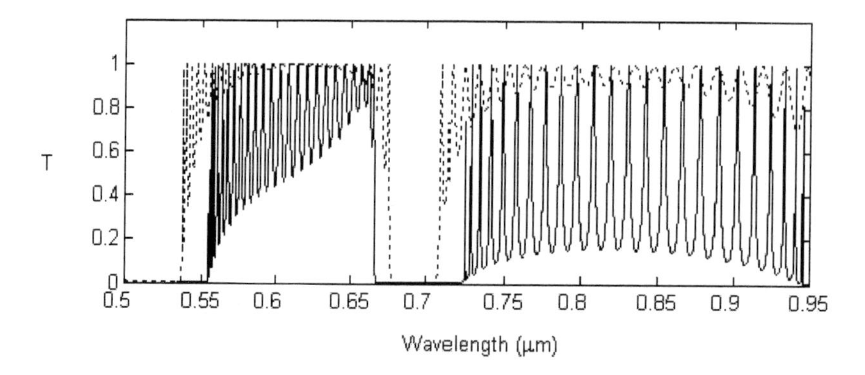

Figure 4(d). Transmission curve for Na_3AlF_6 / Tin Sulfide structure for η=0.6 at 60 degree.

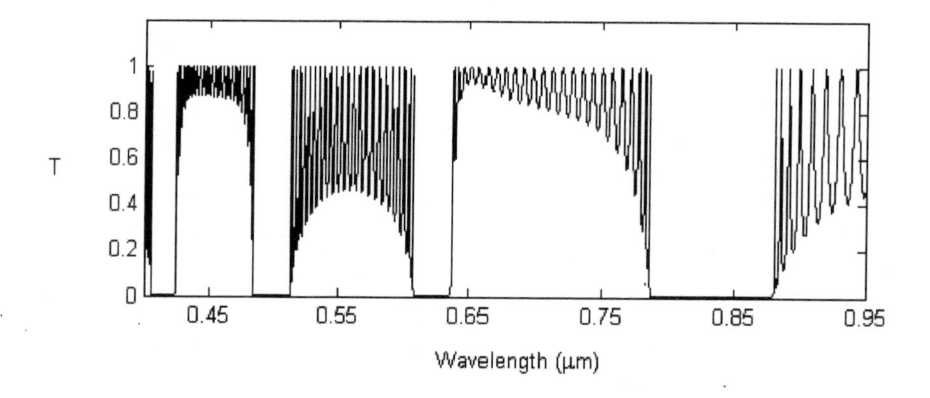

Figure 4(e). Transmission curve for SiO_2/ Tin Sulfide structure for η=0.4 at 0 degree.

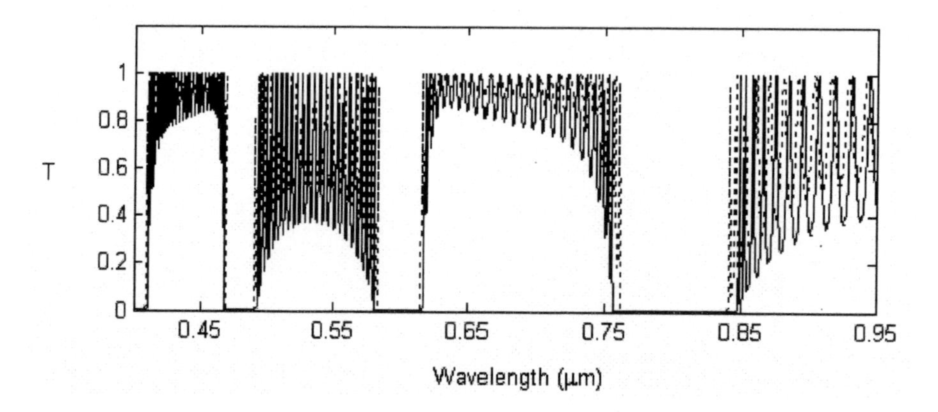

Figure 4(f). Transmission curve for SiO_2/ Tin Sulfide structure for η=0.4 at 30 degree.

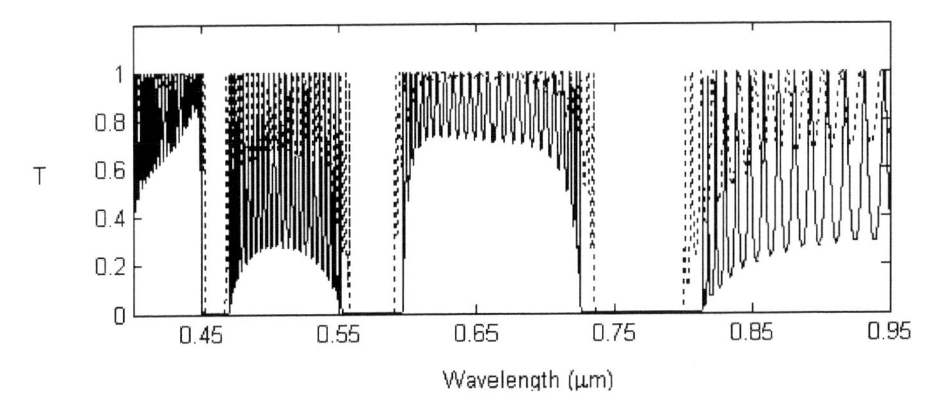

Figure 4(g). Transmission curve for SiO_2/ Tin Sulfide structure for η=0.4 at 45 degree.

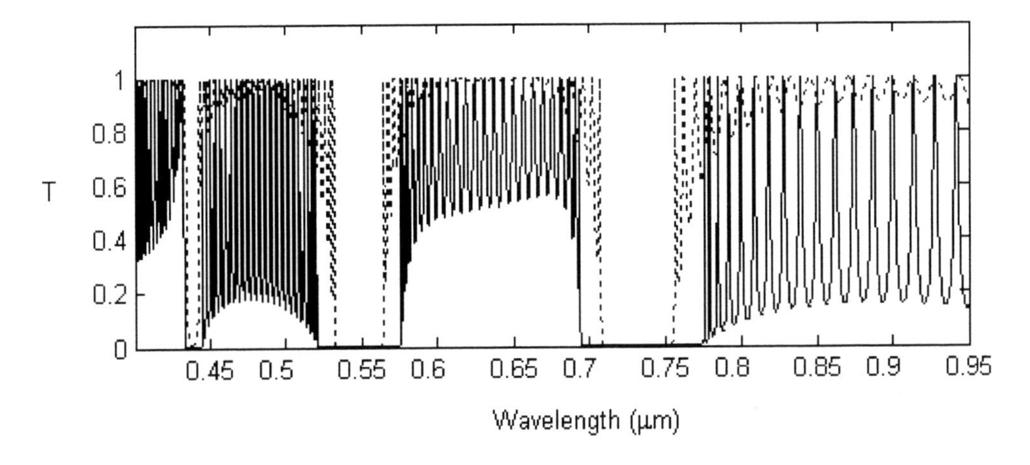

Figure 4(h). Transmission curve for SiO_2/ Tin Sulfide structure for η=0.4 at 60 degree.

Table 3. Allowed ranges for Na_3AlF_6 / Tin Sulfide structure for η=0.6.

Angle (degree)	Transmission bands for s-polarization (µm)	Transmission bands for p-polarization (µm)	Common pass bands (µm)	Width of pass bands (µm)
0	0.500-0.576	0.500-0.576	0.500-0.576	0.076
	0.621-0.758	0.621-0.758	0.621-0.758	0.137
	0.848-1.150	0.848-1.150	0.848-1.150	0.302
30	0.500-0.549	0.500-0.553	0.500-0.549	0.049
	0.600-0.727	0.595-0.734	0.600-0.727	0.127
	0.810-1.090	0.804-1.100	0.810-1.090	0.280
45	0.500-0.518	0.500-0.527	0.500-0.518	0.018
	0.578-0.697	0.568-0.708	0.578-0.697	0.119
	0.770-1.020	0.755-1.050	0.770-1.020	0.250
60	0.555-0.668	0.538-0.680	0.555-0.668	0.113
	0.724-0.950	0.700-0.990	0.724-0.950	0.226

Table 4.Allowed bands for SiO_2/ Tin Sulfide structure for η=0.4.

Angle (degree)	Transmission bands for s-polarization (μm)	Transmission bands for p-polarization (μm)	Common pass bands (μm)	Width of pass bands (μm)
0	0.400-0.407 0.423-0.485 0.512-0.610 0.635-0.790 0.880-1.240	0.400-0.407 0.423-0.485 0.512-0.610 0.635-0.790 0.880-1.240	0.400-0.407 0.423-0.485 0.512-0.610 0.635-0.790 0.880-1.240	0.007 0.062 0.098 0.155 0.360
30	0.410-0.469 0.492-0.582 0.614-0.758 0.845-1.170	0.408-0.471 0.490-0.585 0.612-0.764 0.840-1.180	0.410-0.469 0.492-0.582 0.614-0.758 0.845-1.170	0.059 0.090 0.144 0.325
45	0.400-0.452 0.468-0.555 0.596-0.728 0.812-1.100	0.400-0.456 0.465-0.560 0.590-0.738 0.795-1.130	0.400-0.452 0.468-0.555 0.596-0.728 0.812-1.100	0.052 0.087 0.132 0.288
60	0.400-0.436 0.444-0.522 0.575-0.688 0.775-1.040	0.400-0.438 0.441-0.534 0.562-0.710 0.750-1.070	0.400-0.436 0.444-0.522 0.575-0.688 0.775-1.040	0.036 0.078 0.113 0.265

4. Conclusions

In this chapter, an exclusive study of effect of number of periods, refractive index contrast, filling fraction and angles of incidence of light on the structure on the output spectra of the 1D photonic crystals has been done. It is found that an increase in the number of periods increases the reflectance, while an increase in refractive index contrast and angle of incidence, increases the reflectance as well as the width of the reflection band i.e. transmission bands get narrower. The effect of filling fraction is also dominant in deciding the transmission or reflection band width. For a given pair of materials i.e. for a fixed refractive index contrast it is the filling fraction, which has a decisive role in determining pass band width of the structure. For any structure with given materials and given lattice parameters it is only angle of incidence of light on the structure that allows to tune transmission or reflection bands as well as their widths.

By knowing the effect of all these parameters, one can control these parameters to engineer the output spectra of these photonic band gap structures and to utilize them as optical filters, reflectors or other devices etc. in a desired and optimum way.

Acknowledgements

I am grateful to Sri. Aseem Chauhan, Maj. Gen. K. K. Ohri, Prof. S. T. H. Abidi, Prof. N. Ram and Brig. U.K. Chopra of Amity University, Lucknow, India for their constant encouragement and support during this work. I am also thankful to Prof. U. Malaviya of Lucknow University, Lucknow, India for her valuable suggestions in this work.

References

[1] Yablonovitch, E. *Phys. Rev. Lett*, 1987, vol. 58, 2059-2062.
[2] John, S. *Phys. Rev. Lett*, 1987, vol. 58, 2486-2489.
[3] Rayleigh, JWS. *Phil. Mag*, 1888, vol. 26, 256-265.
[4] Bykov, VP. Sov. *J. Exp. Th. Phys.*, 1972, vol. 35, 269-273.
[5] Bykov, VP. *Quant. Elec.*, 1975, vol. 4, 861-871.
[6] Ohtaka, K. *Phys. Rev. B*, 1979, vol. 19, 5057-5067.
[7] Wu, CJ; Gwo, SJ. of Electromagn. *Waves and Appl.*, 2007, vol. 21, 821-827.
[8] Born, M; Wolf, E. *Principles of Optics*, Cambridge University Press: London, U.K., 1980, 66-67.

In: Photonic Crystals
Editor: Venla E. Laine, pp. 149-171
ISBN 978-1-61668-953-7

Chapter 7

TRANSMISSION THROUGH KERR MEDIA BARRIERS WITHIN WAVEGUIDES: DEVICE APPLICATIONS

A.R. McGurn
Department of Physics
Western Michigan University
Kalamazoo, Michigan 49008-5252

Abstract

The transmission properties of guided waves in photonic crystal waveguides containing barriers formed from Kerr nonlinear media are studied theoretically. The photonic crystal waveguides are formed in a two-dimensional (square lattice) photonic crystal of linear dielectric cylinders by cylinder replacement, with replacement cylinders made of linear dielectric media. The channel of the waveguide is along the x-axis of the photonic crystal. Barriers formed of Kerr nonlinear media are introduced into the waveguide by cylinder replacement of waveguide cylinders in the barrier region by cylinders containing Kerr nonlinear media. The transmission of guided modes incident on the Kerr media barriers, from the linear media waveguides, is computed and studied as a function of the parameters characterizing the Kerr nonlinear barrier media. A focus of the study is on the excitation of intrinsic localized modes within the barrier media and on the effects on these modes from additional off-channel and in-channel features that are coupled to the barrier media. Systems treated are a simple Kerr nonlinear media barrier formed by replacing waveguide sites by cylinders containing Kerr nonlinear media, Kerr barriers that couple to off-channel impurity features formed by cylinder replacement of off-channel sites of the photonic crystal which are adjacent to the barrier, and Kerr barriers containing an in-channel impurity site. The off-channel and in-channel impurity features may be formed of linear media and/or Kerr media replacement cylinders and are found to give rise to multiple bands of intrinsic localized modes. Suggestions are made for possible technological applications of these types of systems.

1. Introduction

There has been much recent interest in photonic crystal waveguides[1, 2, 3, 4, 5, 6, 7, 8]. These are structures formed in photonic crystals for the purpose of channeling the flow of electromagnetic energy. In two-dimensional photonic crystals, which are designed as an array of parallel axes dielectric cylinders with axes arranged in a two-dimensional Bravais lattice, waveguides are created by either removing or replacing a row of the photonic crystal cylinders. By choosing the dielectric properties of the photonic crystal and/or the replacement cylinders correctly, electromagnetic guided wave modes become trapped in the waveguide channel where they then move along the channel in a direction perpendicular to the cylinder axes. The energy flow carried by electromagnetic waves along the waveguide channel is in many respects similar to the energy flow accompanying the motion of electrons in electronic circuits. This has suggested the investigation of the properties of photonic crystal based optical systems that give analogies of electronic circuit behaviors. An important feature, in such a comparison of the optical and electronic systems, is that in some cases the response of the energy flow in photonic crystal systems to time variations can be much faster than that found in analogous electronic systems. This may be of considerable interest in computational and information related processes because, just as in electronics, circuits of photonic crystal waveguides can be made by connecting waveguides together into complex flow patterns for the manipulation and modification of energy signals[5, 6, 7, 8, 9, 10, 11]. Consequently, it has been proposed, as part of the study of Opto-Electronics, that circuits formed of photonic crystal waveguides be used to complement and perhaps improve upon their electronic counterparts in various types of technological applications[12, 13, 14, 15, 16, 17, 18, 19, 20, 21].

Many of the important properties of electrical circuits involve the application of nonlinear responses that occur in electronic materials[22]. Such applications are seen, for example, in transistors and diodes. As a result, in order to arrive at complete optical analogies of electronic circuits it is necessary to develop the study of optical nonlinearities in photonic crystal waveguide structures[11, 21, 23, 24, 25, 26, 27, 28, 29, 30, 31]. This will be one of the foci of the studies presented in this chapter. The nonlinearities in optical systems arise from the dependence of the dielectric constants of some of the materials in the systems on the electric fields that are applied to them[32, 33, 34, 35]. Numerous suggestions have been made for the development of photonic crystals containing nonlinear components in designs which include, for example, switches, transistors, etc. These are based on the application of Kerr nonlinearity properties of various optical materials. In addition, an important problem in optics is the development of devices which change the frequency of light through the generation of second harmonics[32, 33, 34]. These are based on the second harmonic generation properties possessed by specific types of materials, and ideas of photonic crystals and photonic crystal waveguides have been successfully applied to improve the generation of second harmonics for use in various optical applications[36, 37, 38].

In this Chapter we will look at some general properties of nonlinear systems exhibited in simple designs based on photonic crystal waveguides. The focus will be on those systems having Kerr nonlinear components. We will look at designs based on modifications of photonic crystals and photonic crystal waveguides formed of linear dielectric media. Into the photonic crystal waveguide we introduce barriers formed of Kerr nonlinear media, barriers formed of Kerr nonlinear media that contain impurity sites within the barrier, and barriers formed of Kerr nonlinear media which have side couplings formed by cylinder replacement within the photonic crystal adjacent to the barrier[23, 27, 39, 40, 41, 42, 43]. The object of our studies is to look at the scattering of guided modes from these types of barriers of nonlinear Kerr media and how the scattering can be used in device applications.

The types of problems addressed are generalizations of the barrier transmission problem studied in quantum mechanics[44]. The difference between our systems and the well known quantum mechanical system, however, is that the properties of the barrier media in our problems depends on the intensity of the field of the wave propagating within it[39, 40]. This greatly changes the scattering and transmission from the barrier and leads to completely new types of excitations which are resonantly excited within the barrier media. When the barrier is composed only of linear dielectric media, the system exhibits a series of transmission resonances, as a function of the dielectric properties of the barrier, arising from the resonant excitation of Fabry-Perot modes within the barrier media. When the barrier contains Kerr nonlinear dielectric media, however, the system exhibits a series of transmission resonances, as a function of the dielectric properties of the barrier, arising from the resonant excitation of Fabry-Perot modes, intrinsic localized modes, dark soliton modes, etc.[32, 21, 23, 25, 26, 27, 45, 46]. The nonlinearity leads to a new, enlarged, set of barrier modes with unusual and potentially important technological properties. The later discussions will center on the study of these new types of modes arising from the nonlinearity of the media, their classification, and the determination of the conditions for their existence[39, 40, 42].

Recently, we have presented general discussions of the type outlined above for the modes resonantly excited within certain forms of barriers and junctions composed of nonlinear optical media that are contained within or connect a number of photonic crystal waveguides formed of linear dielectric media[39, 40, 42]. (In the following presentation, however, discussions will only involve barriers of a new type not previously treated by us, and junctions will not be discussed.) It was shown there, and will be explained later, that the dielectric properties of the Kerr material are characterized by two parameters and that these two parameters can be used to develop a classification scheme for the barrier modes of the nonlinear system. One parameter gives the dielectric constant of the Kerr material in the limit of zero applied electric field and the other gives the dependence of the Kerr dielectric constant on the intensity of the applied electric field. It was shown that by studying the transmission characteristics of guided modes through the Kerr barrier and junctions as functions of the two parameters characterizing the Kerr media, a series of transmission maxima or resonances are located as points in the two-dimensional space defined by the two parameters characterizing the Kerr media of the barrier or junction. These transmission resonances were shown to be associated with the excitation of resonant modes within the barrier and junctions. The types of modes found included: Fabry-Perot modes, intrinsic localized modes[23, 46], dark soliton modes[21, 39], etc. A mapping of the transmission maxima within the two-dimensional parameter space of the Kerr parameters allowed for the association of mode types resonantly excited within the barrier and junctions (i.e., Fabry-Perot, intrinsic localized, etc.) with features (i.e., lines or ridges) occurring in the pattern of transmission resonances in the two-dimensional Kerr parameter space. This offers a useful scheme for understanding the conditions needed for the observation of resonant transmission in barriers and junctions and a classification of the excitations found within the barriers and junctions under specific resonant conditions. A visualization of the solutions of the nonlinear system results, allowing for an understanding of the origin of the various types of modes present in the system. The visualization is similar to other types of visualizations that have been employed successfully in the study of the dynamics of a variety of nonlinear dynamical systems, e.g., biological populations, economic models, nonlinear oscillators, etc.[39].

Ideas of generating graphical representations of the possible solutions of the dynamics of nonlinear systems are not new to us but have a long history and are common to the study of a large class of nonlinear systems[47]. The solutions of nonlinear dynamical equations often exhibit a high degree of complexity so that methods of graphical visualization are not only useful but are most natural in the representation and classification of the solutions. These ideas have their origins in the Poincare mappings[47], which attempt to give simplified representations in phase space of the solution trajectories of nonlinear systems, and have evolved into a long history of the study and classification of patterns of solutions generated from nonlinear equations and mappings. Well known examples are found in the study of chaos where the understanding of the nature of solutions and their stability with respect to various fixed points and attractors in phase space are used to characterize the dynamics of chaotic nonlinear systems[48]. (A famous example of this is found in the study of the logistic mapping[49].) Here, relatively simple types of nonlinearity are often found to lead to complex patterns of solutions within such phase space representations. Another set of complex patterns are the Turing patterns observed in chemical and biological systems[50, 51]. These are found as types of self sustaining and generating patterns that have a complex evolution in space with time and arise from the interplay of the rate of diffusion and reactive transitions initiated in various inhomogeneous processes. In a more recent work, Wolfram[52] has studied by graphical means programs and algorithms, showing that amazingly complex patterns are generated from small sets of short, compact, recursive rules. Complexity is found to follow from the straightforward application of a simple rule over and over again. In all of these studies fundamental, complex, changes in the dynamics come from the introduction of very simple nonlinearities (even in some cases from nonlinear perturbations) and are understood in terms of graphic visualizations of the solution sets of the system. The graphic presentations given by us later are similar to all of those mentioned above in giving a simple pictorial presentation of solution sets that are complex in nature and which, unlike the solutions of linear systems of equations, do not obey simple laws of composition.

The goal of our discussions is to extend the visualizations methods we have developed to classify the solutions of the resonantly excited modes within our simple Kerr waveguide barrier system to treat more complex barrier geometries[39, 40, 42]. The more complex barriers that are discussed are barriers with couplings to off-channel and in-channel impurity features. These include: single off-channel impurity sites of linear media coupled to the barrier, single off-channel impurity sites of nonlinear media coupled to the barrier, and single nonlinear or linear impurity sites occurring within the barrier of Kerr nonlinear sites. A focus is on the effects of the off-channel and in-channel impurities on the intrinsic localized (discrete soliton-like) modes that are resonantly excited within the barriers of the systems. Intrinsic localized modes, as we show, only exist within Kerr media barriers and are absent in linear media barriers.

In Section 2, a general discussion of our model and of the difference equation method used in our presentation is given. This is followed by a discussion of the mapping in the two-dimensional Kerr parameters space for a simple barrier and then for barriers with off-channel and in-channel feature. In Section 3, the conclusions are presented.

2. The Model, a Review of the Difference Equation Approach, and Results

In this section we introduce and discuss the models of a two-dimensional photonic crystal and a two-dimensional photonic crystal containing photonic crystal waveguides with barriers of Kerr nonlinear dielectric media[39, 40, 42]. The photonic crystal is composed as an array of parallel axes dielectric cylinders on a Bravais lattice and the waveguide and barrier structures are formed by cylinder replacement in the photonic crystal[9, 10, 11, 23, 27]. The electromagnetic modes that are of interest to us propagate in the plane of the Bravais lattice with electric fields polarized along the direction of the cylinder axes. Specifically, the focus is on guided modes of the waveguide and their scattering properties from the barrier of Kerr nonlinear dielectric media. To study the guided modes of this system we use a difference equation approach that has been developed by us in a number of previous publications[9, 10, 11, 25, 26]. It is applicable to systems of a particular type of replacement cylinder geometry which will be explained later. The difference equation approach is briefly reviewed and employed in the discussion presented in this section[9, 10, 11, 25, 26].

We first discuss the treatment of a simple barrier formed by the replacement of waveguide cylinders by cylinders containing Kerr nonlinear media. This is followed by a treatment of Kerr nonlinear barriers that have additional couplings to off-channel features formed by off-channel cylinder replacements within the bulk photonic crystal and a treatment of Kerr nonlinear barriers containing in-channel Kerr nonlinear and linear media impurity sites.

2.1. Simple Kerr Barrier within the Waveguide

In Fig. 1 schematic drawings are presented of a two-dimensional photonic crystal and a two-dimensional photonic crystal modified to include a photonic crystal waveguide containing a barrier of Kerr nonlinear media. Figure 1a represents a two-dimensional square lattice photonic crystal formed of infinite, parallel axes, dielectric cylinders. The plane perpendicular to the axes of the cylinders is represented, and for our considerations the photonic crystal is composed of linear dielectric media. Figure 1b shows the waveguide and barrier of Kerr nonlinear media within the photonic crystal in the plane perpendicular to the axes of the dielectric cylinders. The waveguide is formed by cylinder replacement in the photonic crystal of a single row of cylinders along the x-axis, and the replacement cylinders are made of linear dielectric media. The barrier is formed in the waveguide by cylinder replacement of waveguide cylinders with dielectric cylinders containing Kerr nonlinear dielectric media. The barrier is formed by replacing seven of the waveguide cylinders such that one of the original waveguide cylinders is left remaining between each of the Kerr media replacement cylinders forming the barrier. The reason for alternating in the barrier between original waveguide cylinders and cylinders containing Kerr media is that it simplifies the mathematical treatment of the scattering of guided modes from the barrier while giving qualitatively similar results to those of a barrier of consecutive Kerr media containing sites[53]. Our previous discussions of waveguide barriers and junctions treated systems with consecutive Kerr replacement cylinders[39, 40, 42]. These are more difficult to treat theoretically.

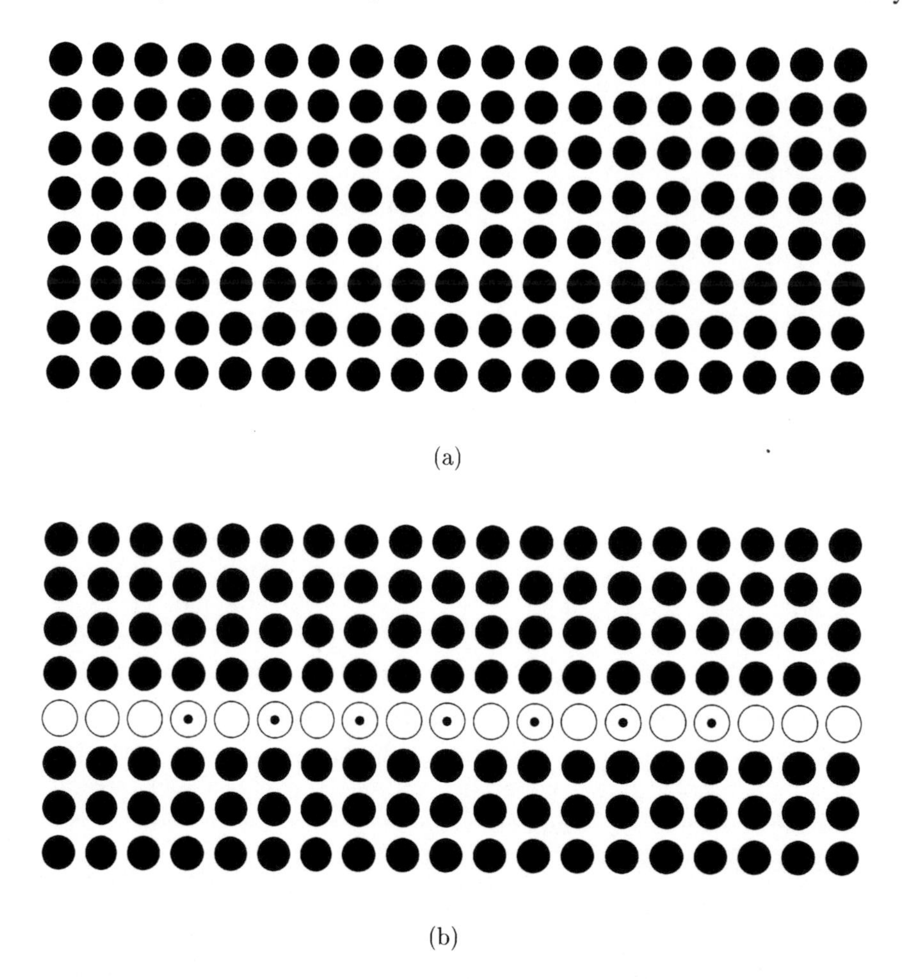

Figure 1. Schematic of: a) A two-dimensional photonic crystal on a square Bravais lattice. Parallel axes cylinders containing one type of linear dielectric material are surrounded by a background of another type of linear dielectric material (vacuum). Electromagnetic waves move in the plane of the Bravais lattice. b) A waveguide in a two-dimensional photonic crystal containing a barrier consisting of seven replacement cylinders containing Kerr nonlinear media. Electromagnetic guided waves move along the waveguide channel within the plane of the Bravais lattice. In both drawings the closed dark circles represent the cylinders of the bulk photonic crystal, the open circles represent the waveguide replacement cylinders composed of linear dielectric media, and the open circles with the center dot are the replacement cylinders of the barrier containing Kerr nonlinear media.

In forming the waveguide and barrier, the replacement cylinders are chosen of a specific form so as to simplify the mathematics and allow for the application of a difference equation approach to the study of the waveguide and barrier modes. The replacement cylinders are formed by adding to the center axis of the cylinders of the photonic crystal an amount of impurity dielectric media such that replacement cylinders are maintained as translationally invariant along their axes[9]. In addition, the radius of the area containing the impurity media in the plane of the photonic crystal Bravais lattice is small so that the electric field of the guided modes changes little over the area containing the impurity media. Under these conditions it can be shown that the equations of electromagnetism relating the electric fields in the impurity media of the dielectric cylinders of the waveguide and barrier to one another are reduced to a set of difference equations.

The difference equation relationships are essentially obtained by using Green's functions methods to rewrite Maxwell's equations for the photonic crystal and waveguide system into the form of an integral equation and applying to this equation the condition that electric fields of the guided modes change slowly over the cross sectional area of the impurity materials of any given replacement cylinder forming the waveguide and barrier[9, 10, 11, 25]. The generation of the set of difference equations, by this method, from the Maxwell's equations for waveguides and barriers has been extensively discussed by us in a number of publications[9, 10, 11, 25] and the reader is referred there for an extensive treatment of these equations. In addition, the reader is also referred to the Appendix of this Chapter where a brief review is given of the derivation of the difference equations formulation from the Maxwell's equations applied to the photonic crystal, waveguide, and barrier system. We shall now turn to a description of the resulting difference equations for our waveguide system and then discuss how they are used to treat guided modes. This is followed by applications to various waveguide-barrier systems for the determination of the transmission characteristics of guided modes through the barrier.

The modes of the waveguide and barrier are characterized by a set of difference equations relating the electric fields at the center of the waveguide and barrier cylinders to one another. The square Bravais lattice of the photonic crystal has lattice constant a_c with lattice sites located in the x-y plane and denoted, in units of the photonic crystal lattice constant, by (l, m) for l, m integers. The cylinders of the photonic crystal are located at and centered about the positions of the sites in the Bravais lattice. For the waveguide, shown in Fig. 1b, along the x- axis, the equations relating the electric fields at the centers of the replacement cylinders to one another are given by[9, 10, 25]

$$E_{n,0} = g_l[E_{n,0} + b(E_{n+1,0} + E_{n-1,0})] \quad (1)$$

for $|n| \geq 8$. In the barrier of seven cylinders containing Kerr nonlinear media with six linear media sites of the waveguide alternating between them, the electric fields are related to one another by

$$E_{2n,0} = g(1 + \lambda|E_{2n,0}|^2)E_{2n,0} + g_l b[E_{2n+1,0} + E_{2n-1,0}] \quad (2)$$

for $|n| < 3$, and

$$E_{2n+1,0} = g_l E_{2n+1,0} + gb[(1 + \lambda|E_{2n,0}|^2)E_{2n,0} + (1 + \lambda|E_{2n+2,0}|^2)E_{2n+2,0}] \quad (3)$$

for $-3 \leq n \leq 2$. Finally, the electric fields at the connections between the waveguide and barrier are related to one another and those in the rest of the system by

$$E_{\pm 6,0} = g(1 + \lambda|E_{\pm 6,0}|^2)E_{\pm 6,0} + g_l b[E_{\pm 7,0} + E_{\pm 5,0}] \quad (4)$$

and

$$E_{\pm 7,0} = g_l[E_{\pm 7,0} + bE_{\pm 8,0}] + gb(1 + \lambda|E_{\pm 6,0}|^2)E_{\pm 6,0}. \quad (5)$$

In these equations $E_{m,j}$ is the electric field at the (m, j) photonic crystal lattice site, g_l characterizes the properties of the dielectric cylinders in the linear media waveguide, g and $\lambda \neq 0$ characterize the properties of the dielectric cylinders in the barrier of Kerr media, and b is the coupling between the electric fields on nearest neighbor sites of the waveguide and barrier as mediated by the materials forming the photonic crystal.

The parameters g and λ are simply related to the dielectric constant, ϵ_{PC}, of the cylinders of the photonic crystal and to the parameters of the dielectric constant of the Kerr nonlinear medium[25]. The general form of the dielectric constant of a Kerr nonlinear medium is given by $\epsilon = \epsilon_0[1 + \lambda'|E|^2]$ where E is the electric field applied to the Kerr medium. In the case that $E = 0$ we find that $g = g(\epsilon_0, \epsilon_{PC}) \propto \epsilon_0 - \epsilon_{PC}$ gives a complete description of the barrier media. For the $E \neq 0$ case, however, additional terms containing λ' characterize the contribution of the field intensity in the Kerr media to its dielectric constant. These addition terms show up in our difference equations as terms involving λ where $\lambda = (\epsilon_0\lambda')/(\epsilon_0 - \epsilon_{PC})$. Notice that as λ' goes to zero in the linear media limit, so does λ.

For additional discussions and details about how the difference equations are developed and how the parameters g_l, g, λ, and b are related to the dielectric and geometric properties of the photonic crystal, waveguide, and barrier the reader is referred to a discussion of these relationships in the Appendix. Further details are also found by consulting Refs. [9],[10],[11], [25].

For the discussions in this chapter, the lattice constant of the photonic crystal and photonic crystal waveguide are taken to be the same. This does not need to be the case, and the theory can be easily generalized to treat systems in which the waveguide and photonic crystals have different lattice constants. By changing the lattice constant of the photonic crystal waveguide, the dispersion properties of the waveguide may be changed as well as the strengths of the couplings between the fields coupled by the waveguide difference equations[23]. In general, it is found that increasing the lattice constant of the waveguide decreases the couplings between the fields at neighboring sites[23]. This comes from the fact that the waveguide modes are at stop band frequencies of the photonic crystal so that their field intensities decay with increasing spatial separation from the waveguide channel into the bulk material of the photonic crystal. In this sense the motion of the guided modes along the waveguide channel can be viewed as a tunneling of the electromagnetic field between waveguide replacement cylinders through the photonic crystal in which they are embedded. This is a feature held in common by many of the physical systems which have descriptions based on tight binding models, of which Eqs. (1) through (5) for the photonic crystal waveguide-barriers are representative.

The difference equations in Eqs. (1) through (5) can be solved analytically. In the incident waveguide channel to the left of the barrier, the solution is composed of a sum of incident and reflected waves having the form[27]

$$E_{n,0} = ue^{ikn} + ve^{-ikn} \quad (6)$$

for $n < -6$, and the transmitted mode in the waveguide channel to the right of the barrier is of the form

$$E_{n,0} = rxe^{ikn} \tag{7}$$

for $n > 6$. Here r and x are real with xr giving the amplitude of the transmitted wave. For incident, reflected, and transmitted waves described by Eqs. (6) and (7), the transmission coefficient, T, is given by $T = |rx/u|^2$ and ranges between 0 and 1.

In the discussions later, T is studied as a function of g and r for fixed values of λx^2 and the parameters b and g_l which characterized the geometric and dielectric properties of the photonic crystal and waveguide. For the results we take numerical values of $b = 0.0869$ and characterize the linear media of the waveguide by g_l given by $g_l = 1/[1 + 2b\cos(k)]$ with $k = 2.9$. These parameters have been used by us in a number of studies of photonic crystal waveguides and circuits formed of linear and nonlinear dielectric media[9, 10, 11, 23, 24, 25, 26, 27]. They are generated from a specific realization of a square lattice photonic crystal that was original studied in Ref. [9, 54] and for the guided modes, originally studied in Ref. [9, 10], of waveguides introduced into that square lattice photonic crystal. The guided modes are modes with a frequency $\frac{\omega a_c}{2\pi c} = 0.440$ at the center of the second frequency stop band of the square lattice photonic crystal in Ref. [9, 54] and have $g_l = 1/[1+2b\cos(k)]$ as the relationship between the wavenumber of the guided modes and the g_l characterizing the dielectric and geometric properties of the waveguide replacement cylinders. Consequently, for a fixed frequency the wavenumber of the incident and scattered guided modes is changed by changing the dielectric properties of the waveguide. In a previous paper we have treated the scattering of guide waves with different wave vector and frequencies and shown that the patterns of solutions in the two parameter space of the Kerr media of the barrier are relatively stable with changes in both frequency and wave vector[42]. In this chapter only solutions with a frequency at the center of the stop band and with $k = 2.9$ near the edge of the waveguide Brillouin zone are treated so that the effects of waveguide dispersion are not considered.

Equations (1) through (5) are solved for the transmission coefficient T by specifying g characterizing the dielectric of the cylinders containing Kerr nonlinear media in the absence of an applied electric field, $\lambda x^2 = 0.001$ which sets the strength of the nonlinearity of the Kerr nonlinear dielectric, and r characterizing the intensity of the wave transmitted through the barrier[27]. For a fixed value of r, T as a function of g is a series of resonant peaks such that each transmission maximum corresponds to a resonantly excited mode in the barrier, i.e., a Fabry-Perot mode, an intrinsic localized mode, a dark soliton-like mode, etc. As an illustration of this, in Fig. 2 a plot is given of T versus g for the linear limit of the system given by $r \to 0+$. (Note: Alternatively one could also take $\lambda = 0$ for the linear limit.) The transmission is displayed as a series of peaks which indicate the Fabry-Perot resonances within the barrier. Associated with each of the transmission resonances is an excited Fabry-Perot mode within the barrier media. As r increases from zero, additional modes (i.e., intrinsic localized modes, dark soliton modes and multiple sets of these types of modes) appear as new resonantly excited modes within the barrier media. The positions of the various resonant transmission peaks in the two-dimensional (r, g) space are determined in this way and a plot of the positions at which they occur in (r, g) space made. The various features in the resulting map of resonances in (r, g) are then associated with different types of resonantly excited barrier modes. In this way, a pattern of ridges and lines is found with each ridge or line associated with a different type of mode, i.e., Fabry-Perot mode, intrinsic localized mode, dark soliton mode, etc.

In Fig. 3 results are presented of such a plot for transmission peaks with $T \geq 0.60$ for the barrier in Fig. 1b. The pattern formed by the resonances is of a complicated appearance[39, 40, 42]. In the $r \to 0$ limit the barrier dielectric becomes linear and exhibits transmission resonances corresponding to a series of Fabry-Perot modes excited within the now linear media barrier. Starting at $r = 0$, the Fabry-Perot modes occur at $g = 1.2039$, 1.2031, 1.2029, 1.1700, 1.1229, 1.0748, 1.0349, and 1.0089. As r increases, the nonlinearity of the barrier media and the effects from its dependence on the field intensity increases. The lowest g Fabry-Perot mode for $1.0 < g < 1.01$ is found to terminate with increasing r before the left hand side of the plot is reached, but the other Fabry-Perot modes of larger $g > 1.01$ continue across the entire plot. Along the various lines shown in the figure, the basic shape of the wave functions of the modes excited within the barrier is preserved while the basic shapes of the wave functions of modes corresponding to different lines in the plot are different. This is true for the Fabry-Perot modes as well as the other types of modes arising from the Kerr nonlinearity of the system.

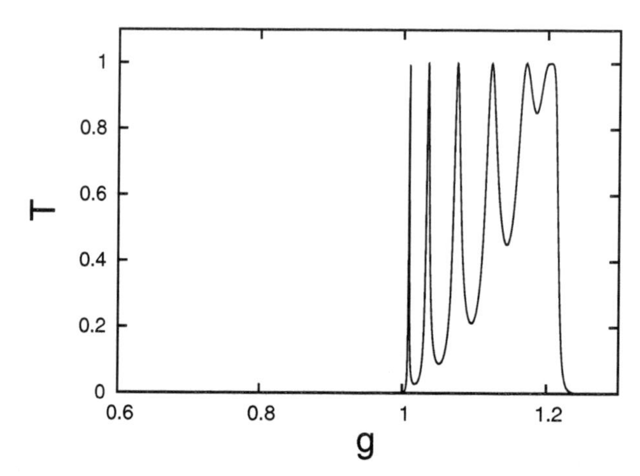

Figure 2. Plot of T versus g for the linear dielectric limit of the simple barrier in Fig. 1b.

A new feature in (r, g) space arising from the Kerr nonlinearity is a series of intrinsic localized mode solutions. The intrinsic localized mode solutions are found in (r, g) for $0.1 < r < 0.6$ and $0.92 < g < 1.0$, and in Fig. 3 to the lower right of the ridge of intrinsic localized modes solutions in this region a notation "ILM" has been added to aid the reader. The line of intrinsic localized modes appears at a value of $r > 0$ and does not evolve from Fabry-Perot modes resonantly excited within the linear limit of the barrier media. In Table 1 a listing of $\lambda|E_{n,0}|^2$ for n = 6, 4, 2, 0, -2, -4, -6 in the seven Kerr sites of the barrier is given for some of the points in Fig. 3 that are intrinsic localized modes resonances. The wave function intensities are highly peaked at the center of the barrier media and decay rapidly to small values at the edges of the barrier. This is a characteristic of intrinsic localized modes whose large peak intensities modify the Kerr dielectric constant of the barrier media so as to maintain and support the localized mode. It is a self-consistent arrangement with the mode modifying the dielectric which in turn allows the mode to exist within the dielectric. The most highly peaked intrinsic localized modes occur at low values of g, while with increasing g the peak heights of the intrinsic localized modes decrease. Eventually the ridge of intrinsic localized mode solutions rises in (r, g) space to meet the lowest branch of Fabry-Perot modes. As r is increased further, regions of multiple intrinsic localized modes (for $0.8 < g < 1.0$ and $r > 1.0$) and dark soliton modes (off the r scale of the figure) are found. These have been discussed by us elsewhere[39, 40, 42, 53]. In this Chapter the focus will be on intrinsic localized modes and their properties in the simple barrier and more complicated barriers considered later.

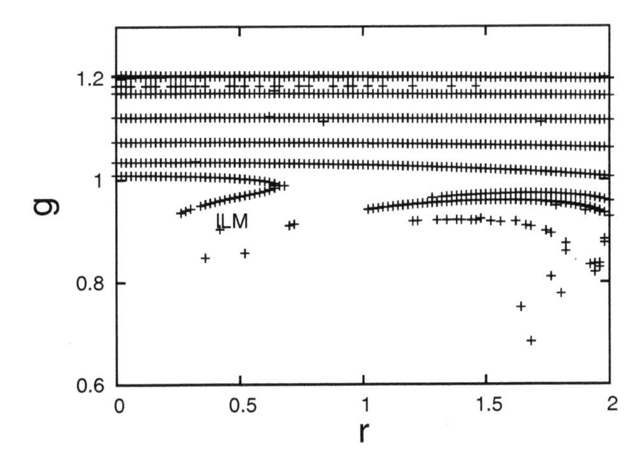

Figure 3. Plot in the space of (r, g) of the locations of the resonant transmission peaks from the simple barrier of seven Kerr sites shown in Fig. 1b. The Fabry-Perot modes are contained within lines beginning at $r = 0$ and continuing across the $r > 0$ plot. The intrinsic localized modes start at $r \neq 0$ at the lowest values of g and, as r increases, rise to meet the lowest line of Fabry-Perot modes. Next to the lower right of the region of intrinsic localized modes the notation ILM has been added in the plot. The Fabry-Perot modes corresponding to different lines in the plot have different types of electromagnetic intensity profiles within the barrier media while those corresponding to the same line have similar electromagnetic intensity profiles within the barrier media.

The intrinsic localized modes, of the type discussed above, show a large center peak intensity created by a small incident wave field intensity[45, 46]. In device applications this offers for a way of providing for the creation of a large field intensity within a dielectric medium through the modulation of a low intensity guided wave, i.e, it offers an amplification function for the incident modulated field. This is similar to the amplification of of an electronic signal provided by certain types of electronic circuits. In addition, the conditions on the system for the excitation of such modes are restrictive so that small variations in system parameters allow for a switching effect. For example, varying the intensity of the incident guided wave can substantially change the amount of the incident wave transmitted or reflected by the barrier. This affect may be elaborated upon, and for an intensity modulated guided mode the Kerr barrier can also act as a type of half-wave rectifier. This is done by off-tuning the average field intensity of the guided mode from the transmission peak of a barrier intrinsic localized mode so that only during part of the modulation cycle would the intrinsic localized mode be resonantly excited with an associated resonant transmission of the guided mode.

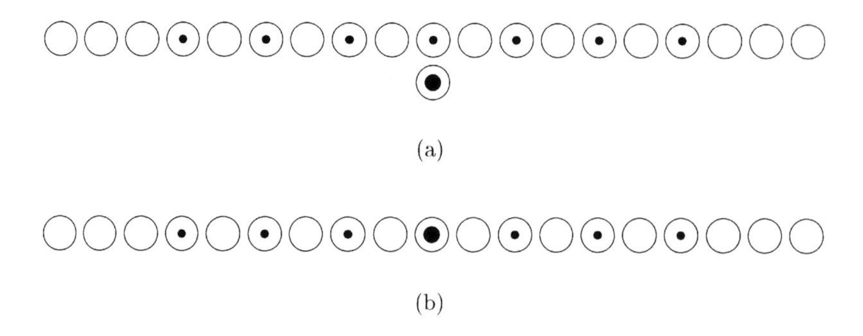

Figure 4. Schematic drawing of: a) a barrier coupled at its center site to an off-channel linear or Kerr nonlinear impurity site and b) a barrier containing a linear or Kerr nonlinear impurity site at its center. In the drawing only the waveguide and impurity sites are shown and the sites of the bulk photonic crystal are suppressed. The open circles are linear media waveguide sites, the open circles with the small dots in the waveguide channel are Kerr nonlinear media sites, and the off-channel or center barrier site open circle with the larger dot is a linear or Kerr nonlinear media impurity. The lattice constant of the waveguide is the same as that of the bulk photonic crystal. For simplicity the closed dark circle sites of the bulk photonic crystal are omitted.

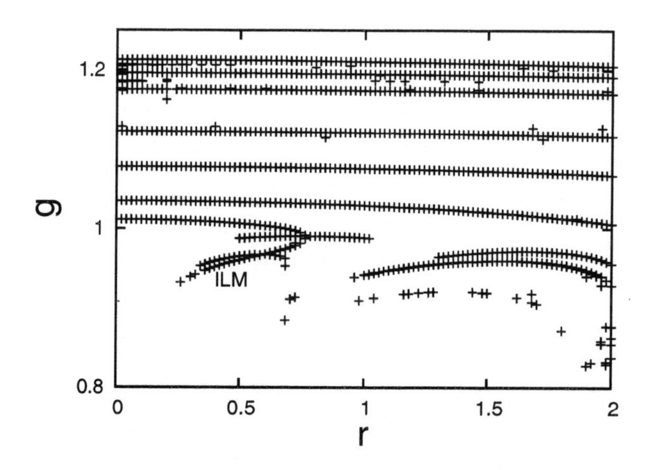

Figure 5. Plot in the space of (r, g) of the locations of the resonant transmission peaks from a simple barrier of seven Kerr sites coupled at the center nonlinear site to an off-channel site of linear dielectric media described by $g_i = 5$. Aside from the off-channel site, the system is the same as in Fig. 3. The intrinsic localized modes present in Fig. 3 are seen to be split by the off-channel coupling. The notation ILM has been added at the lower right of the region of the intrinsic localized modes.

2.2. Kerr Media Barrier with Linear Dielectric Off-Channel Feature

A generalization of the Kerr media barrier problem is to couple the center site of the barrier to an off-channel impurity site formed by cylinder replacement of a photonic crystal site positioned next to the center barrier site and off the waveguide channel. A schematic

Table 1. Intrinsic localized mode field intensities $\lambda|E_{n,0}|^2$ on the simple barrier sites for $n =$ 6, 4, 2, 0,-2, -4,-6 for selected values of (r, g) and for $k = 2.9$.

Intrinsic Localized Modes							
(r, g)	$n = 6$	$n = 4$	$n = 2$	$n = 0$	$n = -2$	$n = -4$	$n = -6$
$(0.3, 0.9433)$	0.0002	0.0026	0.0234	0.1197	0.0356	0.0027	0.0002
$(0.4, 0.9570)$	0.0003	0.0038	0.0260	0.0948	0.0261	0.0039	0.0003

drawing of this case is given in Fig. 4a, and we shall take the off-channel impurity site to be of linear dielectric media. The equations for this geometry are the same as those in Eqs. (1) through (5) with the modifications that now the equation

$$E_{0,0} = g(1 + \lambda|E_{0,0}|^2)E_{0,0} + g_l b(E_{-1,0} + E_{1,0}) \tag{8}$$

from Eq. (2) is replaced by

$$E_{0,0} = g(1 + \lambda|E_{0.0}|^2)E_{0,0} + g_l b(E_{-1,0} + E_{1,0}) + g_i b E_{0,-1} \tag{9}$$

and an equation for the off-channel site given by

$$E_{0,-1} = g_i E_{0,-1} + gb(1 + \lambda|E_{0,0}|^2)E_{0,0} \tag{10}$$

is added to complete the system of equations for the case described by Fig. 4a. In Eqs. (9) and (10) g_i is related to the dielectric media of the off-channel replacement cylinder and may differ from the g_l of the replacement cylinders forming the waveguide channel. In the limit that $g_i \to 0$ the off-channel replacement cylinder reduces to a cylinder of the bulk photonic crystal so that the equations of the barrier with an off-channel site reduce to those of the simple barrier shown in Fig. 1b.

The modified Eqs. (1) through (5) and Eqs. (8) through (10) are solved for the same incident, reflected, and transmitted wave boundary conditions given in Eqs. (6) and (7), using the same values of b, λx^2 and wavenumber k from which the results in Figs. 2 and 3 were obtained. In Fig. 5 the resulting plot of locations of the resonant transmission peaks in (r, g) space is shown for the case in which $g_i = 5.0$. Only peaks for which $T > 0.6$ are indicated in the plot. As in Fig. 3 a series of seven lines of Fabry-Perot modes are observed. These lines begin at $r = 0$ for $g > 1.0$, and the six top lines continue across the entire plot while the lowest line of Fabry-Perot modes ends near $r = 0.7$. In addition, a series of lines of intrinsic localized modes is also found in the plot. These occur for $g < 1.0$ and begin at $r > 0$. The notation "ILM" is added to the lower right of the series of lines of intrinsic localized mode solutions occurring in the region of (r, g) for $0.2 < r < 1.0$ and $0.9 < g < 1.0$. It is seen that the off-channel impurity splits the original branch of intrinsic localized modes of the simple barrier in Fig. 3 into three branches of intrinsic localized modes. Again the branches of intrinsic localized modes are in the lower left corner of the plot, with the branches appearing at values of $r > 0.20$. As r increases, one of the branches rises in g to meet the lowest valued branch in g of Fabry-Perot modes. The other two branches are of a more horizontal nature as r increases, and both cross the rising branch

Table 2. Listing for the barrier with coupling to an off-channel linear media site of the intrinsic localized mode field intensities. Valves are shown for $\lambda|E_{n,0}|^2$ in the barrier sites for $n = 6, 4, 2, 0, -2, -4, -6$ for selected values of (r, g) and g_i for $k = 2.9$ and $\lambda x^2 = 0.001$.

Intrinsic Localized Modes for $g_i = 5.0$							
(r, g)	$n = 6$	$n = 4$	$n = 2$	$n = 0$	$n = -2$	$n = -4$	$n = -6$
$(0.4, 0.9519)$	0.0003	0.0041	0.0304	0.1126	0.0306	0.0042	0.0003
$(0.4, 0.9593)$	0.0003	0.0037	0.0242	0.0871	0.0582	0.0107	0.0007
$(0.6, 0.9682)$	0.0006	0.0072	0.0361	0.0749	0.0360	0.0072	0.0006
$(0.6, 0.9897)$	0.0005	0.0052	0.0165	0.0277	0.0316	0.0121	0.0012
Intrinsic Localized Modes for $g_i = -5.0$							
(r, g)	$n = 6$	$n = 4$	$n = 2$	$n = 0$	$n = -2$	$n = -4$	$n = -6$
$(0.6, 0.9711)$	0.0006	0.0069	0.0328	0.0677	0.0328	0.0070	0.0006
$(0.6, 0.9739)$	0.0005	0.0067	0.0298	0.0609	0.0353	0.0082	0.0007
$(0.6, 0.9846)$	0.0005	0.0056	0.0202	0.0371	0.0339	0.0109	0.0010

of intrinsic localized modes with increasing r. Again, directly to the right of the region of intrinsic localized modes is a region of multiple intrinsic localized mode solutions (i.e., for $r > 1.0$ and $0.9 < g < 1.0$) which will be discussed elsewhere.

In Table 2 the intensity profiles (i.e., $\lambda|E_{n,0}|^2$ versus $n = 6, 4, 2, 0, -2, -4, -6$) are presented for intrinsic localized modes at various (r, g) shown in Fig. 5. Representative solutions for all of the three branches of intrinsic localized modes are given in the Table. In general, the intensity profiles show an intensity peak near the center of the barrier that decreases rapidly to the edges of the barrier. This is a characteristic of the intrinsic localized modes. The intensity of the modes near the center of the barrier are great enough to impact the dielectric properties of the barrier media so that the modes are self-consistently created by and then sustain the dielectric properties of the barrier needed for the modes' existence. From the results in Table 2, it is seen that, for a given value of r, modes with increasing values of g tend to exhibit intensity peaks that are broader within the barrier. For a comparison in Table 2 results are presented both for $g_i = 5.0$ and $g_i = -5.0$. The general features of the mode intensity plots are similar for both of these cases.

The effects on the intrinsic localized modes of changing the value of g_i are generally small. This is seen in the results presented in Fig. 6 which shows a plot of the g at which intrinsic localized modes occur as a function of g_i for fixed $r = 0.4$ (denoted in the figures by the lower +'s) and $r = 0.6$ (denoted in the figures by the upper x's). A mild change in the value of g is found for changing g_i at a given fixed value of r. For some values of g_i further splittings of the intrinsic localized mode branches are found, but these splittings are small.

In regards to possible device applications, an interesting feature of of the plot in Fig. 5 is the presence of a set of intrinsics localized mode branches that cross each other in (r, g) space. These crossings allow for the possibility of a complex series of switchings within these systems. Moving on a circular path centered about the point of crossing of two branches of intrinsic localized modes, a series of four transmission resonances are encountered for small changes in r and g as one circulates around the crossing point of the two branches of modes. The changes in r are obtained by altering the field intensity of the guided modes whereas the changes in g may be mediated by mechanical or chemical stimuli applied to the system.

2.3. Kerr Media Barrier with Non-Linear Off-Channel Feature

A further generalization of the off-channel impurity problem is to have the off-channel site in Fig. 4a be formed of Kerr media impurity material. In this generalization Eqs. (9) and (10) become

$$E_{0,0} = g(1+\lambda|E_{0,0}|^2)E_{0,0} + g_l b(E_{-1,0} + E_{1,0}) + g_i b(1+\lambda|E_{0,-1}|^2)E_{0,-1} \quad (11)$$

and

$$E_{0,-1} = g_i(1+\lambda|E_{0,-1}|^2)E_{0,-1} + gb(1+\lambda|E_{0,0}|^2)E_{0,0}, \quad (12)$$

respectively. Whereas the linear off-channel site was characterized by g_i, the nonlinear site is characterized by g_i and λ.

In Fig. 7 the transmission resonances for $T > 0.6$ are plotted in (r, g) space for the parameters g, k, b, g_i, and λx^2 used in Fig. 5. Note that the parameter λx^2 for the barrier and impurity Kerr sites, for simplicity, are taken to be the same. The results in Fig. 7 are found to differ very little from those in Fig. 5. The reason for this is explained by the results for the field intensities in the barrier and off-channel sites given in Table 3. The off-channel site has, in general, a small field intensity so that its nonlinearity has little effect on the fields in the barrier sites. As a result the fields within the barrier sites are essentially the same as those in Table 2 for the off-channel site of linear dielectric media. This is found to be the case for $1.0 < g_i < 5.0$ and $-5.0 < g_i < -1.0$ which have been investigated by us.

2.4. Kerr Media Barrier with a Kerr Media or a Linear Media Impurity at the Center Site of the Barrier

Another generalization of the Kerr media barrier problem is to replace the Kerr site at the center of the barrier by an impurity site formed of a different Kerr medium. A schematic drawing of this case is given in Fig. 4b. The equations for this geometry are the same as those in Eqs. (1) through (5) with the following modifications: In Eq. (2), the equation for $n = 0$ is replaced by

$$E_{0,0} = g_i(1+\lambda|E_{0,0}|^2)E_{0,0} + g_l b[E_{1,0} + E_{-1,0}] \quad (13)$$

where $g_i \neq g$ is the parameterization for the Kerr impurity site, and for simplicity we retain the same parameter λ for both the Kerr barrier sites and the impurity site at the center of the

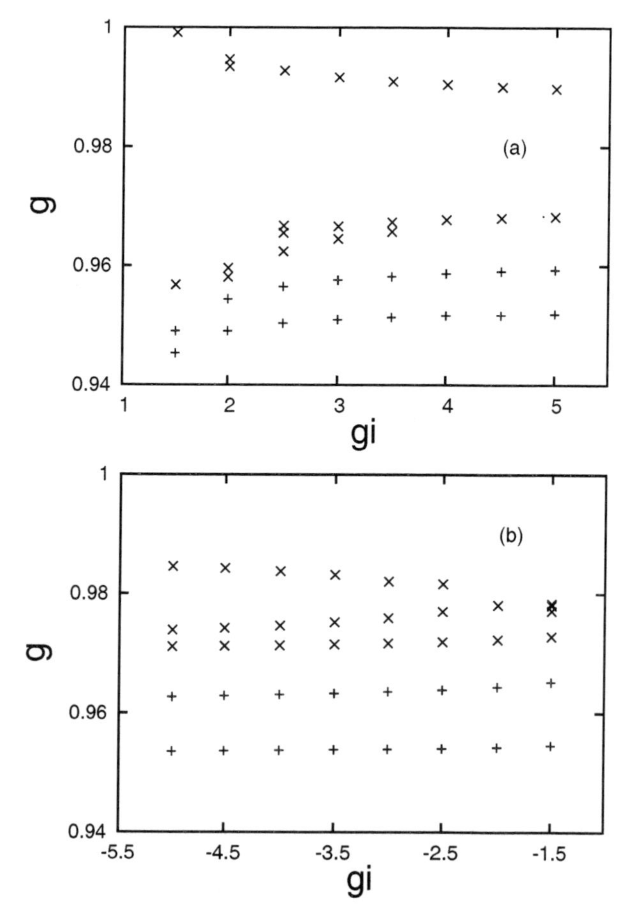

Figure 6. Plots of g versus g_i for the Kerr barrier coupled to a single off-channel linear media site as in Fig. 4a. In a) results are for an off-channel site formed of media with a positive g_i, and in b) results are for an off-channel site formed of media with a negative g_i. The upper x's are for the $r = 0.6$ case and the lower +'s are for the $r = 0.4$ case. Aside from the values of g_i, the values of parameters used in this plot are the same as those in Fig. 3.

barrier. In Eq. (3), the equation for $n = 0$ is replaced by

$$E_{1,0} = g_l E_{1,0} + gb(1 + \lambda|E_{2,0})|^2)E_{2,0} + g_i b(1 + \lambda|E_{0,0}|^2)E_{0,0} \quad (14)$$

and the equation for $n = -1$ is replaced by

$$E_{-1,0} = g_l E_{-1,0} + g_i b(1 + \lambda|E_{0,0}|^2)E_{0,0} + gb(1 + \lambda|E_{-2,0}|^2)E_{-2,0}. \quad (15)$$

In Fig. 8a a plot is given of the resonant transmission peaks in (r, g) space with $T > 0.6$, using the same parameters for g, k, λx^2 as in Fig. 3 and taking $g_i = 1.0$. Again a series of Fabry-Perot modes beginning at $r = 0$ and extending across the figure are observed in the region $1.0 < g < 1.2$. For this system, however, two different regions of intrinsic localized modes are observed. One is found below the Fabry-Perot modes (i.e., for $g < 1.0$ and $0.0 < r < 0.6$) and one above the Fabry-Perot modes (i.e., for $g > 1.2$ and $r > 0.2$).

Table 3. Listing for the barrier with coupling to an off-channel Kerr media site of the intrinsic localized mode field intensities. Valves are shown for $\lambda|E_{n,0}|^2$ in the barrier sites for n = 6, 4, 2, 0, -2, -4, -6 and $\lambda|E_{0,-1}|^2$ of the off-channel site (denoted by "off" in the Table) for selected values of (r, g) and g_i for $k = 2.9$ and $\lambda x^2 = 0.001$.

Intrinsic Localized Modes for $g_i = 5.0$								
(r, g)	$n = 6$	$n = 4$	$n = 2$	$n = 0$	$n = -2$	$n = -4$	$n = -6$	off
$(0.4, 0.9519)$	0.0003	0.0041	0.0304	0.1126	0.0306	0.0042	0.0003	0.00006
$(0.4, 0.9593)$	0.0003	0.0037	0.0242	0.0871	0.0582	0.0107	0.0007	0.00004
$(0.6, 0.9683)$	0.0006	0.0072	0.0360	0.0746	0.0361	0.0073	0.0006	0.00004
$(0.6, 0.9688)$	0.0006	0.0072	0.0354	0.0734	0.0368	0.0075	0.0006	0.00004
$(0.6, 0.9897)$	0.0005	0.0052	0.0165	0.0277	0.0316	0.0121	0.0012	0.00001
Intrinsic Localized Modes for $g_i = -5.0$								
(r, g)	$n = 6$	$n = 4$	$n = 2$	$n = 0$	$n = -2$	$n = -4$	$n = -6$	off
$(0.6, 0.9711)$	0.0006	0.0069	0.0328	0.0677	0.0328	0.0070	0.0006	0.00002
$(0.6, 0.9739)$	0.0005	0.0067	0.0298	0.0609	0.0353	0.0082	0.0007	0.0001
$(0.6, 0.9846)$	0.0005	0.0056	0.0202	0.0371	0.0339	0.0109	0.0010	0.00001

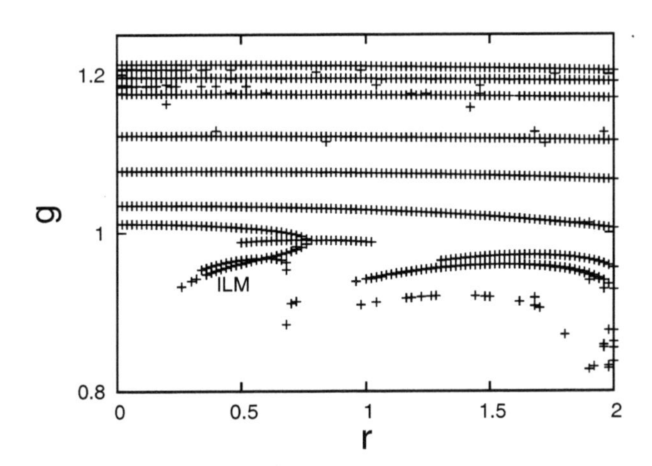

Figure 7. Plot of the resonant transmission peaks in (r, g) space for the case of a single off-channel site with $g_i = 5.0$ and $\lambda x^2 = 0.001$. The results show little difference from the $\lambda x^2 = 0.0$, linear impurity media, results in Fig. 5. The notation ILM has been added at the lower right of the region of the intrinsic localized modes.

As with the previous system studied, the lines of intrinsic localized modes begin at some $r > 0$ and continue with increasing r into the plot. The lowest Fabry-Perot branch meets the lower ridge of intrinsic localized modes at $r \approx 0.6$, while the upper branch extends across the figure. In Table 4 results are shown for the intensity profiles within the barrier of modes within the two branches of intrinsic localized modes.

It is interesting to compare the problem of a Kerr nonlinear impurity at the center of the barrier with the case in which the center site barrier impurity is made of linear impurity media. For this case, Eq. (13) becomes

$$E_{0,0} = g_i E_{0,0} + g_l b[E_{1,0} + E_{-1,0}], \quad (16)$$

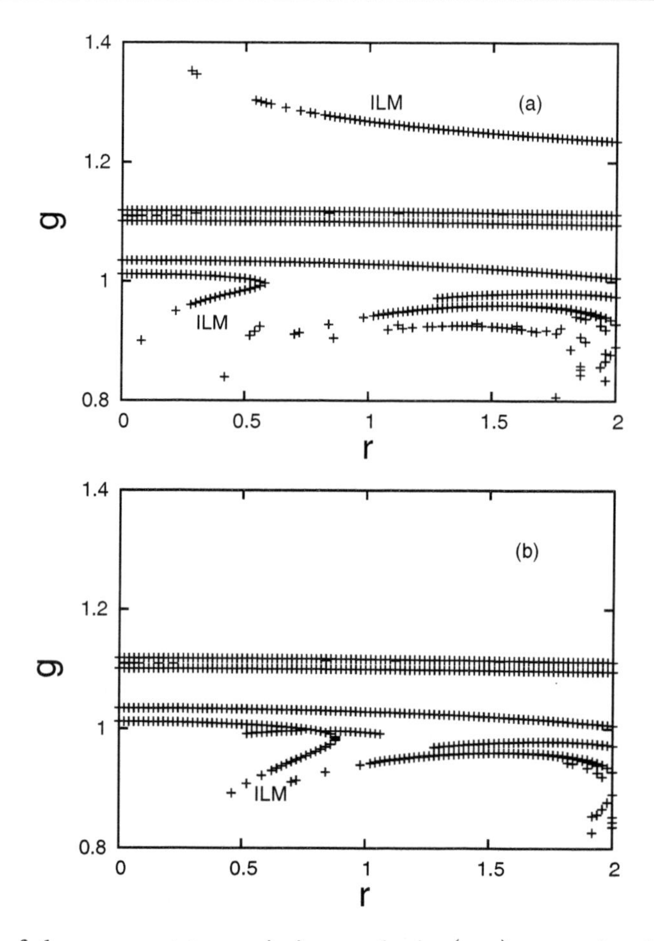

Figure 8. Plot of the resonant transmission peaks in (r, g) space for the case of a Kerr barrier containing an impurity site at it center: a) Kerr impurity site with $g_i = 1.0$ and $\lambda x^2 = 0.001$. The notation ILM has been added at the lower right of the $g < 1.0$ region of the intrinsic localized modes and above and adjacent to the $g > 1.2$ region of intrinsic localized modes. b) A linear media impurity site with $g_i = 1.0$. The notation ILM has been added to the lower right of the $g < 1.0$ region of intrinsic localized modes.

Eq. (14) becomes

$$E_{1,0} = g_l E_{1,0} + gb(1 + \lambda|E_{2,0}|^2)E_{2,0} + g_i b E_{0,0}, \tag{17}$$

and Eq. (15) becomes

$$E_{-1,0} = g_l E_{-1,0} + g_i b E_{0,0} + gb(1 + \lambda|E_{-2,0}|^2)E_{-2,0}. \tag{18}$$

This amounts to taking $\lambda = 0$ at the center site, but $\lambda \neq 0$ on the other Kerr sites of the barrier.

Table 4. Listing of the intrinsic localized mode field intensities for the barrier with a Kerr impurity site at the center of the barrier, i.e., at $(0,0)$. Valves are shown for $\lambda|E_{n,0}|^2$ in the barrier sites for $n = 6$, 4, 2, 0, -2, -4, -6 for selected values of (r,g) and for $k = 2.9$ and $\lambda x^2 = 0.001$. The impurity site has $g_i = 1.0$ and $\lambda x^2 = 0.001$.

Intrinsic Localized Modes							
(r,g)	$n=6$	$n=4$	$n=2$	$n=0$	$n=-2$	$n=-4$	$n=-6$
$(0.3, 0.9634)$	0.0001	0.0020	0.0125	0.0506	0.0125	0.0020	0.0002
$(0.4, 0.9750)$	0.0002	0.0029	0.0142	0.0415	0.0141	0.0029	0.0002
$(0.5, 0.9850)$	0.0004	0.0039	0.0145	0.0308	0.0146	0.0039	0.0004
$(0.6, 1.2972)$	0.0003	0.0016	0.0117	0.1362	0.0118	0.00176	0.0005
$(1.0, 1.2684)$	0.0009	0.0033	0.0167	0.1416	0.0163	0.0027	0.0003

In Fig. 8b results are presented for the $g_i = 1.0$ linear impurity and the usual values of g, b for $k = 2.9$ and $\lambda x^2 = 0.001$. The results are similar to those in Fig. 8a, with a renormalization of the the plotted ridges for $g < 1.2$. For $g > 1.2$, however, the branch of intrinsic localized modes found in Fig. 8a is no longer present in the results of Fig. 8b. This indicates the importance of the nonlinear media at the center site in developing intrinsic localized mode excitations. This is due to the fact that the mode intensity is greatest at the center site so that the dielectric nonlinearity exhibits the largest response at the center site. In addition to the renormalization of the lines and ridges for $g < 1.2$ a new branch of intrinsic localized modes is found that develops near horizontally with increasing r.

3. Conclusions

The properties of a number of simple photonic crystal circuit elements containing a barrier of Kerr nonlinear media and various linear media and nonlinear media impurities have been discussed. The possible modes that can be resonantly excited by guided waves incident on the barriers have been determined and listed in an (r,g) space plot of the transmission resonances of the barrier. This is a type of spectroscopy which allows for the modes of the barrier to be classified and their behaviors under changes in the nonlinear barrier media shown. Emphasis was placed on the determination of the Fabry-Perot and intrinsic localized modes. While the Fabry-Perot modes exist in the linear limit of the barrier media and exist in renormalized forms as the barrier nonlinearity is increased, the intrinsic localized modes only are found in the case of nonlinear barriers as they only exist in nonlinear systems. The various ridges and lines of resonances in (r,g) are such that barrier modes on the same ridge or line have similar wavefunctions whereas modes on different ridges or lines have distinctly different types of wavefunctions. This gives a very useful classification scheme for the possible modes that are excited within the barrier.

The nonlinear barriers display characteristics important in various types of amplification, switching, and rectification applications. In addition, the crossing of different branches of intrinsic localized modes can be used to formulate an intricate series of switching processes that are provided for by a number of different mechanisms.

Appendix A

A brief summary of the origin of the difference equation formulation is given here for the case in which the electric field of the modes is polarized with the electric field parallel to the axes of the dielectric cylinders. For more details the reader is referred to Refs. [9], [10] and [25].

The photonic crystal is described by a periodic dielectric function $\epsilon(\vec{x}_{||})$ where $\vec{x}_{||}$ is in the plane of the Bravais lattice, perpendicular to the axes of the dielectric cylinders. A waveguide is defined by introducing a change in the dielectric constant, $\delta\epsilon(\vec{x}_{||})$, in a row of dielectric cylinders. Using the Helmholtz equation for the electric fields of the modes and exact methods of Green's functions[9, 10, 11, 25, 26], the electric fields of the waveguide modes satisfy [9, 10, 25]

$$E(\vec{x}_{||}) = \frac{\omega^2}{c^2}\int d^2x'_{||}G(\vec{x}_{||},\vec{x'}_{||})\delta\epsilon(\vec{x'}_{||})E(\vec{x'}_{||}). \tag{19}$$

Here $G(\vec{x}_{||},\vec{x'}_{||})$ is the Green's function of the Helmholtz equations for the photonic crystal described by $\epsilon(\vec{x}_{||})$. The evaluation of the Green's functions for the system addressed in this paper is discussed in Refs.[9, 54] where it is shown that the Green's functions for frequencies within the photonic crystal stop band exhibit an overall rapid decay with increasing separation of $\vec{x}_{||}$ and $\vec{x'}_{||}$ within the system, particularly for multiple separations of the lattice constant[23]. The stop band decay is due to the non- propagating nature of the wave functions for frequencies within the stop band, and is different from the behavior shown at pass band frequencies.

If $\delta\epsilon(\vec{x}_{||})$ is non-zero only in a small region about the axes of cylinders forming the waveguide channel and $E(\vec{x}_{||})$ changes slowly over $\delta\epsilon(\vec{x}_{||})$ in each such cylinder, then Eq. (19) reduces (for a waveguide along the $x-$axis for which only on site and nearest neighbor site interactions are significant) to a set of difference equations given by

$$E_{n,0} = g_p[f_{n,0}E_{n,0} + b(f_{n+1,0}E_{n+1,0} + f_{n-1,0}E_{n-1,0})]. \tag{20}$$

In Eq. (20), $f_{n,0} = 1 + \lambda|E_{n,0}|^2$ with $\lambda = 0$ for linear dielectric media, $g_p = \frac{\omega^2}{c^2}\int d^2x'_{||}\, G(0,\vec{x'}_{||})\delta\epsilon(\vec{x'}_{||})$ where the integral is over the cylinder centered at $(0,0)$, and $b = \int d^2x'_{||}G(a_0\hat{i},\vec{x'}_{||})\delta\epsilon(\vec{x'}_{||})/\int d^2x'_{||}G(0,\vec{x'}_{||})\delta\epsilon(\vec{x'}_{||})$ where the integrals are computed as in the case of g_p and a_0 is the lattice constant of the waveguide. In the evaluation of all of the integrals in the definitions of g_p and b, $\lambda = 0$. Both the Green's functions and their space integrals are evaluated from the eigenvalues and eigenvectors of the electromagnetic equations of motion for the photonic crystal[9, 54] using methods that are common in the study of ionic impurities in metals[55].

The parameters of the photonic crystal used in this paper are the same as those used in a number of previous publications, and we refer the read to these for a more detailed discussion[9, 10, 11, 25]. The two-dimensional photonic crystal is a square lattice with cylinders of dielectric constant $\epsilon = 9$ and radius $R = 0.37796a_c$ where a_c is the lattice constant of the square lattice, and waveguide and barrier impurity materials occur in a square cross section of side $0.02a_c$. The waveguide is taken along the $x-$ axis of the square lattice such that the lattice constant of the waveguide is $a_0 = a_c$. The dielectric constant of the Kerr media is taken of a generic isotropic form so that a semi-quantitative indication of the effects of nonlinearity are given. This is not an uncommon thing to do in theoretical treatments in nonlinear optics, and it is hoped that the results presented will stimulate studies on specific systems in which more detailed considerations of the form of the dielectric constants may be made.

References

[1] Pollaock, C.; Lipson, M. "Integrated Photonics", Kluwer Academic Publishers: Boston, 2003.

[2] Prasad, P. N. "Nanophotonics", John Wiley and Sons: Hoboken, 2004.

[3] McGurn, A. R. In "Survey of Semiconductor Physics"; Boer, K. W.; Ed.: John Wiley and Sons, Inc.: New York, 2002, Chp. 33.

[4] Sakoda, K. "Optical Properties of Photonic Crystals"; Springer: Berlin, 2001.

[5] Joannopoulos, J. D.; Meade, R. D.; Winn J. N. "Photonic Crystals"; Princeton University Press: Princeton, 1995.

[6] Joannopoulos, J. D.; Vilenueve, P. R., Fan, S. *Nature* **386**, 1997, pp. 143-149.

[7] Londergan, J. T.; Carini, J. P.; Murdock, D. P. "Binding and Scattering in Two-Dimensional Systems: Applications to Quantum Wires, Waveguides, and Photonic Crystals"; Springer: Berlin, 1999.

[8] Zolla, F.; Renversez, G.; Nicolet, A.; Kuhlmey, B.; Guenneau, S.; Felbacq, D. "Foundations of Photonic Crystal Fibers"; World Scientific: New Jersey, 2005.

[9] McGurn, A. R. *Phys. Rev. B* **53**, 1996, pp. 7059-7064.

[10] McGurn, A. R. *Phys. Rev. B* **61**, 2000, pp. 13235-13249.

[11] McGurn, A. R. *Phys. Rev. B* **65**, 2002, pp. 75406-1 - 75406-11.

[12] Hunsperger, R. G. "Integrated Optics: Theory and Technology", 4th Ed. Springer: Berlin, 1984.

[13] Yeh, C. "Applied Photonics". Academic Press: San Diego, 1990 Chps. 2 and 12.

[14] Abdonovic, I.; Uttamchandarri, D. "Principles of Modern Optical Systems", Artech House: Norwood, 1989.

[15] Tanida, J.; Ichioka, Y. In "Progress in Optics"; Wolf, E.; Ed.; Elsevier: Amsterdam, 2000, Vol. XL, pp 77-114.

[16] Kilin, S. Ya. In "Progreess in Optics", Vol. 42, E. Wolf, E.; Ed.; Elsevier: Amsterdam, 2001, Vol. 42, pp. 1-92.

[17] Glesk, I.; Wang, B.C.; Xu, L.; Baby, V.; Prucnal, P. R. In "Progress in Optics"; Wolf, E.; Ed.; Elsevier: Amsterdam, 2003, Vol. 45, pp. 53-118.

[18] Feitelson, D. G. "Optical Computing"; MIT Press: Cambridge, 1988.

[19] "Optical Computing"; Wherrett, B. S.; Tooley, F. A. P.; Ed.; Edinburgh University Press: Edinburgh, 1989.

[20] Mansuripur, M. "Classical Optics and Its Applications"; Cambridge University Press: Cambridge, 2002.

[21] Kivshar, Y. S.; Agrawal, G. P. "Optical Solitons"; Academic Press: Amsterdam, 2003.

[22] Ibach, H.; Luth, H. "Solid-State Physics"; Springer: Berlin, 2003.

[23] McGurn, A. R. *Chaos* **13**, 2003, pp. 754-765.

[24] McGurn, A. R. *J. Phys. CM* **14**, 2004, pp. S5243-S5252.

[25] McGurn, A. R. *Phys. Lett. A* **251**, 1999, pp. 322-335.

[26] McGurn, A. R. *Phys. Lett. A* **260**, 1999, pp. 314-321.

[27] McGurn, A. R.; Birkok, G. *Phys. Rev. B* **69**, 2004, pp. 235105-1 - 235105-13.

[28] Cowan, A. R.; Young, J. F. *Phys. Rev. E* **68**, 2003, pp. 46606-1 - 46606-16.

[29] Yanik, M. F., Fan, S., Soljacic, M., and Joannoupoulis, J. D., *Opt. Lett.* **28**, 2003, pp. 2506-2510.

[30] Soljacic, M.; Joannopoulos, J. M. *Nat. Mater.* **3**, 2004, pp 211,215.

[31] Xu, Z.; Maes, B.; Jeang X.; Joannopoulos, J. D.; Torner, L.; Soljacic, M. *Optics Letters* **33**, 2008, 1763-1767.

[32] Mills, D. L. "Nonlinear Optics"; Springer-Verlag: Berlin, 1998.

[33] Banerjee, P. P. "Nonlinear Optics"; Marcel Kekker, Inc.: New York, 2004.

[34] Boyd, R. W. "Nonlinear Optics 2nd Ed."; Academic Press: Amsterdam, 2003.

[35] Kittel, C. "Introduction to Solid State Physics"; Wiley: New York, 1976.

[36] Berger, V. *Phys. Rev. Lett.* **81**, pp. 4136-4139.

[37] Shi, B.; Jiang, Z. M.; Wang, X. *Opt. Lett.* **26**, 2001, pp. 1194-1196.

[38] Ren, F. F.; Li, R.; Cheng, C.; Wang, H. T.; Qui, J. R.; Si, J. H.; Hirao, K. *Phys. Rev. B* **70**, 2004, pp. 245109-245113.

[39] McGurn, A. R. *Phys Rev B* **27**, 2008, pp. 115105-115115.

[40] McGurn, A. R. *J Phys: Condens. Matter* **20**, 2008, 025202-025212.

[41] McGurn, A R. in "Nonlinear Research Perspectives" Ed. Wang, 2007, Novascience Publishers.

[42] McGurn, A. R. *J Phys: Condes. Matter* **21**, 2009, pp.485302- 485312.

[43] McGurn, A. R. *Compexity* **12**, 2007, pp.18-40. 39

[44] Merzbacher, E. "Quantum Mechanics", 3rd Edition, 1998, Hamilton Printing Company, New York.

[45] Sievers, A, J.; Takeno, S. *Phys. Rev. Lett.* **61**, 1988, pp. 970-974.

[46] Sievers, A. J.; Page, J. B. In "Dynamical Properties of Solids"; Horton, G. K.; Maradudin, A. A.; Ed.; Elsevier: Amsterdam, 1995, Vol. 7, pp. 137-255.

[47] Lam L. (ed.) "introduction to Nonlinear Physics". 1996 Springer, New York

[48] Pau, T. "Attractors, Bifurcations, and Chaos", 2000, Springer, New York

[49] Rasband, S. N. "Chaotic Dynamics of Nonlinear Systems" 1990, Wiley, New York.

[50] Turing, A. M., *Phil. Trans. R. Soc. B* **237**, 1952, 37-55.

[51] Rabinovich, M. I.,; Ezershy, A. B.; Weidman, P. D. "The Dynamics of Patterns" 2000, World Scientific, Singapore.

[52] Wolfram, S. "A New Kind of Science", 2002, Wolfram Media, Urbana.

[53] McGurn, A. SPIE Proceedings, Vol. **7395**, 2009, Ed. by E. S. Kawata, V. M. Shalaev, and D. P. Tsai, 7395 1T-1 to -11.

[54] Maradudin, A. A.; McGurn, A. R. *J Opt. Soc. Am. B* **10** pp. 307-316.

[55] Callaway, J. "Energy Band Theory", 1963, Academic, New York.

In: Photonic Crystals
Editor: Venla E. Laine, pp. 173-189
ISBN: 978-1-61668-953-7

Chapter 8

INFLUENCE OF THE SOLID-LIQUID ADHESION ON THE SELF-ASSEMBLING PROPERTIES OF COLLOIDAL PARTICLES

***Edina Rusen*[1]*, Alexandra Mocanu*[1]*, Bogdan Marculescu*[1]*, Corina Andronescu*[1]*, I.C. Stancu*[1]*, L.M. Butac*[1]*, A.Ioncea*[2] *and I. Antoniac*[3]**

[1]University Politehnica Bucharest, Depart. of Polymer Science, Bucharest, Romania
[2] S.C. Metav S.A. Bucharest,C.A., Bucharest, Romania
[3]University Politehnica Bucharest, Faculty of Materials Science and Engineering, Bucharest, Romania

Abstract

Colloidal dispersions with self-assembling properties have been prepared by soap-free emulsion copolymerization of styrene (St) with hydroxyethylmethacrylate (HEMA), using various St/HEMA molar ratios. Investigations by SEM, AFM and reflexion spectra have shown that copolymer particles give colloidal crystals while monodisperse particles of polystyrene, with no HEMA units, do not crystallize. XPS analysis of the copolymer particles has shown that the HEMA hydrophilic units tend to concentrate on the surface of the copolymer particles, while the core is richer in St. The copolymer particles have been separated and redispersed in other liquids than water (ethylene glycol, formamide, ethanol, hexane, and toluene) to investigate the conditions that allow self-assembling after the diluent evaporation. For each case, contact angles have been measured and the adhesion work has been computed. The results have shown that the contact angle does not influence the self-assembling process, while the adhesion work is an important parameter. Only dispersions characterized by a value of the adhesion work superior to 87 mJ/m^2 did crystallize after the diluent evaporation.

Keywords: soap-free polymerization; self-assembling; contact angle; adhesion work

1. Introduction

Spontaneous formation of photonic crystals from colloidal dispersions of either organic polymers or inorganic materials has raised a great scientific interest materialized in a large number of papers published in the last decade.

Although many alternative methods for obtaining colloidal crystals by deposing spherical, monodisperse particles, on plane surface have been reported [1,2], the auto-assembling process always involves three distinctive stages:

- a concentration of the colloidal particles on the substrate surface, phenomenon based on the reciprocal adherence;
- a compact disposition of the particles in a planar 2D network due to the lateral inter-particles attractions;
- an increase of the thickness of the layer by successive deposition of new crystalline planes on the initial planar area, thus leading to a 3D – spatial network, with a compact hexagonal (hcp) or centered cubic (ccp) geometry.

Since none of the above stages has been proven as dominant for the entire process, it is reasonable to accept that the relative importance of the stages is the same as their chronological order. One of the major problems is to define the nature of the motrix force of the auto-assembling process. An exhaustive and pertinent analysis of the complex physical interaction phenomena involved during concentration of the disperse particles [3,4] revealed the fact that the capillary forces play a decisive role, both due to their intensity and their interaction radius.

Consequently, most of the theoretical studies focus on obtaining precise methods for computing the capillary force, based mainly on the geometry of the physical system and less on the chemical nature of the solid surface or the liquid medium [5-21]. The only accessible parameter (by experimental measurements) that reflects the specifics of these interactions is the contact angle (**θ**) between the liquid and the solid surface. The most relevant results from the above-mentioned studies lead to some conclusions that are considered relevant premises for the present work:

a. Several theoretical studies are oriented towards obtaining mathematical models describing physical systems consisting of pairs of solid bodies separated by a liquid film. The two bodies have well defined shape and size, but their chemical nature is ignored. The chemical structure of the liquid is not involved either and it's only physical property taken into account is the superficial tension (**γ**) [5-12]. In other works, where experimental data is used to validate the theoretical models proposed, the liquid involved is always water and the dispersed solid phase is composed by monodisperse particles of polystyrene, SiO_2 [1,2,16,18,20] or submicronic sand particles [21]. As for the plane, solid surfaces used as the support for the deposition and the spontaneous auto-assembling of the particles, glass (as such or with the surface covered by hydrophobic films of perfluor-derivatives or organo-siloxane compounds [2,18]), Au, Ag, mica [16] were used. A notable exception is the study of the auto-aggregation of ice or tetrahydrofurane hydrate from n-dodecane [15]. Note that the attraction resulted from capillary forces acts on any particle, whatever the nature and shape of the dispersed particles but the result is (or may be) a crystalline network only if these are spherical and with a strictly uniform size.

b. Many works assume explicitly (even if only qualitatively) that the contact angle has a crucial influence on the attractive forces and consequently on the auto-assembling capacity of the colloidal particles [13,14,16,20]. This is obviously due to the decisive role of the interfacial compatibility between the solid and the liquid. The shape of the liquid meniscus (concave or convex), the value and the sign of the capillary force depend on the interfacial compatibility. The role played by the contact angle is explicitly or implicitly considered in more complex mathematical models [9,16,18]. Moreover, the relations used to compute the capillary force used in most studies [11,12,15,17,19] are those given by equations (1) and (2) below:

$$F_c \sim \gamma \, (\cos \theta_1 + \cos \theta_2) \qquad (1)$$

$$F_c \sim \gamma \cos \theta \qquad (2)$$

Relation (1) applies for attractions between surfaces of different chemical composition (S_1 and S_2) separated by a liquid film with the superficial tension $\boldsymbol{\gamma}$. Relation (2) is valid for the capillary attraction between two spherical surfaces having the same nature. Obviously, relation (1) reduces to (2) for two surfaces with the same chemical composition ($\boldsymbol{\theta_1 = \theta_2}$). For two different surfaces, from which one (S_2) has a strong liophilic character (therefore the liquid spreads on the respective surface) ($\theta < 10^0$ leads to $\boldsymbol{\cos \theta \approx 1}$), relation (1) reduces to equation (3):

$$F_c \sim \gamma \, (\cos \theta_1 + 1) \qquad (3)$$

c. Many papers use the term "adhesive forces" instead of "capillary forces" and replace the term "attraction" with "adhesion". The use of such notions in works of a high scientific level is not figurative but reflects a real phenomenon: the liquid film that mediate the attraction between the surfaces of two solid bodies acts as an adhesive layer that connects two identical or different substrates. If the liquid (the adhesive) is superficially compatible with the solid surface ($\theta < 90^0$ leading to $\boldsymbol{\cos \theta > 0}$), then the capillary forces are attractive, while for an incompatibility between the liquid and the solid ($\theta > 90^0$ leading to $\boldsymbol{\cos \theta < 0}$) the same forces are repulsive. Meanwhile, considering the capillary force as derivated from an attraction potential **W**, gives an alternative definition to the classical one (Laplace) [5,6]:

$$F_c = -\frac{dW}{dL} \qquad (4)$$

Even if the physical nature of this potential is not clearly defined, obviously it must be an electrostatic potential, the same that is considered as the basis for the intermolecular attractive forces (dispersion, orientation and induction forces). In fact the energetic approach of any adhesive assembling results form the interpenetration of the hybrid force fields localized at the solid-liquid interface (adhesion) and the homogeneous forces acting in the bulk of the liquid film (cohesion).

Therefore, the auto-assembling capacity of the colloidal particles should take into account not only the capillary forces but also the adhesion work (**W_a**), a thermodynamic parameter used frequently to characterize the interfacial compatibility between a liquid and a solid. W_a may be obtained from experimental data using the classical relation Young-Dupré:

$$W_a = \gamma_L(1+\cos\theta) \quad (5)$$

Note also that from the above relations it results that $\mathbf{F_c \approx W_a}$.

Using the adhesion work makes possible not only the interpretation of the experimental data already available but also it allows predicting the behavior of a solid-liquid system by using analytical relations such as $W_a = W_a(\gamma_S^D, \gamma_S^P, \gamma_L^D, \gamma_L^P)$, proposed by Fowkes or Wu [22,23]. The necessary and sufficient condition for using such equations is to know the values of the dispersive and respectively polar components of the superficial tensions involved in the adhesion phenomenon.

The aim of the present work is to validate the above hypothesis. Colloidal dispersions of various chemical compositions have been obtained and the behavior of the particles in a variety of liquids, with different chemical composition, has been examined. The investigation supposes measuring, for every solid-liquid pair, of the adhesion work, coupled with evaluating the aptitude of crystallization of the same systems on a single plane substrate (hydrophilic glass). Obviously, the selected colloidal suspensions must have a size monodispersity, since this is the most important condition for self-assembling. The disperse particles are styrene (St) – hydroxyethylmethacrylate (HEMA) copolymers with various compositions while liquids used have been selected according to both their superficial tensions and wide range of polarities.

2. Materials and Methods

2.1. Materials

The St purchased from Merck has been purified through distillation under vacuum. HEMA (Aldrich) has been passed through separation columns filled with Al_2O_3 to remove inhibitors. The initiator used, potassium persulphate (KPS) (Merck) has been recrystalized from a mixture ethanol/water and then vacuum dried.

For the precipitant homopolymerisation of St the diluent was methanol (M) (Merk) previously purified by distillations. The initiator was azo-iso-butyro-di-nitrile (AIBN) (Merk) purified by recrystallisation from M.

For redispersing monodisperse particles and respectively for measuring contact angles several liquids have been used: formamide (Merck), ethylene-glycol (Aldrich), ethanol (Aldrich), hexane (Fluka) and toluene (T) (Aldrich).

2.2. Soap-Free Emulsion Polymerization

A total mass of 1.55 g of monomer mixture (with a different molar ratio St:HEMA for each test) has been added to 20 ml distilled water together with a constant concentration of KPS ($2.3 \cdot 10^{-3}$ mol/l). The reaction mixture was nitrogen bubbled and then maintained for 48 h at 75°C under continuous stirring. In test **A** the monomer was only styrene while tests **B**, **C**, **D** and **E** consisted of mixtures St:HEMA with the molar ratios 17:1, 8:1, 5:1 sand 3.3:1 respectively. The final lattices have been dialyzed in distilled water for 7 days, using cellulose dialysis membranes (molecular weight cutoff: 12.000-14.000).

Latexes have been precipitated in M, filtered and dried under vacuum before measuring the final conversions (that have been in all cases above 99%) and to obtain the cylinders used for measuring the contact angles.

2.3. Precipitant Polymerization

2 mol/l St have been added to 10 ml M together with an AIBN concentration of $2 \cdot 10^{-3}$ mol/l. After nitrogen purging the polymerization vial has been kept for 4 hors at 60^0C, under stirring. The resulted polymer was separated by filtration and the unreacted monomer was eliminated by Soxhlet extraction and dried under vacuum until a constant mass has been reached.

2.4. Obtaining the Colloidal Crystals

The colloidal dispersions (in water or other liquids) have been deposed by dropping on lamellae of hydrophilised glass (the hydrophilic character was obtained by immersing the glass in a 30% NaOH solution for 24 hours) degreased immediately before use. Specimens (colloidal dispersions) have been maintained at constant temperature (40^0C) until complete drying.

2.5. Redispersing of the Monodisperse Particles in Liquid Media

The opal was detached from the glass sheets with a metallic blade, manually grinded in a mortar and redispersed by ultrasounds in various liquid media, at concentrations similar to the initial latexes (1.55 g/20 ml).

2.6. Characterization

The morphologies of polymer particles were observed using a Zeiss Evo 50 XVP Scanning Electron Microscope (SEM). The probes were sputtered with a thin layer of gold prior to imaging. Structural characterizations of the opals were performed using the AFM NT-MDT P47H apparatus. The reflectance has been recorded using a UV-VIS spectrophotometer Jasco V550. A XPS K-ALPHA spectrometer has been used to establish the specimen composition. Contact angles have been measured with a EW-59780-20 Contact Angle Meter 110 VAC, 50/60Hz. The method used for obtaining the specimens was different from the one used for copolymers with similar composition reported in the literature [24,25]; in the present work, solid particles have been compressed under a pressure of 15MPa using a PerkinElmer mini-press.

3. Results and Discussion

Since there is a well-known difference of solubility in aqueous media of the two monomers (St, respectively HEMA) a first stage of the research was to synthesize copolymers with various compositions by soap-free emulsion polymerization. The differences in the solubilities of the monomers, macroradicals and respectively polymers will result in differences in the size of the final particles [26,27], since the particle size is directly influenced by the mechanism of their generation and growth. Therefore, the films obtained from the latexes thus obtained have been investigated using the SEM technique to asses their morphology (figure 1).

The analysis of the images in figure 1 shows that the particles size diminishes when the molar ration St/HEMA is lower. Since both the initiator concentration and the total mass of the monomers have been kept constant in all the synthesis processes, the only possible explanation for the increase in the number of particles when the St/HEMA molar ratio diminishe resides in the change of the topochemistry of the process resulted from the different concentrations of the hydrophilic (water-soluble) monomer in the reaction mixture. A reasonable supposition is that St/HEMA ratio influences the kinetics of the process both in the nucleation stage and in what concerns the subsequent behavior of the generated particles. This remark is valid for several other binary systems consisting of a hydrophobic and a hydrophilic monomer [28,29].

A comparison between the morphology of the particles from latexes **A-E** (from the SEM data, figure 1) shows that for the particles containing HEMA units the monodispersity allows the self-assembling leading to colloidal crystals. Supplementary investigation methods, such as AFM (figure 2) and reflexion spectra have been used to give more information on the colloidal crystals.

From figure 2, a well-ordered, close packed structure can be noticed with a triangular arrangement, that can be attributed to planes <111> of a face centered cubic (fcc) system [30,31]. Moreover FTT's image displays sharp peaks confirming the existence of long range crystalline order.

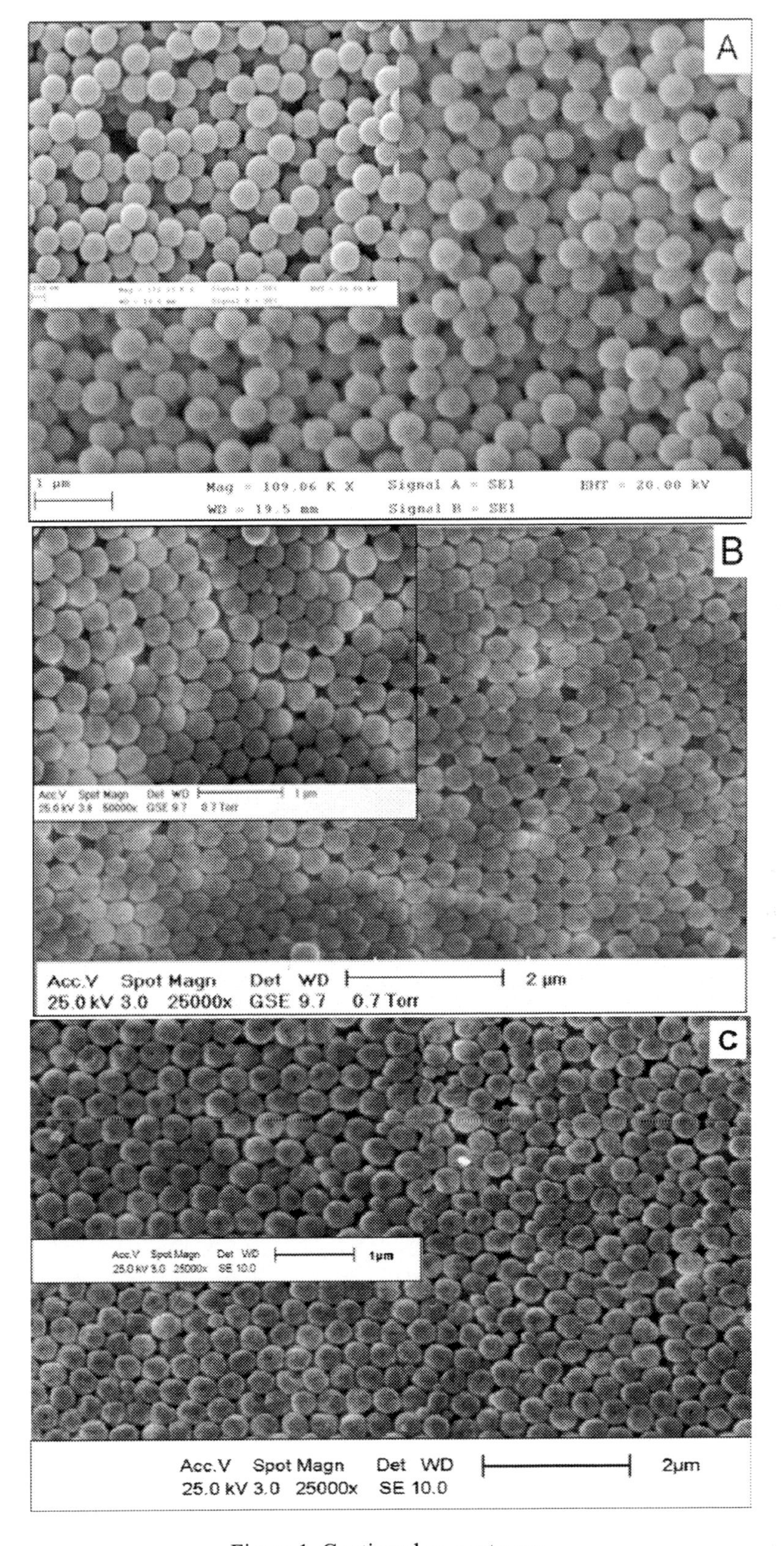

Figure 1. Continued on next page.

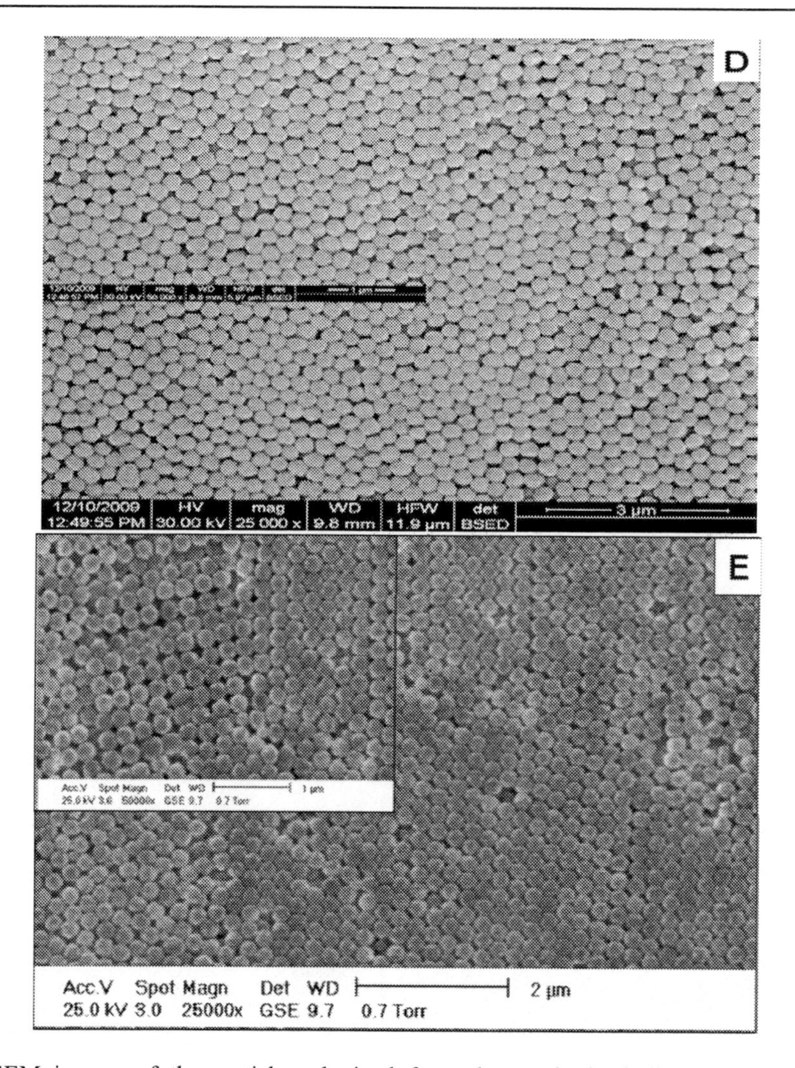

Figure 1. SEM images of the particles obtained from the synthesized dispersions (in water): A (d=430nm) monodisperse, B (d=380nm) monodisperse, C (d=330nm) monodisperse, D (d=290nm) monodisperse and E (d=254) monodisperse

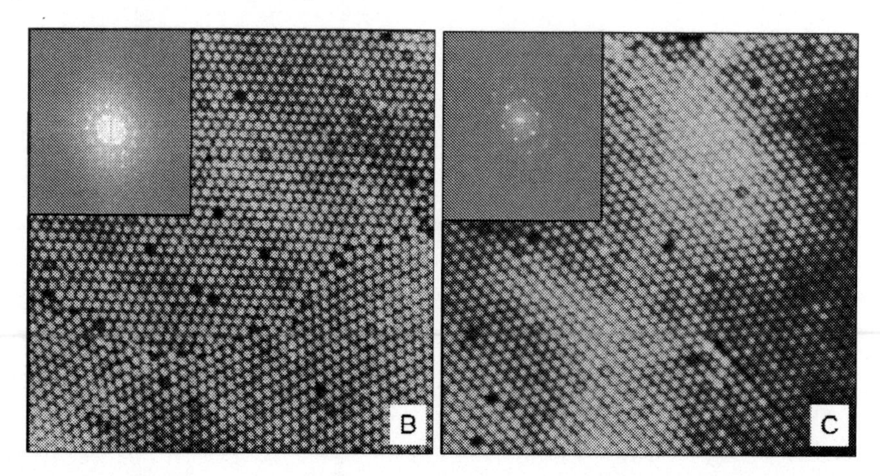

Figure 2. AFM image of the top surface of the opal and FFT's image from the samples B and C.

In figure 3, the reflection peaks located at 448 nm and 618 nm, are due to the Bragg reflection of periodic arrangement of spheres in the solid layer. A sharp peak indicates a regular, periodic structure.

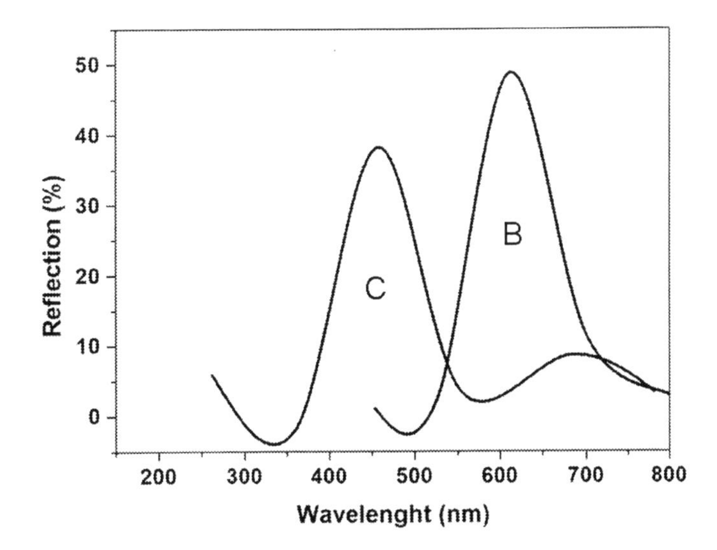

Figure 3. Reflection spectra of the opal films from the systems B and C.

Other works concerning the analysis of the St/HEMA soap-free copolymerization have shown a core-shell morphology of the colloidal particles [24,25]. Several characterization methods have been used to prove that the superficial layer is richer in HEMA as compared to the global composition of the copolymer. Since in the present work the superficial behavior of the particles generated from the St/HEMA system is investigated, a more precise characterization of the chemical composition of the superficial layer has been obtained by XPS.

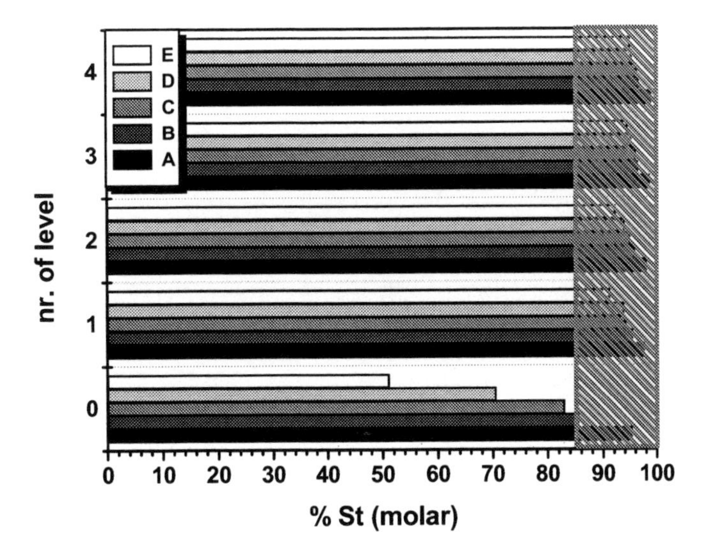

Figure 4. XPS results for depths between 0 and 160 nm (the shaded area gives the limits of the ST fraction from the initial composition of the monomers

Since the nucleation is homogeneous [32-35], the macroradicals precipitated during this stage will be found on the surface of the final particles; their composition and molar mass (depending on the initial ratio of the monomers) are important factors acting on the size of the nuclei at the moment when colloidal stability is achieved (at the end of the nucleation stage). Obviously, the colloidal stability of the particles is given both by the ionic groups provided by the initiator and the hydroxyl, water-soluble groups from the HEMA units. The XPS analysis gave the composition of the particles obtained from latexes **A-E** both on the surface and for depths reaching up to 160 nm (which corresponds roughly with the center of a particle); scannings were performed every 40 nm (figure 4).

For system **A** the surface contains 95% polystyrene (PSt) while the rest consists of oxygen and sulphur, constituents of the SO_4^- groups resulted from the initiator. In all the other specimens, the percentage of PSt in the superficial layer of the particles is much lower than the corresponding fraction of St in the initial monomer mixture. This is another argument towards the homogeneous nucleation. A more important remark is that the PSt percentage on the surface decreases with the St/HEMA ratio in the initial monomer composition [25], which confirms that the composition of the macroradicals that reach the critical precipitation size is different. Therefore, the size of the nuclei when the colloidal stability achieved is different. If the initial composition of the monomer mixture is richer in HEMA, the critical precipitation size increases, the size of the auto-stabilized nuclei is lower and therefore the final size of the particles is lower (meaning that more particles will be generated). When the analysis is made at various distances from the surface of the particles the content of St increases, up to values higher than the St content in the initial monomer mixture, thus balancing the superficial lower amount of the same monomer. To reduce the level of error, sample **E** has been re-analyzed, with a step of 3 nm. Results are shown in figure 5.

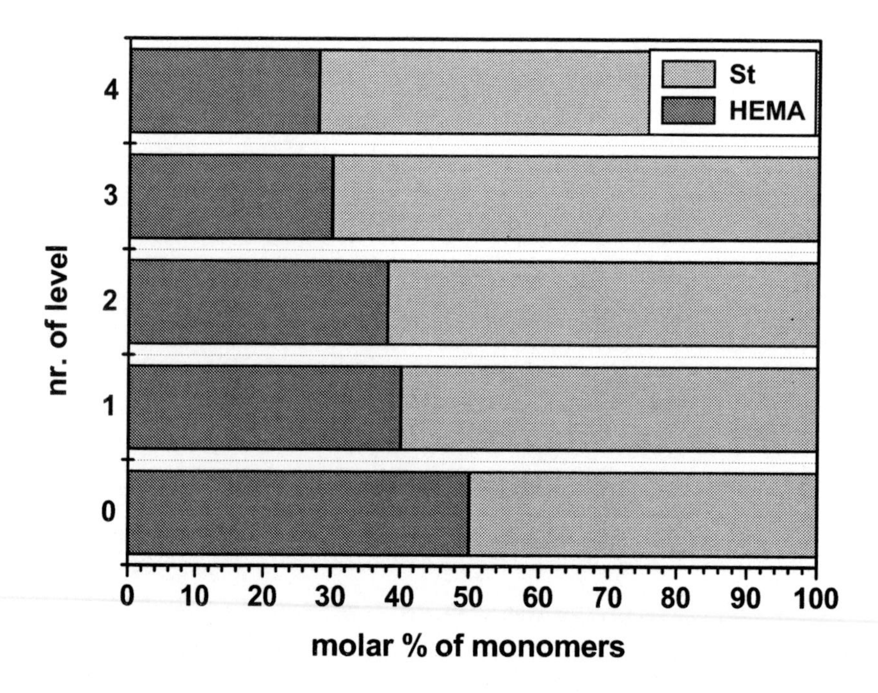

Figure 5. XPS analysis of specimen E with a step of 3 nm.

Figure 5 also shows a slight decrease of the HEMA percentage with the distance from the surface, which confirms the accuracy of the data previously obtained. To confirm the results, specimen B was dissolved in N-methyl pyrrolidone and then deposed on a glass sheet. After solvent evaporation the transparent layer has been analyzed. XPS data (figure 6) showed no difference between the results at different depths and the copolymer composition was identical with the one of the initial monomer mixture.

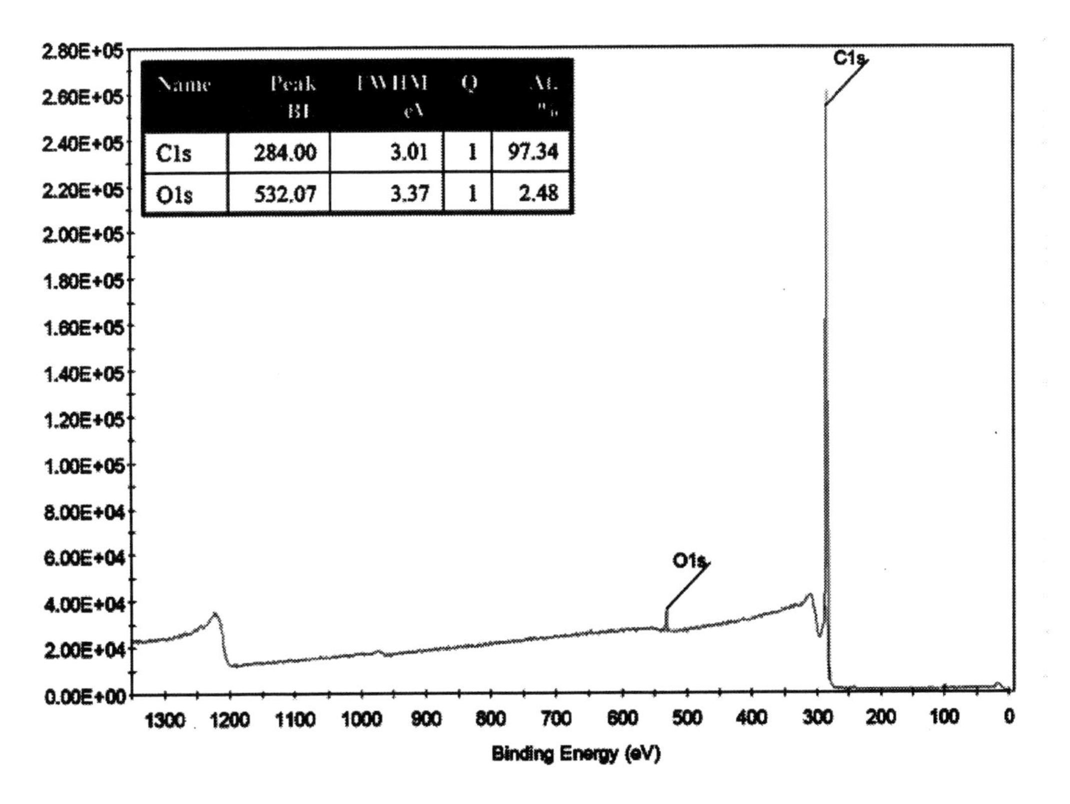

Figure 6. XPS analysis for the film resulted from specimen B after dissolution in NMP and evaporation of the solvent.

The results (figure 4-5) show clearly that the HEMA units are concentrated in the superficial layer of the particles, thus increasing the hydrophilicity above the value corresponding to the global composition of the copolymers. This result is valid for all the initial compositions of the monomer mixture. Since the superficial roughness has a strong influence on the lipophilicity (and consequently on the values of the contact angle) a characterization of the surfaces microgeometry (by AFM) was considered as mandatory.

Figure 7 shows the surface of the particles before (**a**) and after (**b**) pressure was applied. Case (**b**) shows a significant lower surface roughness (which should give relevant data resulting from the contact angle measurement) and the details show a change of shape of the initially spherical particles, which confirms that the diminished roughness is not accompanied by a change in the chemical nature of the surface.

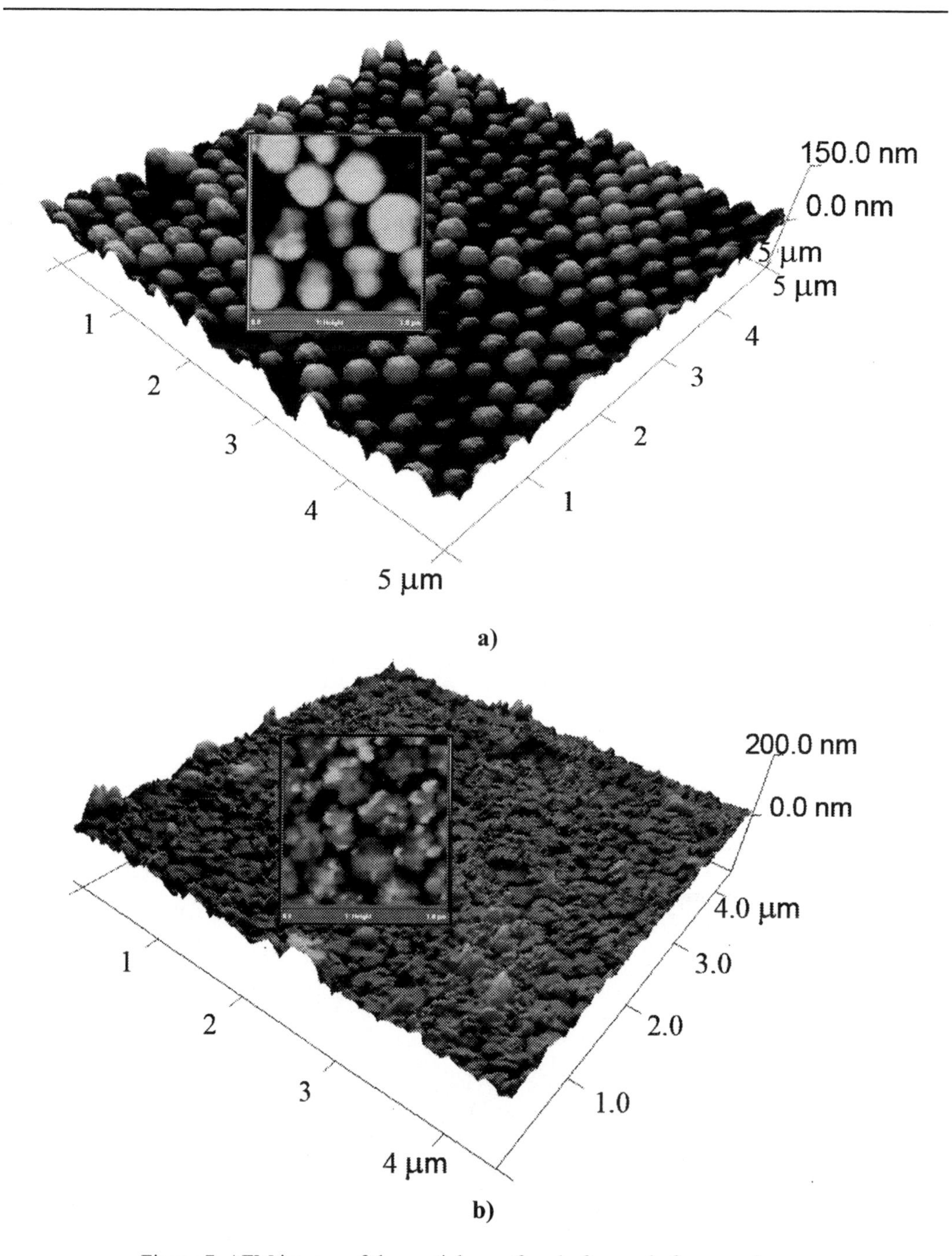

Figure 7. AFM images of the particles surface before and after pressing.

An initial validation of the method was performed by measuring the contact angle of water on PSt, that do not contain any ionic, hydrophilic groups. The polymer was obtained by precipitant polymerization in M, with AIBN as initiator. Fine particles (d=0.1-1 µm) have been obtained, dried and then pressed into cylinder, using the same technique as for the products of soap-free emulsion polymerization. The measured contact angle, $\theta = 94.59\pm2^0$, was practically identical with the theoretical value. The theoretical value, $\theta = 92^0$, was obtained by computation, using the relation:

$$\gamma_L(1+\cos\theta) = 2\left(\sqrt{\gamma_L^{LW}\gamma_S^{LW}} + \sqrt{\gamma_L^{+}\gamma_S^{-}} + \sqrt{\gamma_L^{-}\gamma_S^{+}}\right)$$

where L:H_2O, S:PSt and :γ_{H2O} = 72.8 mN (γ^{LW} = 21.8; $\gamma^+ = \gamma^-$ = 25.5), γ_{PSt} = 40.7 mN/m (γ^{LW}=42; γ^+=0; γ^-=1.1) [23].

If the value of the adhesion work would determine the nucleation and growth of the colloidal crystals, then one could anticipate that the spherical monodisperse particles should self-assemble also from other liquids than water. The self-assembling condition should be that the value of the adhesion work must exceed a given threshold value, decided by the contact angle of the liquid on the solid surface and the superficial tension of the liquid. Both the adhesion work (computed according to the Young- Dupré relation) and the term $\boldsymbol{\gamma\cos\theta}$ (which decides the value of the capillary force) increase with the polarity of the liquids, for the same solid surface (constant γ_S^P). Reciprocally, for a given liquid (a constant γ_L^P) the behavior towards the solid surface is decided by the polarity of the solid, i.e. by the St/HEMA ratio in the copolymer.

Therefore, the copolymers **A-E** have been colloidally dispersed in ethylene-glycol, formamide, ethanol, hexane, toluene and then deposed on hydrophilic glass lamellae, to examine the possible tendency towards self-assembling. For all cases, the contact angles of the same liquids on the surfaces of the particles obtained by copolymerization have been measured (table 1).

The data in table 1 lead to some interesting conclusions:

- The self-assembling capacity was first evaluated by the apparition of the iridescence characteristic for opal and subsequently proved by SEM investigation. For all specimens that have shown selective diffraction the SEM images revealed a colloidal crystal structure while the specimens with a white-opaque aspect have shown a non-ordered morphology (similar to specimen **A** from figure 1);
- The value of the contact angle by itself does not influence the self-assembling process. Even if the surface has been perfectly wetted ($\boldsymbol{\theta<10^0}$), this was not a sufficient condition for crystallization, as shown by the particles (with various chemical compositions) dispersed in ethanol, hexane or toluene. Experiments have shown that even if $\cos\theta \approx 1$, opal was not formed by evaporation of the liquid.
- The value of the adhesion work seems a more relevant parameter for the process, as compared to the term $\boldsymbol{\gamma_L\cos\theta}$. The least value of the adhesion work for which self-assembling occurs is in the range of **87-88 mJ/m²**. All systems characterized by W_a<87 mJ/m^2 do not have self-assembling properties. As an example, specimens **D** in formamide, respectively **A, B** and **C** in ethylene glycol are characterized by a higher value of $\boldsymbol{\gamma_L\cos\theta}$ than the specimens **B** and **C** in water. The first group does not crystallize (W_a<87 mJ/m^2), while the second group does (W_a>87 mJ/m^2). Such is the case of specimen **D** in ethylene glycol; in figure 8,

the SEM image of the particles after ethylene glycol evaporation, clearly shows a crystalline structure (W_a=88.11 mJ/m^2).

Table 1. Behavior of the St/HEMA particles after evaporation of the liquid from various colloidal dispersions

Specimen	Liquid (mN/m)	$W_a=\gamma_L(1+\cos\theta)$ mJ/m^2	$\gamma_L\cos\theta= W_a-\gamma_L$ mJ/m^2	Spontaneous self-assembling capacity
A	WATER (γ_L=72.80)	85.32	12.52	No
B		93.45	20.65	Yes
C		97.20	24.40	Yes
D		103.18	30.38	Yes
E		109.57	36.77	Yes
A	FORMAMIDE (γ_L=58.20)	71.19	12.99	No
B		73.85	15.65	No
C		79.43	21.23	No
D		83.80	25.60	No
E		87	26.80	Yes
A	ETHYLENE GLYCOL (γ_L=47.70)	73.59	25.89	No
B		78.36	30.66	No
C		82.58	34.88	No
D		88.11	40.41	Yes
E		91.09	44.02	Yes
A	ETHANOL (γ_L=22.10)	$\theta<10^0$ $W_a\approx44.20$	≈22.10	No
B				No
C				No
D				No
E				No
A	HEXANE (γ_L=18.43)	$\theta<10^0$ $W_a\approx36.86$	≈18.43	No
B				No
C				No
D				No
E				No
A	TOLUENE (γ_L=28.40)	$\theta<10^0$ $W_a\approx56.80$	≈28.40	No
B				No
C				No
D				No
E				No

Replacing water as a dispersion medium (respectively as the "adhesive" for the colloidal particles) rises some complex problems, most of them practical and not theoretical. First, it is difficult to find liquids with increased polarity and high superficial energy (γ_L>40 mJ/m^2) that will not dissolve or swell the particles. The self-assembling condition is that the liquid should act strictly at the level of the solid-liquid interface, without affecting the shape or the physical state of the solid particles. Second, the surface of the particles should have an increased polarity, therefore a high free superficial energy. In the present case, this would mean a higher HEMA fraction in the copolymer. However, if the HEMA amount exceeds the concentration in copolymer **E**, particles swell in water and by evaporation compact layers are obtained. The same phenomenon happens when dimethylformamide, N-methylpyrollydone, nitrobenzene or aniline are used as dispersion media for the copolymers **B-E.**

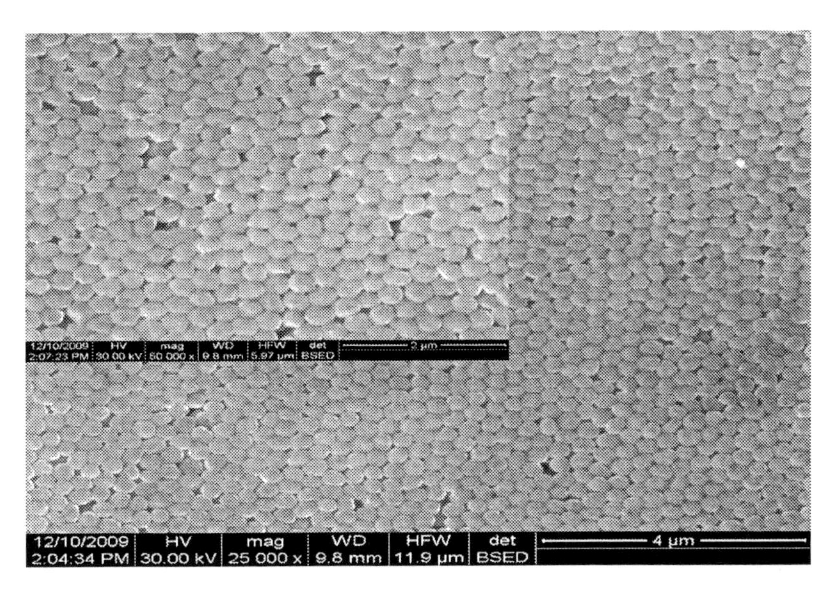

Figure 8. SEM image of specimen D after ethylene glycol evaporation.

4. Conclusions

Several colloidal dispersions of copolymer particles with various St/HEMA ratios have been obtained by soap-free polymerization. The particles monodispersity is the main condition for self-assembling of the particles to obtain colloidal crystals. The shell-core morphology of the dispersed particles has been confirmed by XPS investigation. The HEMA units are concentrated in the superficial layer while the core is richer in St units. Experimental data have proven that the colloidal monodisperse particles can self-assemble in crystalline networks if dispersed in other liquids than the water. The reciprocal attractions between the particles, mediated by the liquid layer, depend both on the chemical nature of the liquid and the chemical composition of the superficial layer of the particles. The adhesion mechanical work, parameter that measures the interaction energy, is relevant enough to explain the behavior of the particles during the concentration stage, until final drying. Polar liquids with high superficial energy favor crystallization of the particles with high-polarity surfaces (increased HEMA fraction). Liquids that are non polar and/or with low surface energies do not allow self-assembling of colloidal particles, whatever the chemical composition of the solid.

Acknowledgements

The National Authority for Scientific Research from The Ministry of Education, Research and Youth of Romania is gratefully acknowledged for the financial support through the exploratory project "Polymeric Biomaterials For Bone Repair. Biomimetism Through Nanostructured Surface", PN-II-ID-2008-2, number 729/2009.

References

[1] Kralchevsky, P. A.; Denkov, N. D. *Curr. Opin. Colloid Interface Sci.* 2001, 6, 383-401.
[2] Velev, O. D.; Lenhoff, A. M.; Kaler, E. W. *Science* 2000, 287, 2240-2243.
[3] Israelachvili, J. N. *Intermolecular and Surface Forces*, 2nd ed.; Academic Press: New York, 1992.
[4] Li, Q.; Jonas, U.; Zhao, X. S.; Kappl, M. *Asia-Pac. J. Chem. Eng.* 2008, 3, 255-268.
[5] Kralchevsky, P. A.; Nagayama, K. *Langmuir* 1994, 10, 23-36.
[6] Kralchevsky, P. A.; Paunov, V. N.; Denkov, N. D.; Ivanov, I. B.; Nagayama, K. *J. Colloid Interface Sci.* 1993, 155, 420-437.
[7] Di Leonardo, R.; Saglimbeni, F.; Ruocco, G. *Phys. Rev. Lett.* 2008, 100, 106103–106106.
[8] Lechman, J.; Lu, N. *J. Eng. Mech.* 2008, 134, 374-384.
[9] Megias-Alguacil, D.; Gauckler, L. J. *AIChE Journal* 2009, 55, 1103-1109.
[10] Han, G.; Dusseault, M. B.; Cook, J. Proceedings of the 6th North America Rock Mechanics Symposium (NARMS), Houston, Texas, 2004.
[11] Lambert, P.; Chau, A.; Delchambre, A.; Régnier, S. *Langmuir* 2008, 24 (7), 3157-3163.
[12] Pakarinen, O. H.; Foster, A. S.; Paajanen, M.; Kalinainen, T.; Katainen, J.; Makkonen, I.; Lahtinen, J.; Nieminen, R.M. *Modell. Simul. Mater. Sci. Eng.* 2005, 13, 1175-1186.
[13] Peyrade, D.; Gordon, M.; Heyvert, G.; Berton, K.; Tallal, J. *Microelectron. Eng.* 2006, 83, 1521-1525.
[14] Cui, Y.; Björk, M. T.; Liddle, A.; Sönnichsen, C.; Boussert, B.; Alivisatos, A. P. *Nano Lett.* 2004, 4 (6), 1093-1098.
[15] Yang, S.; Kleehammer, D. M.; Huo, Z.; Sloan, E.D.; Miller, K. T. *Colloid Interface Sci.* 2004, 277, 335–341.
[16] Ulmeanu, M.; Zamfirescu, M.; Medianu, R. *Colloids Surf., A* 2009, 338, 87-92.
[17] Willet, C. D.; Adams, M. J.; Johnson, S. A.; Seville, J. P. K. *Langmuir* 2000, 16, 9396-9405.
[18] Mori, Y. *Colloids Surf., A* 2007, 311, 61-66.
[19] Butt, H. J.; Kappl, M. *Adv. Colloid Interface Sci.* 2009, 146, 48-60.
[20] Yan, Q.; Wong, C. C.; Chiang, Y. M. *Thin Solid Films* 2008, 516, 5632-5636.
[21] Kolesnikov, Y. I.; Mednykh, D. A. Proceedings of the 18th Session of the Russian Acoustical Society, Taganrog, 2006.
[22] Patrick, R. L. *Treatese on Adhesion and Adhesives*, Dekker, M.; New York 1967, Vol. 1, 1969, Vol. 2.
[23] Packman, D. E. *Handbook of Adhesion*, 2nd ed.; J. Wiley and Sons, 2005.
[24] Martín-Rodríguez, A.; Cabrerizo-Vílchez, M. A.; Hidalgo-Álvarez, R. *Colloids Surf., A* 1996, 108, 263-271.
[25] Wang, F.; Hu, J.; Yang, W.; Wang, C. *J. Polym. Sci.* 2007, 45, 4552-4563.
[26] Rusen, E.; Mocanu, A.; Corobea,C.; Marculescu, B. *Colloids and Surfaces A: Physicochemical and Engineering Aspects*, doi:10.1016/j.colsurfa.2009.11.020.
[27] Rusen, E.; Mocanu, A.; Marculescu, B. *Colloid and Polymer Science,* in press.

[28] Preda, N.; Matei, E.; Enculescu, M.; Rusen, E.; Mocanu, A.; Marculescu, B.; Enculescu, I. *Journal of Polym. Research,* in press.
[29] Chen, Y.; Sajjadi, S. *Polymer*, 2009, 50, 357-365.
[30] He, X.; Thomann, Y.; Leyrer, R. J.; Rieger, J. *Polym. Bull*. 2006, 57, 785–796.
[31] Waterhouse, G.; Waterland, M. *Polyhedron,* 2007, 26, 356–368.
[32] Arai, M.; Arai, K.; Saito, S. *J. Polym. Sci., Part A: Polym. Chem*, 1979, 17, 3655-3665.
[33] Fitch, R. M.; Prenosil, M. P.; Sprick, K. J. *J. Polym. Sci., Part C : Polym. Symposia*, 1969, 27, 95-118.
[34] Chern, C. S. *Prog. Polym. Sci*, 2006, 31, 443-486.
[35] Fitch, R.M. *Polym.* J., 1973, 5, 467-483.

In: Photonic Crystals
Editor: Venla E. Laine, pp. 191-207
ISBN: 978-1-61668-953-7

Chapter 9

OPTICAL LOGIC GATE BASED IN A PHOTONIC CRYSTAL

A. Wirth L. Jr*[*] *and A.S.B. Sombra
Laboratory of Telecommunications and Science of Materials LOCEM, Department of Physics, Federal University of Ceará (U.F.C.) - Fortaleza, Ceará, Brazil.
Department of Teleinformatics Engineering, DETI, Center of Technology, Federal University of Ceará (U.F.C.) - Fortaleza, Ceará, Brazil.

Abstract

Taking into account that the current development of the world requires Increasingly devices operating at higher frequencies and bandwidths, and that these requirements are no longer possible to be obtained by electronic, we propose and analyze an optical logic gate based on nonlinear photonic crystal (PhC). This device uses an optical directional coupler. In our simulations we used the following methods: PWE (Plane Wave Expansion), FDTD (Finite-Difference Time-Domain), and COMSOL, which is integrated to our own BiPM (Binary Propagation Method).

Keywords: Optical Device; Optical Logic Gate; Photonic Crystal.

1. Introduction

Although optical fibers allow the use of light in modern telecommunications systems, these systems must use the electronics, in order to process the signals. Furthermore, electronic devices, such as, for example, modern computers are increasingly subjected to hardware heating, due the high operational frequency, which is a serious problem. The current physical limitations of the electronics encourage researchers around the world to obtain optical devices, which can overcome these limitations [1].

PhCs offer unprecedented opportunities for molding the flow of light [2,3,4]. We can make the following analogy between Semiconductors and PhCs: "The Semiconductors make

*E-mail address: awljeng@gmail.com. (Corresponding author)

the electronics and PhCs enable the integrated optical." Furthermore, PhCs work with high efficiency in systems using high frequency and large bandwidth.

The logic optical gate is based on a directional coupler, which is driven by a command signal. The coupling region works as a periodic waveguide between the W1 waveguides. When the command signal is inserted in the coupling region, the refractive index decreases (due to the non-linear effects) and causes the increase of the coupling coefficient value. If we consider that the coupler was designed to operate in the bar state, the increase of the coupling coefficient should be sufficient to bring the coupler to work in the cross state. Thus, the devices changes from bar to cross state, and the data signal is switched from the input port to the output port.

The only way to obtain the output signal is inserting an optical signal in both the inputs ports. So, any other combination of input signals on these ports will not produce the output optical signal and the output level is zero, ie, the device acts as an optical logic gate (AND gate).

Therefore, we have analyzed and proposed an optical logic gate embedded in nonlinear photonic crystal, based on a nonlinear directional coupler, which acts as a switch.

In Section 2 we begin with a description of the nonlinear photonic crystal used to obtainment of the optical logic gate. In Section 3 the detailing of our optical logic gate are presented, and we show the structure, as well as the calculations and simulations developed in the device. Finally, in section 4 we show our conclusions concerning the optical logic gate, which is embedded in a nonlinear photonic crystal.

2. Nonlinear Optical Directional Coupler Working as Switch

Our optical logic gate is based in a nonlinear optical directional coupler embedded in a two-dimensional (2D) triangular lattice of air holes with radius $r_b = 0.31a$, where "a" is the lattice constant. The vertical layer of the PhC is formed by an InP cladding (n = 3.17), an InGaAsP core (n = 3.35), and an InP substrate. This structure can be represented accurately by an effective refractive index $n_{eff} = 3.258$ ($\lambda = 1550$nm).

The directional coupler has two defect line waveguides (W1) and the coupling region is a line of air holes (Figure 1).

The switching of the data signal is possible due to the different propagation constants of the two split modes that arise in the system [5]. The switching mechanism is based on the very small proximity of the two waveguides (W1) so that the mode inserted within the input waveguide is split up due to the perturbation. Therefore, the even mode splits into an even-even (e-e) and an even-odd (e-o) mode and the odd mode splits into an odd-even and an odd-odd mode. Indeed, the modes in the coupler (supermodes) have even and odd symmetries with respect to the plane equidistant from the axis of the waveguides. The area where only the (e-e) and the (e-o) modes are present is called "quasi-single mode region".

If we decrease de radius of the air holes between the waveguides (r_c), the difference between the propagations constants become smaller and the coupling length is reduced. However, the bandwidth is reduced too. Moreover, with the increasing of the outer border air holes (r_e) of the two coupled W1, the change in the coupling length is not strong, but the bandwidth is not negatively affected.

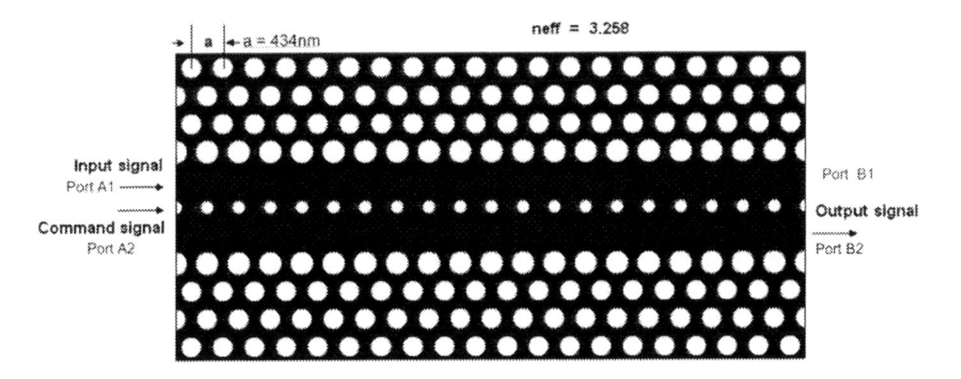

Figure 1. Directional coupler structure.

The photonic band gap (PBG) for the coupler with $r_b = 0.31a$, $r_c = 0.19a$ and $r_e = 0.35a$ covers the normalized frequencies ($u = a/\lambda$) from $u = 0.274$ to $u = 0.286$. We adopted a = 434nm ($u = 0.28$ for $\lambda = 1.55\mu m$). Hence, the minimal coupling length regarding the bar state is $L_c = 23.436\mu m$ for the whole normalized frequencies band [6].

Figure 2 shows the dispersion relation of the coupler for TE polarized light (E_x and E_y is inside the plane (x,y) and H_z is perpendicular to the plane (x,y)). The dispersion relation of the modes e-e, e-o, o-e and o-o were obtained by PWE method. The inset at the top of the graph sketches the "quasi-single" mode region.

According to the supermodes method when the supermodes are traveling within the coupler, they possess different propagation constants (β_{even} e β_{odd}) and the coupling length of a symmetrical directional coupler is:

$$L_c = \frac{1}{2} L_B = \frac{\pi}{\beta_{odd} - \beta_{even}} \text{ (Cross state)} \tag{1}$$

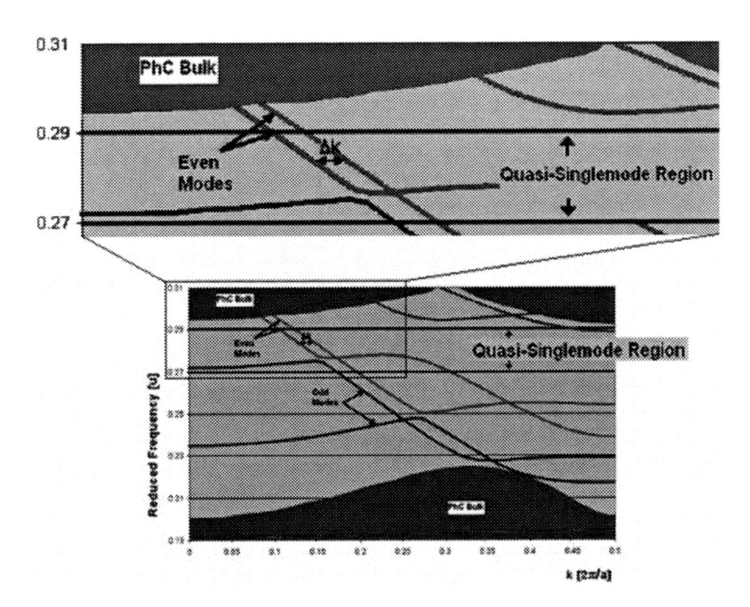

Figure 2. Dispersion relation of the PhC coupler (even mode and odd mode).

From the coupled mode method, the coupling length of a symmetrical directional coupler is:

$$L_c = \frac{\pi}{2k} \text{ (Cross state)}, \tag{2}$$

where k is the coupling coefficient.

Indeed, the directional coupler is driven by the input signal 2 (IS_2), which is inserted in the coupling region, which works as a periodic waveguide between the W1 waveguides. If the IS_2 signal is working in the coupling region the refractive index decreases (due to the non-linear effects), which causes the increase of the coupling coefficient value. Taking into account that the power of the IS_2 signal for obtaining the needed difference in the refractive index to bring the coupler of the bar state to the cross state is a function of the Kerr coefficient, of the group velocity and of the coupler length, we can use these three parameters to determine the lowest possible power of this IS_2 signal.

However, due to great difficulty in obtaining the Kerr coefficient change, the most feasible is to increase the coupler length by the power periodic transfer characteristic of this device.

3. Optical Logic Gate Detailing

Figure 3 shows the structure of the optical logic gate.

The IS_2 signal (command signal) is inserted in the coupling regiôn, which operates as a waveguide. Hence, the coupling region acts as a periodic waveguide [7]. When the command signal is working in the coupling region the refractive index decreases (due to the non-linear effects), which causes the increase of the coupling coefficient value. If we consider that the coupler was designed to operate in the bar state, the increase of the coupling coefficient should be sufficient to bring the coupler to work in the cross state.

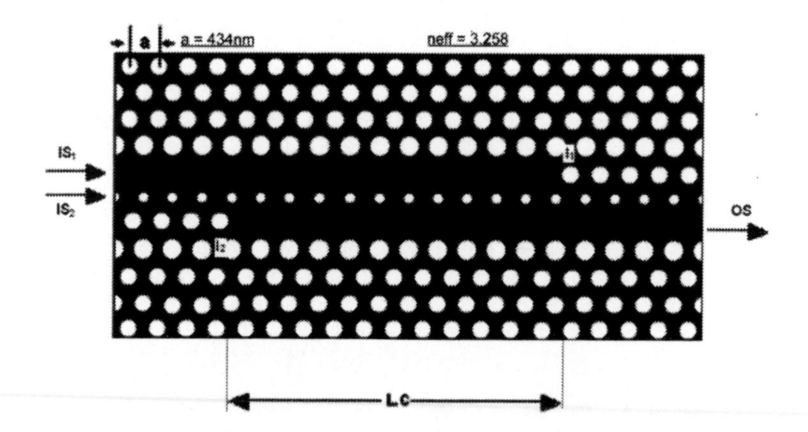

Figure 3. Structure of the optical logic gate.

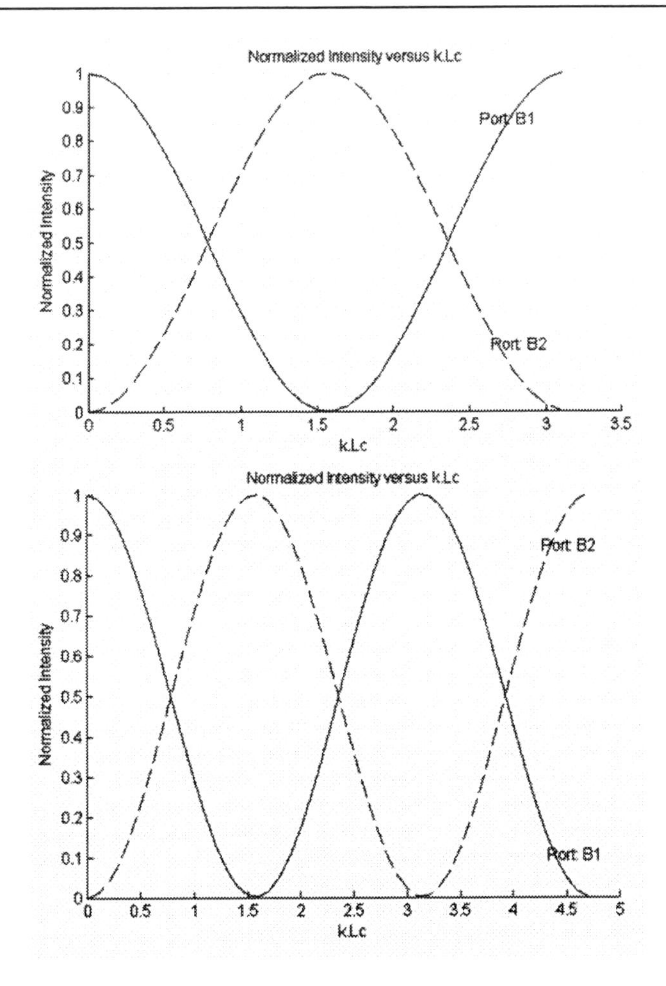

Figure 4. (a): Before the insertion of the IS_2 signal, all entry optical power exits through the B1 port. (b): After the IS_2 signal insertion all entry optical power now exits through the B2 port.

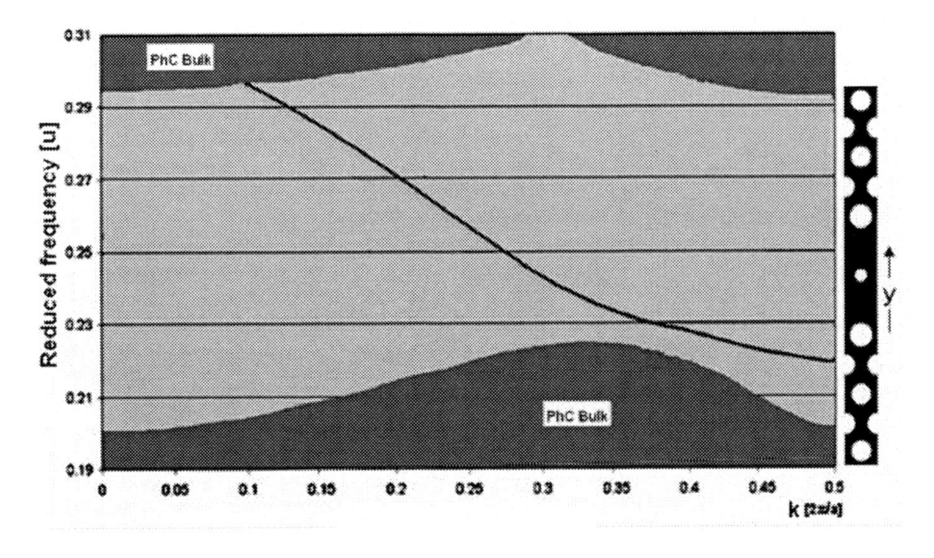

Figure 5. Dispersion relation of the IS_2 signal.

Figure 4 shows that for a given coupler length, the difference between the propagation constants for the even and odd supermodes, referring to the cross state ($\Delta\beta_c$ - after the insertion of the command signal) is equal to $1.5x\Delta\beta_d$ where $\Delta\beta_d$ is the value of the difference between the propagation constants for the even and odd supermodes, referring to the bar state (before the insertion of the command signal).

Figure 5 shows the dispersion relation of the command signal obtained by means of PWE. The inset of the graph sketches the supercell used in the plane wave expansion calculation.

The normalized frequency of the command signal is u = 0.2194. The wave number of this command signal is located very near of the limit of the irreducible Brillouin zone. The three signals inside the device could cause four-wave mixing. However, for this to occur would require the matching of the frequencies as well as of the wave vectors (phase-matching). The phase matching requirement is that $\Delta k_j = 0$ ($k_j = n_j w/c$), j = 1 to 4. It is difficult to satisfy phase-matching in our device as result of the variations in the structure of the PhC.

The calculation of the necessary power should take into account the increase of the refractive index in the coupling region (Δn) [8], which depend on the used PhC structure, and on the normalized frequency of the command signal.

$$\Delta n = 3n_2|E|^2 \frac{V_{gcr}}{V_{gcw}} = 3n^2 I \frac{V_{gcr}}{V_{gcw}} = 3\frac{n_2 P}{A_{eff}}\frac{V_{gcr}}{V_{gcw}} \rightarrow P = \frac{(\Delta n)A_{eff}}{3n_2\left(\frac{V_{gcw}}{V_{gcr}}\right)} \tag{3}$$

In equation (3) P is the desired optical power of the command signal, n_2 is the non-linear refractive index (Indium Gallium Arsenide Phosphide (InGaAsP) has $n_2 \approx -5.9x10^{-16} m^2/W$ at $\lambda = 1.55\ \mu m$)) [9], E is the electric field intensity, A_{eff} is the mode effective area, V_{gcw}, is the command signal group velocity in a conventional axial uniform waveguide, and V_{gcr} is the low group velocity of the command signal in the coupling region (the wave number is located very near of the limit of the irreducible Brillouin zone). The factor of 3 is due the cross-phase modulation, which induces an index change twice as strong as self-phase modulation, and because the longitudinal confinement of the mode is not uniform [10].

To get low optical power for the command signal, we need low value of Δn (equation (3)).

Since our coupler was originally designed to works in the bar state, we have that $\Delta\beta_d L_c = 2\pi$, where $\Delta\beta_d$ is the difference between the propagation constants for the even and odd supermodes, referring to the bar state. As we need to modify the coupler to works in the cross state we obtain that $\Delta\beta_c L_c = 3\pi$, where $\Delta\beta_c$ is the difference between the propagation constants for the even and odd supermodes, referring to the cross state. Hence, we must use the command signal to increase the value of $\Delta\beta_d$ to $\Delta\beta_c = (3/2)x\Delta\beta_d$. So, our coupler works now in the cross state.

Therefore, for the coupler with $L = nxL_c$ to change from the bar state to the cross state

$$\Delta\beta_{c(n)} = \frac{(2n+1)}{2n}\Delta\beta_d \tag{4}$$

Figure 6 shows the values of the needed differences between the propagation constants of the two supermodes for the coupler working in the bar state ($\Delta\beta_d$) and in the cross state ($\Delta\beta_c$) as well as the value of the increase of $\Delta\beta$ to change the coupler of the bar state to the cross state depending on the length of the coupler (λ = 1.55μm).

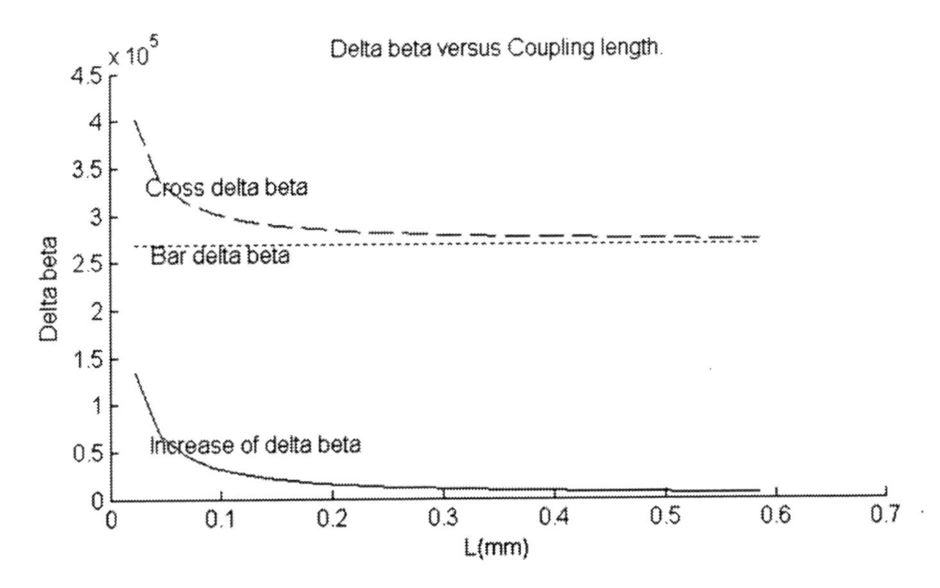

Figure 6. Values of the needed propagation constants.

We know that $n_e(\lambda)=\dfrac{\beta_e c}{w}=\dfrac{\beta_e\lambda}{2\pi}$ (even mode) and $n_o(\lambda)=\dfrac{\beta_o c}{w}=\dfrac{\beta_o\lambda}{2\pi}$ (odd mode), where n_e and n_o are the refractive index for the even mode and odd mode, respectively. The decrease of the refractive index causes slightly shifts of the even mode, whereas the odd mode remains almost unchanged. Hence, to bring the coupler from bar state to cross state we need that $\Delta n=\dfrac{\lambda}{2\pi}(\Delta\beta_e)\approx\dfrac{\lambda}{2\pi}(\Delta\beta_c-\Delta\beta_d)$, that is,

$$\Delta n=\frac{\lambda}{2\pi}\left(\frac{3\pi}{L_c}-\frac{2\pi}{L_c}\right)=\frac{\lambda}{2L_c}.$$

Therefore,

$$\Delta n_{(n)}=\frac{\lambda}{2\pi}\left(\Delta\beta_{c(n)}-\beta_{d(n)}\right)=\frac{\lambda}{2\pi}\left(\frac{(2n+1)\Delta_{d(n)}}{2n}-\Delta d_{(n)}\right)=\frac{\lambda\Delta\beta_d}{4\pi n} \qquad (5)$$

Figure 7(a) shows the required decreases in the refractive index and figure 7(b) shows the values for the optical power of the command signal to change the coupler from the bar state to the cross state depending on the length of the coupler (λ = 1.55μm).

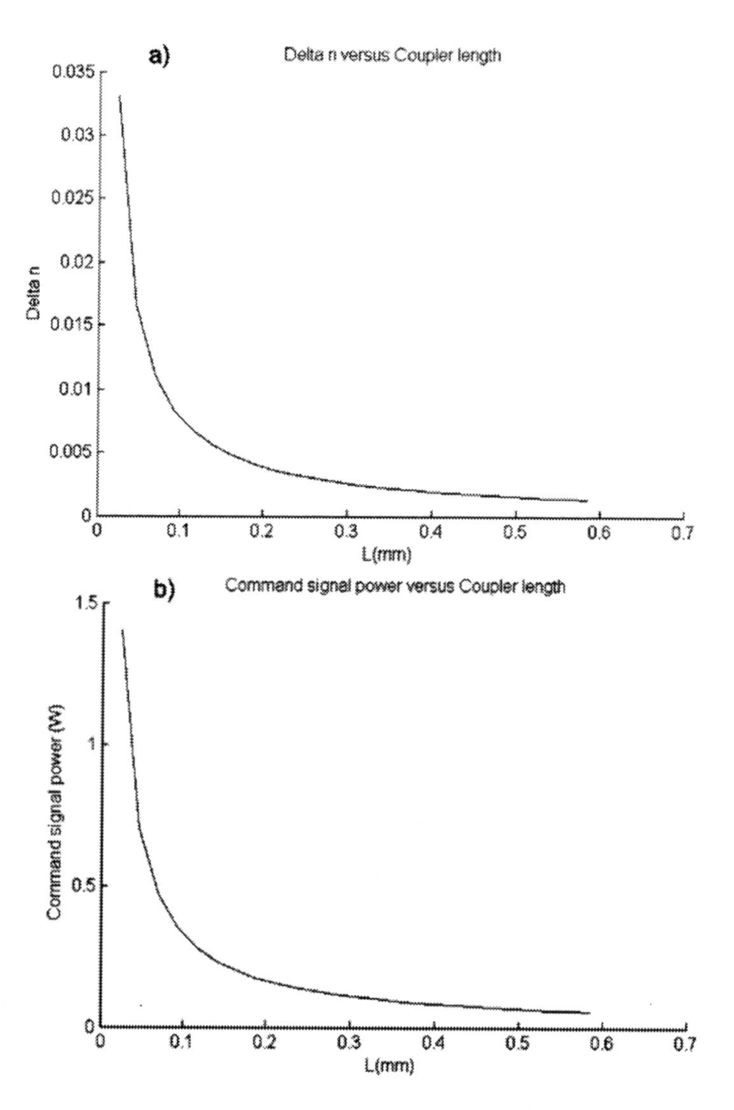

Figure 7. (a): Values of the required decreases in the refractive index. (b): Values for the optical power of the IS_2 signal.

We can note that the Δn and P dependence regarding the coupler length is very large [11, 12].

As the group velocity of the data signal is approximately ten times larger than the group velocity of the command signal, the pulse width of the command signal needs to be ten times greater than the width of the data signal pulse. Hence, the transmission rate of the pulses in the transmission line should be calculated assuming that its width has a value ten times higher than its real value.

To obtain the switching of the data signal, we need at the same time of the command signal.

To analyze the gradual increment of the refractive index using the PWE method, we gradually decrease the dielectric constant of the coupling region keeping unchanged the remainder of the structure. The results of these simulations showed the effects associated with the gradual increase in the power of the command signal.

The values found by equation (4) agree with the results obtained by PWE.

As we can see in equation (5), the desired change in the refractive index depends on the wavelength of the data signal too. Figure 8 shows the graph of the optical power of the command signal vs. data signal wavelength for the coupler length ten times greater than the minimal coupler length.

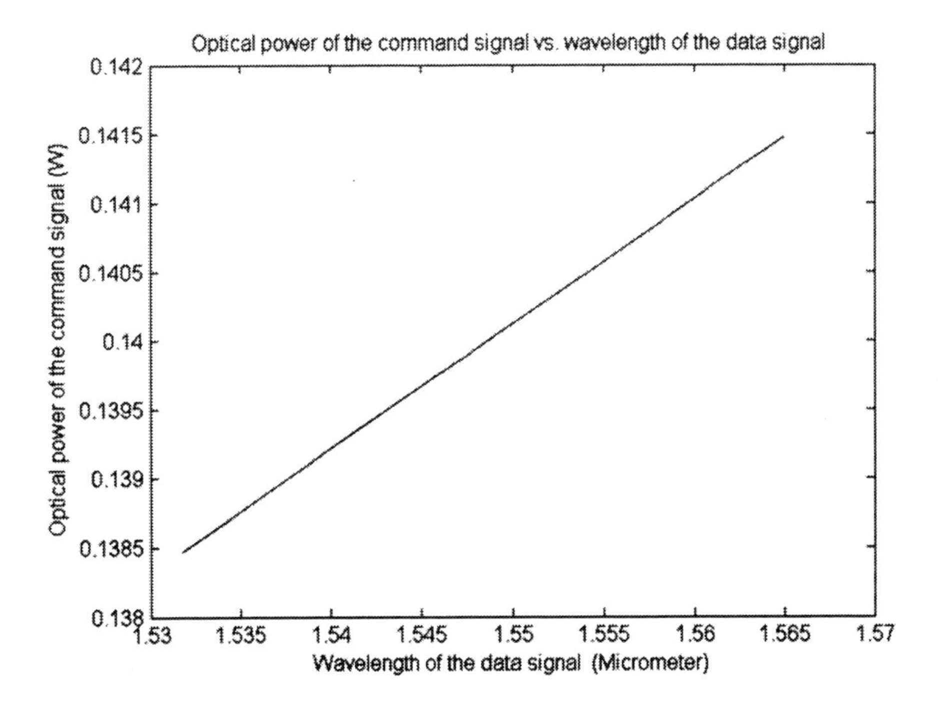

Figure 8. Graph of the optical power of the IS_2 signal vs. data signal wavelength ($L = 10*L_c$).

As expected, the required optical power of the command signal grows linearly with the increase of the wavelength of the data signal.

The periodically corrugated interfaces of the PhCs waveguides can originate coupling between the co-directional fundamental mode and counter-propagation higher-order modes, leading to the formation of Mini-Stop Bands (MSBs). Our simulations took into account the wave vectors within the PBG but outside of MSBs.

Theoretically, the modes below of the light line are lossless for ideal PhCs waveguides with perfect translational invariance. Any propagation loss that does occur in this region is extrinsic scattering loss due to fabrication structural disorder, which destroys the translational invariance. The lowest loss value reported for these PhCs waveguides is 5dB/cm. PhC waveguides are expected to yield a larger propagation loss because of their complicated shapes. The propagation loss reflects underlying physics relating to several different scattering mechanisms, which is significantly more complicated than a simple v_g-scaling expectation. The bandwidth with loss < 10 dB/cm is u = 0.273 to u = 0.284) [13]. Hence, in this region we adopted the loss value equal to 10dB/cm.

High-order dispersion in the slow-light regime of photonic crystal (PhC) waveguides was measured by utilizing integrated Mach-Zehnder interferometer (MZI) structures, and compared with theoretical results obtained from 3D plane-wave calculations. Highly accurate measurements of group-velocity dispersion (GVD), third-order dispersion (TOD) and fourth-order dispersion (FOD) at high group-index (n_g) values were enabled by minimizing external phase-distortions and increasing signal-to-noise ratio in the MZI. The experimental results for GVD (β_2), TOD, and FOD parameters were ~ $10^2 ps^2/mm$, ~ $10^4 ps^3/mm$, and ~ $10^5 ps^3/mm$, respectively [14]. We adopted $\beta_2 = 10^2 ps^2/mm$.

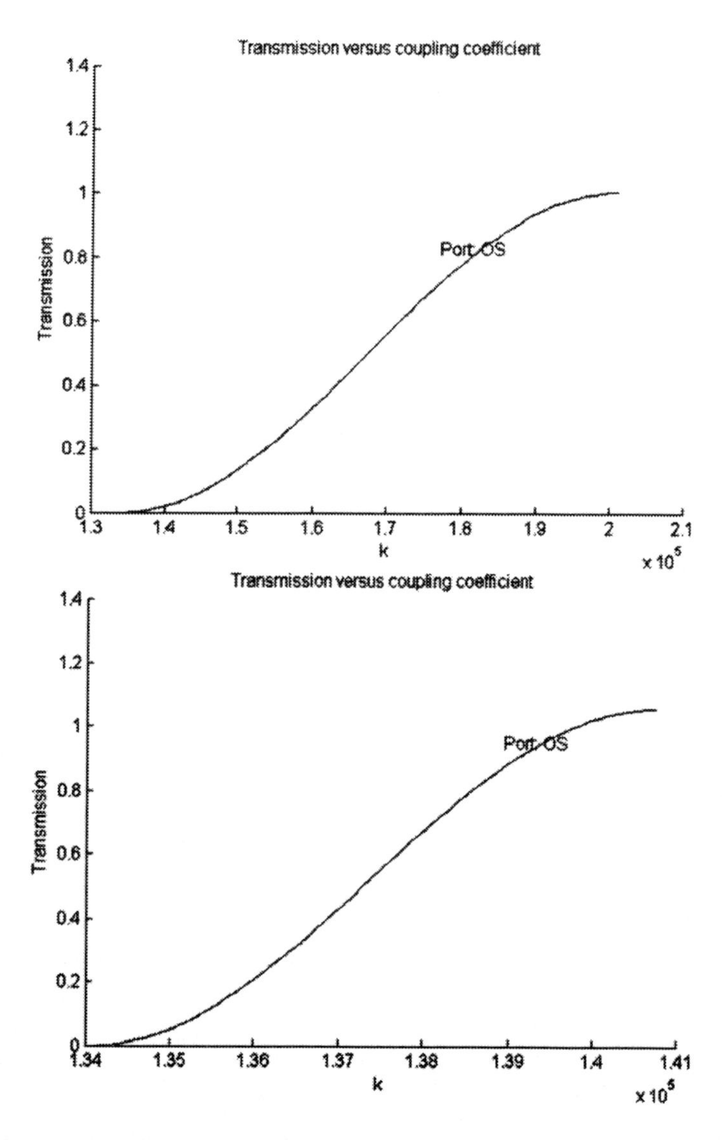

Figure 9. Variance in the transmission depending on the k values. (a): Minimal coupler length. (b): Coupler length ten times greater than the minimal coupler length.

In order to analyze the nonlinear effects in the PhC structure, we developed a numerical method ("Binary Propagation Method" (BiPM)). This method, which uses "Split-Step Fourier" and "fourth-order Range-Kutta", solves the coupled modes equations:

$$\frac{\partial A_1}{\partial z}+\frac{1}{V_g}\frac{\partial A_1}{\partial t}+\frac{i}{2}\beta_2\frac{\partial^2 A_1}{\partial t^2}+\frac{\alpha}{2}A_1=i\gamma\left(\left|A_1\right|^2+B\left|A_2\right|^2\right)A_1+ikA_2 \quad (6)$$

$$\frac{\partial A_2}{\partial z}+\frac{1}{V_g}\frac{\partial A_2}{\partial t}+\frac{i}{2}\beta_2\frac{\partial^2 A_2}{\partial t^2}+\frac{\alpha}{2}A_2=i\gamma\left(\left|A_2\right|^2+B\left|A_1\right|^2\right)A_2+ikA_1\,. \quad (7)$$

V_g is the group velocity, β_2 is the GVD (group-velocity dispersion), α is the loss parameter, γ is nonlinearity coefficient ($n_2.2\pi / \lambda A_{eff}$), B is the parameter which govern the XPM-induced nonlinear coupling, and k is the coupling coefficient [15].

Figure 9 shows the variance in the transmission depending on the k values for the minimal coupler length (Figure (9a)), and for coupler length ten times greater than the minimal coupler length (Figure (9b)). Please, note that the increase in the value of the coupling coefficient, due to the nonlinear effects caused by IS_2 signal, reverses the coupler from the bar state to the cross state.

Figure 10(a) shows the field distribution concerning the data signal (u = 0.2194) inside the coupler with length of $2xL_c$ (cross state) using the FDTD, before the insertion of the command signal (linear regime). At Figure 10(b) we show the field distribution of the data signal inside this coupler, after the insertion of the command signal (bar state at non-linear regime). We used equation (5) to obtain the value of $\Delta n_{(2)}$.

We can note that equations (6) and (7) take into account the final value of n_2 by the value of γ.

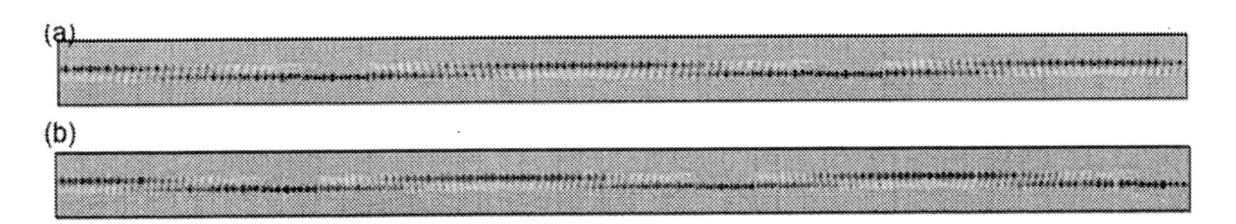

Figure 10. (a): Field distribution concerning the data signal inside the coupler with length of $2xL_c$ (cross state) before the insertion of the IS_2 signal. (b): Field distribution of the data signal inside the coupler, after the insertion of the IS_2 signal (bar state).

Plots of the H_z fields of the lowest-order of the TE modes at k very near of the limit of the irreducible Brillouin zone (u = 0.2194), which are travelling within the periodic guide is shown in figure 11(a). We can see that the electrical field is completely confined in the coupling region. Hence, the increasing of the value of the refractive index due the nonlinear effect arises only in the coupling region, where the periodic waveguide is located and this command signal influences the process of switching only by modifying of the refractive index. Figure 11(b) shows the electric field in the coupling region, which is located in the holes and also in approximately 92% of the area of the dielectric in the coupling region. Hence, we can neglect the consequences it causes for the resultant Δn_{eff} in the coupling region.

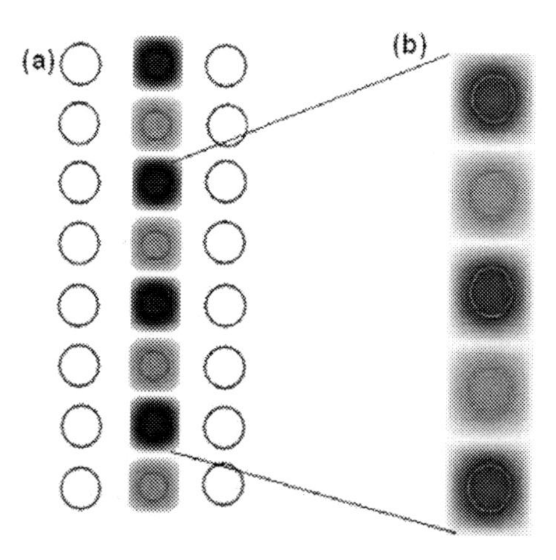

Figure11. Plots of the fields of the first band of the TE modes within the periodic guide.

The Gaussian pulse used in our simulations, which is inserted in port IS_1 of our coupler, is determined by equation (8)

$$I(0,T) = A\exp\left(-\frac{T^2}{2T_0^{\,2}}\right), \tag{8}$$

where $A = (P_0)^{1/2}$ is the normalized amplitude, P_0 is the optical power of the input data signal, T_0 is the half-width at 1/e-intensity point, and T is measured in a frame of reference moving with the pulse at the group velocity V_g (T = t - x/Vg). In our simulations, we used a Gaussian pulse with full width at half maximum (FWHM) equal to 2ps ($T_{FWHM} = 1.665\ T_0$).

Considering the propagation of Gaussian pulses through the coupler with minimum coupling length (L_c), the input and output pulses do not suffer any changes in their formats, temporal and spectral (frequency domain). However, the greater the length of the coupler, the greater change in formats, temporal and spectral, of the output Gaussian pulses. Nevertheless, for all the possible lengths of our coupler, the output pulses behave within the standards for telecommunications systems. Figure 12(a) shows the temporal shape of the input and output pulses for couplers with length equal to $6xL_c$ and figure 12(b) for the length equal to $12xL_c$. We can notice that the greater the length of the coupler is the wider is the bandwidth of the output pulses.

Furthermore, figure 13(a) shows the spectral shape of the input and output pulses for couplers with length equal to $6xL_c$ and Figure 13(b) for length equal to $12xL_c$. We can see that unlike the case of the temporal shape, the greater the length of the coupler is less is the bandwidth of the output pulses.

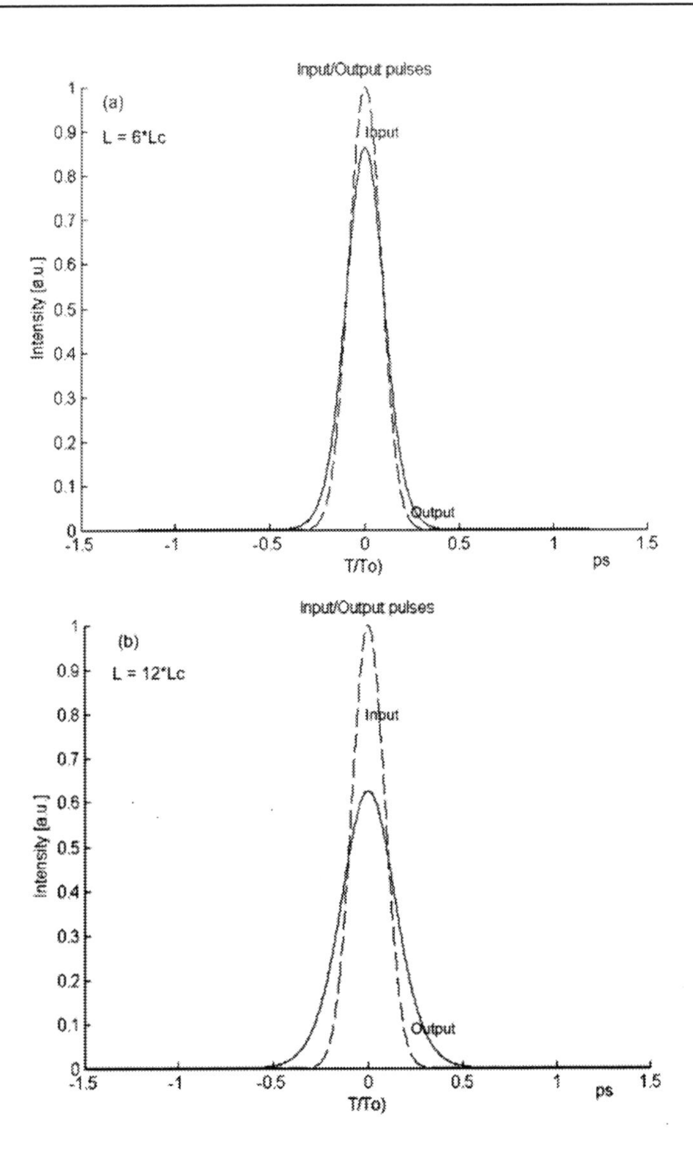

Figure 12. Temporal shape of the input and output pulses. (a): Couplers with length equal to $6xL_c$. (b): Couplers with length equal to to $12xL_c$.

The reflectivity R at the interface I_1 is obtained by:

$$R = \frac{\cos^4(\Delta\beta L)}{\cos^4(\Delta\beta L) + 4sen^2(\Delta\beta L)\cos^2(\Delta\beta L)} \tag{9}$$

For the coupling length L = n*Lc, if we get $\Delta\beta L = 2\pi$, then R = 1. Note that at this time the coupler is operating in the bar state. Therefore, in this case all optical power inserted into the optical P_1 port (input) is reflected at the interface I_1, and returns to that port.

Furthermore, if the coupler is operating in the cross state, we get $\Delta\beta L = 3\pi$, then R = 0. In this case, and all optical power inserted within P_1 port is switched to the adjacent waveguide and exits the device through of its output port.

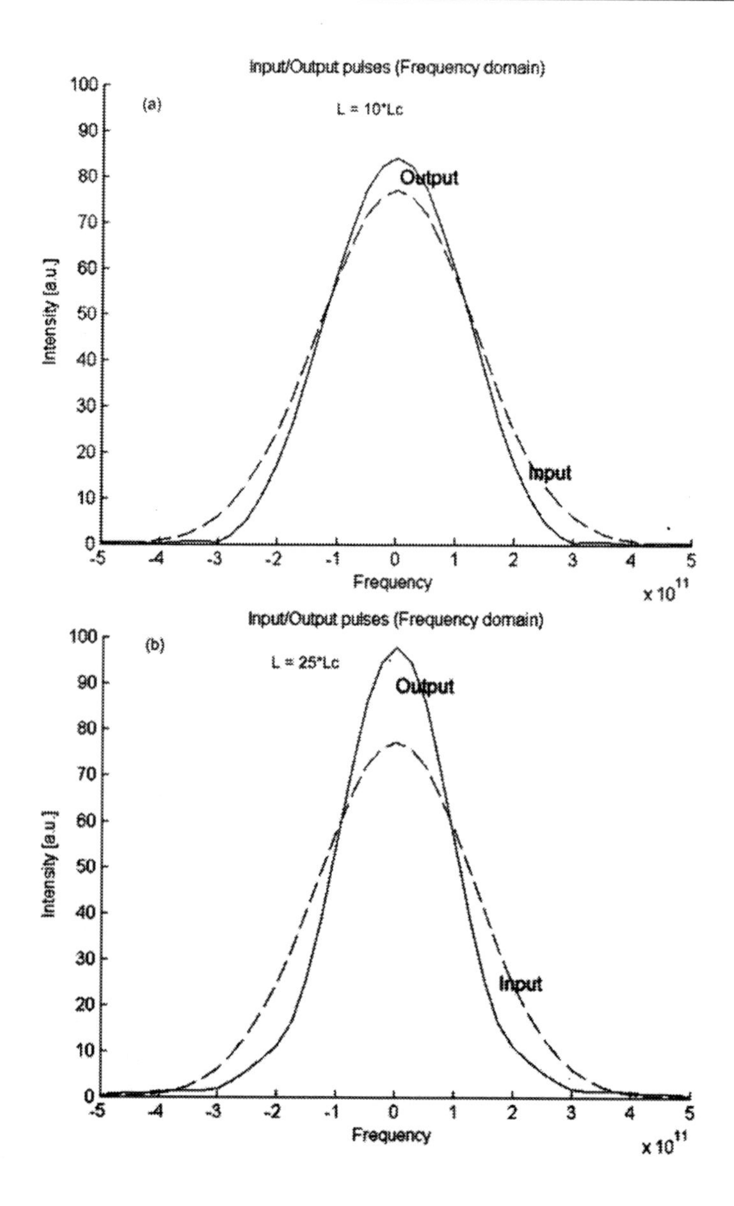

Figure 13. Spectral shape of the input and output pulses. (a): Couplers with length equal to $6xL_c$. (b): Couplers with length equal to to $12xL_c$.

Note that the only way to obtain the output signal is inserting an optical signal in both the inputs ports. So, any other combination of input signals on these ports will not produce the output optical signal (OS) and the output level is zero, ie, the device acts as an optical logic gate (AND gate), as we show below.

IS_1	IS_2	OS
0	0	0
0	1	0
1	0	0
1	1	1

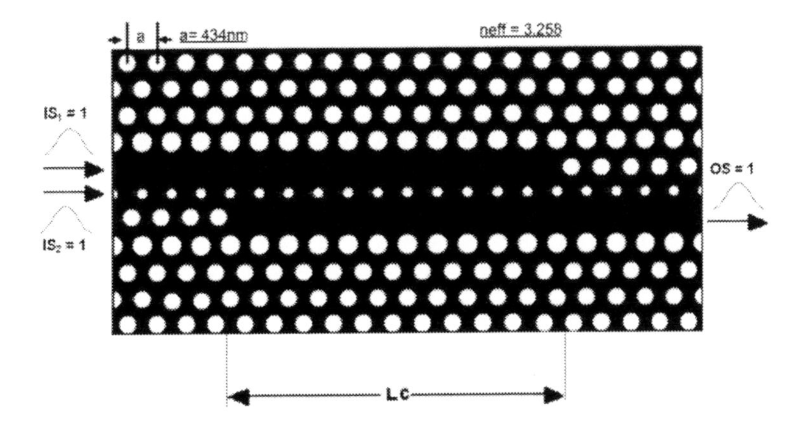

Figure 14. Obtainment of the output signal (OS).

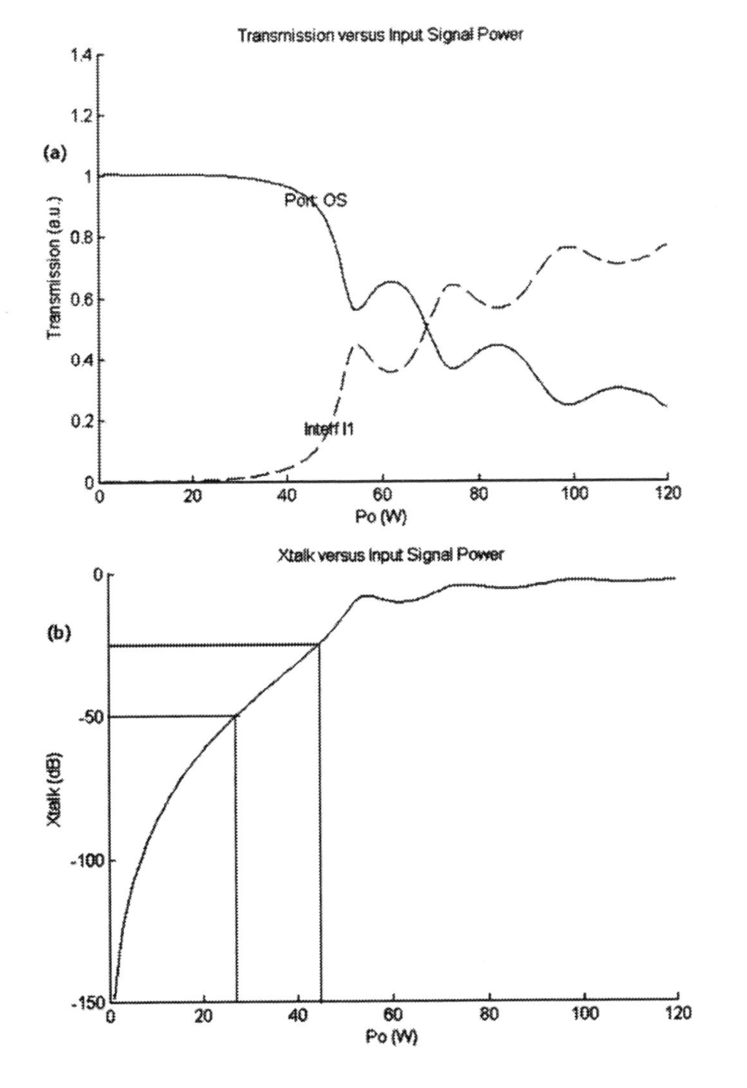

Figure 15. (a): Transmission at interface 1. (b): Far crosstalk (Xtalk) at interface 1.

Note that if the waveguide was formed from conventional dielectric waveguides, most of the light in the closed waveguides would be radiated into the cladding and lost. However, our waveguides are embedded in a photonic crystal. Hence, the light at the end of these waveguide are fully reflected and propagate back.

The optical logic gate works with two different power levels. The power level of the input signal IS_2 depends on the length of the coupler, as shown in Figure 6 (b). Theoretically, we can use any level power to the input signal IS_1, which, may or may not, be switched to output port of the optical logic gate. However, taking into account the coupler in the cross state, for example, the huge increase of the power level of the input signal IS_1 causes the transmission at the output port (OS_1) where is the interface 1 (Figure 15(a)) and hence far crosstalk (Xtalk) at this interface (Figure 15 b)). Figure 15 was obtained by BiPM method for the coupler with length equal to L_c.

XRatio is defined as the ratio between the power level of the signal at the wanted output port (OS) and the power level of the signal at the unwanted output port (OS_1). Considering the coupler in the cross state, the figure 16, obtained via BiPM method, shows the xRatio for the coupler with length equal to L_c.

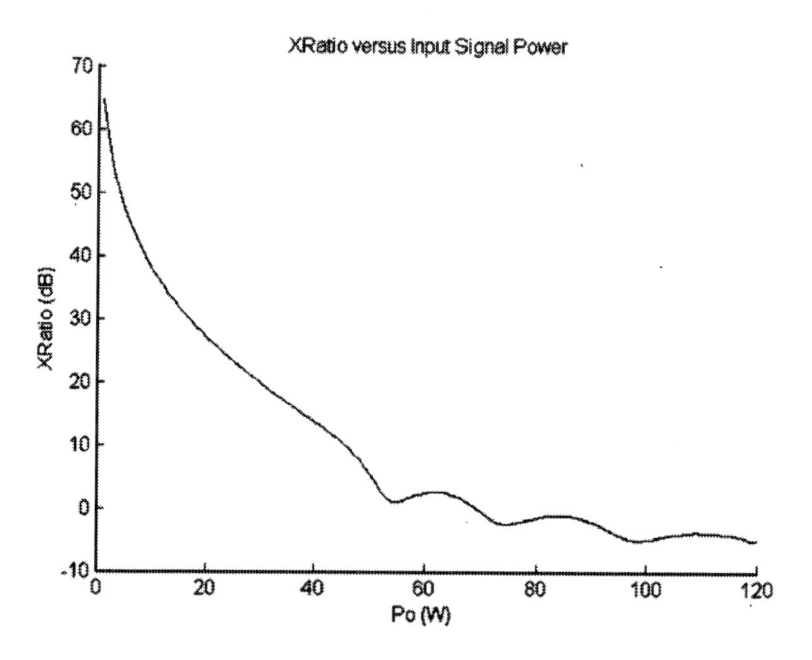

Figure 16. XRatio (OS/OS_1).

4. Conclusions

We have analyzed and proposed an optical logic gate embedded in a nonlinear photonic crystal. This optical logic gate is based in a directional optical coupler, which changes from the bar state to the cross state, when we insert the input signal IS_2. Note that the only way to obtain the output signal is inserting an optical signal in both the inputs ports. So, any other combination of input signals on these ports will not produce the output optical signal (OS) and the output level is zero, ie the device acts as an optical logic gate(AND gate).

Acknowledgments

This work was partly sponsored by National Council of Scientific and Technological Development (Conselho Nacional de Desenvolvimento Científico e Tecnológico – CNPq).

References

[1] Marin Soldjacic and J. D. Joannopoulos, *Enhancement of nonlinear effects using photonic crystals*, Physics Department, Massachusetts Institute of Technology, 211 – 218, 2004.

[2] John D. Joannopoulos, Steven G. Johnson, Joshua N. Winn, and Robert D. Meade, *Photonic Crystals, Molding the Flow of Light*, Princeton University Press, 66 – 93 (2008).

[3] Sakoda, K., *Optical Properties of Photonic Crystals*, Springer, Berlin, 2001.

[4] Johnson, S. G.& Joannopoulos, J. D., *Photonic Crystals: The Road from Theory to Practice*, Kluwer,Norwell, 2002.

[5] Ahmed Sharkawy, Shouyuan Shi and Dennis W. Prather, Electro-optical switching using coupled photonic crystal waveguides, *Optics Express* **1048**, (2002).

[6] Patrick Strasser, Ralf Flückiger, Robert Wüest, Franck Robin, and Heinz Jäckel, InP-based compact photonic crystal directional coupler with large operation range, *Optics Express* **8472**, (2007).

[7] Pi-Gang Luan, Kao-Der Chang, *Transmission characteristics of finite periodic dielectric waveguides*, Optical Society of America, (2006).

[8] Daryl M. Beggs, Thomas P. White, Liam O'Faolain, and Thomas F. Krauss, Ultracompact and low-power optical switch based on silicon photonic crystals, *Optics Letters* Vol. 33 No. 2, (2008).

[9] Ilya Fushman and Jelena Vuckovic, Analysis of a quantum nondemolition measurement scheme based on Kerr nonlinearity in photonic crystal waveguides, *Optics Express*, Vol. 15, No. 9, pp. 5559-5571, (2007).

[10] F. Cuesta-Soto, A. Martinez, J. Garcia, F. Ramos, P. Sanchis, J. Blasco, and J. Marti, All-optical switching structure based on a photonic crystal directional coupler, *Optics Express* **161** (2004).

[11] A. Wirth Lima J. and A. S. B. Sombra, Switching cell embedded in photonic crystal, *Microsystem Technologies* **15**:821–825, (2009).

[12] A. Wirth Lima J., M. Gomes Silva, A. C. Ferreira, and A. S. B. Sombra, All-Optical Nonlinear Switching Cell made of Photonic Crystal, *Journal of the Optical Society of America*. A, Vol. 26, No. 7, pp. 1661-1667, (2009).

[13] E. Kuramochi, M. Notomi, S. Hughes, A. Shinya, T. Watanabe and L. Ramunno3, Disorder-induced scattering loss of line-defect waveguides in photonic crystal slabs, *Physical review, B* **72**, 161318 (R), pp. 161318_1 – 161318_4, (2005).

[14] Solomon Assefa and Yurii A. Vlasov, *High-order dispersion in photonic crystal waveguides*, Optical Society of America, (2007).

[15] Govind P. Agrawal, *Nonlinear fiber optics*, Academic Press, 295 – 301, 414, (1995).

In: Photonic Crystals
Editor: Venla E. Laine, pp. 209-244

ISBN 978-1-61668-953-7

Chapter 10

TWO-DIMENSIONAL PHOTONIC CRYSTALS WITH GLIDE-PLANE SYMMETRY FOR INTEGRATED PHOTONICS APPLICATIONS

*Adam Mock**
Central Michigan University
Ling Lu†
University of Southern California

PACS 42.55.Px, 42.55.Sa, 42.55.Tv, 42.82.Et.

Keywords: Photonic crystal, waveguides, optics, photonics, photonic crystal laser, photonic crystal waveguide, semiconductor laser, microcavity, integrated optics, integrated photonics, finite-difference time-domain method, group theory.

1. Introduction

During the past two decades, there has been considerable attention paid to materials with periodic modulation of the refractive index. Such structures, called photonic crystals, can have periodicity in one, two or three dimensions and typically utilize a period that is approximately one half of the operating wavelength. One dimensional photonic crystals have been used in practical devices inlcuding distributed feedback lasers [28, 50] and vertical cavity surface emitting lasers [64] which are utilized in applications ranging from optical communication systems to laser printing. Devices based on two and three dimensional photonic crystals remain an ongoing research topic. This chapter discusses the use of two dimensional photonic crystals for integrated photonics applications. In particular we focus on the band structure and modal properties of waveguides and cavities that posess non-symmorphic space group symmetry in the form of a glide plane.

Two dimensional photonic crystals are a promising candidate for integrated photonics due to the ability to realize a diverse range of devices by simple rearrangements of the hole pattern. Figure 1 illustrates what a simple photonic crystal based photonic integrated

*E-mail address: mock1ap@cmich.edu
†E-mail address: lingl@usc.edu

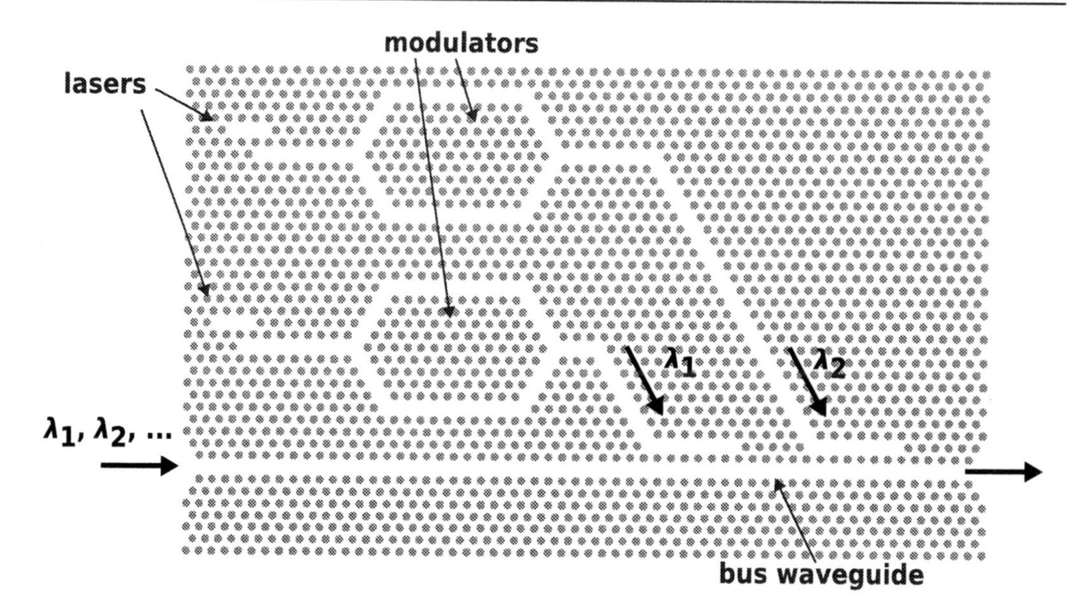

Figure 1. Schematic diagram of a photonic integrated circuit built using two dimensional photonic crystals.

circuit could look like. A line defect extends from left to right and functions as a bus waveguide carrying several wavelengths ($\lambda_1, \lambda_2, \ldots$) appropriate for a wavelength division multiplexed optical communication system. The circuit in Figure 1 includes two defect cavities functioning as on-chip laser sources. The sources feed Mach-Zehnder modulators. The modulated signal is then coupled to the bus waveguide. Such a circuit could function as a signal regenerator or as an on-chip optical data source.

Figure 2(a) illustrates the field behavior of a photonic crystal waveguide mode. The mode propagates from left to right and is confined via Bragg reflection from the lattice of air holes along the y-direction. A resonant mode of the three missing hole cavity ("L3" cavity) is shown in Figure 2(b) demonstrating the two dimensional confinement ability of the lattice. The fields depicted in Figure 2(a) and (b) were obtained numerically using the finite difference time domain method which will described later in this chapter. The photonic crystal geometry is defined by assigning a spatially varying index of refraction. The index in the hole regions were set to 1.0 consistent with air and to 3.4 in the material regions consistent with semiconductor materials such as indium phosphide, gallium arsenide and silicon operating at wavelengths in the near infrared suitable for telecommunication applications. These semiconductor materials are attractive for integrated photonics due to their large refractive index, their direct bandgap (for InP and GaAs) and low absorption loss in the near infrared (for silicon).

Integrated photonic devices have many applications including sensing [5, 14, 15, 36, 63] and high bandwidth, low power data routing [6, 9, 12, 25, 41, 52, 54, 58, 71]. In addition to the previously discussed versatility of photonic crystals, in many cases devices built from photonic crystals are more compact and have lower loss than devices based on index guiding. Waveguide bends with a small bend radius have been demonstrated [61] which

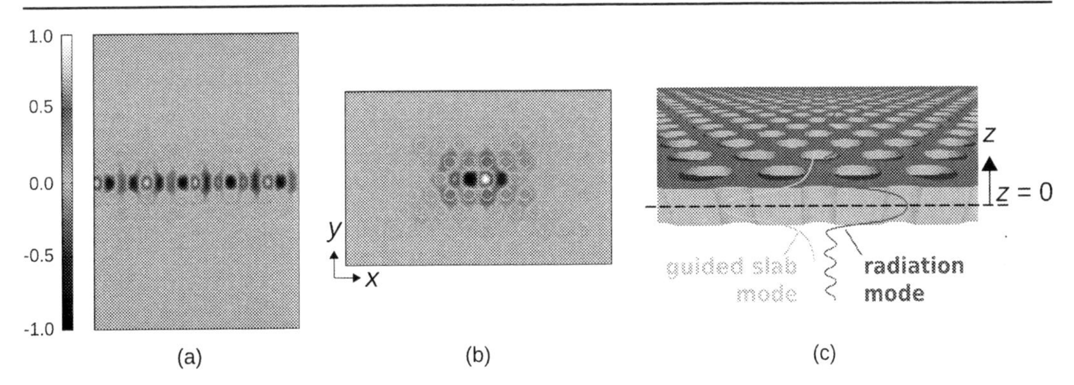

Figure 2. (a) Field distribution ($H_z(x,y,z=0)$) in a photonic crystal linear defect waveguide. (b) Field distribution ($H_z(x,y,z=0)$) in a photonic crystal three missing hole cavity (L3). (c) Schematic diagram showing the vertical field distribution in a finite height slab.

enables light to be routed and moved on chip. Resonant cavities with mode volumes on the order of a cubic wavelength and quality (Q) factors exceeding 10^6 have been fabricated and measured [1]. Such small and high Q factor cavities can function as filters, laser cavities or sensors. Their small mode volume and low loss make these cavities attractive for enhancing nonlinear interactions as well as studying quantum optics [18, 35, 43, 51, 76].

An interesting feature of two dimensional photonic crystal waveguides is their modified dispersion properties. The periodic modulation of the refractive index along the waveguide direction causes coupling between planewaves with propagation constants that are connected by reciprocal lattice vectors of the periodic waveguide structure. When the propagation constant reaches π/a, where a is the periodicity of the waveguide, a standing wave forms due to the presence of forward and backward propagating plane waves with the same propagation constant. In this regime, the waveguide bands flatten and the group velocity tends toward zero. It has been proposed to use this slow light effect for optical buffers or memories [2, 3, 29, 72].

Although three dimensional photonic crystals potentially offer the most degrees of freedom in controlling the properties of chip scale guided waves, they are difficult to fabricate due to the volumetric subwavelength three dimensional patterning required. On the other hand, the fabrication of two dimensional photonic crystals is compatible with layer-by-layer fabrication techniques which have been developed by the semiconductor electronics industry. Defining a two dimensional photonic crystal pattern into a layer of material approximately one half wavelength thick provides confinement in the vertical direction via total internal reflection. The vertical behavior of the guided and radiation modes are shown in Figure 2(c). Electromagnetic energy can be confined in-plane using the bandgap properties of the photonic crystal lattice, and the high index slab allows for confinement in the vertical direction. Thus three dimensional confinement is possible using two dimensional photonic crystals.

Because the vertical confinement mechanism is total internal reflection, only wavevectors that approach the air interface from inside the slab at a sufficiently glancing angle are confined. Wavevectors that are incident at a normal angle will be partially reflected and partially transmitted according to the Fresnel equations. When total internal reflection does not occur, energy from inside the slab radiates vertically into free space. This radiation is depicted in Figure 2(c) and is often the dominant radiative loss mechanism in two dimensional photonic crystal devices. For this reason, photonic crystal devices are often made from free standing thin membranes suspended in air. A free standing membrane provides the best vertical confinement because air has the lowest refractive index leading to the smallest possible critical angle and the fewest lost wavevectors that are not totally internally reflected.

Although the free standing membrane geometry provides the best optical confinement due to the low refractive index of air, the resulting suspended membrane typically has a thickness between 200 and 300 nm. Such a thin membrane presents an increased thermal and electrical resistance to heat and current passing to and from the operating regions of these devices. The lattice of air holes making up the cladding exacerbates this problem. Because of the large thermal resistance introduced by both the thin membrane and air-hole lattice, continuous wave lasing in high Q factor cavities is often difficult. This is because heat builds up in the cavity as carriers are excited by an external energy pump, and the resulting temperature increase degrades the gain properties of the active material. Similarly, the thin membrane and air hole array make it difficult to inject current via electrical contacts on the exterior of the air-hole array for pumping via electrical injection. In order for two dimensional photonic crystal devices to be a viable candidate for integrated photonics, one must show that photonic crystal cavities can be constructed that are suitable for continuous wave and electrically addressed lasing.

One solution to improving the thermal properties of photonic crystal membrane devices is to introduce a substrate with a thermal conductivity greater than that of air which acts as a heat sink. The photonic crystal semiconductor layer can then be grown on or bonded to this heat sinking substrate. This approach has been shown to lower the thermal resistance of photonic crystal defect cavity lasers by a factor of 20 leading to continuous wave lasing [40]. In addition to providing a thermal heat sink, bonding or growing the photonic crystal semiconductor layer on a heat sinking substrate improves the mechanical stability of the devices. If the substrate of choice is silicon dioxide, this approach would make the fabrication of devices more compatible with existing silicon fabrication techniques and potentially facilitate the integration of electronic and photonic components on a single chip. Improved heat sinking would also benefit nonlinear devices which often require a large operating field intensity. The enhanced thermal dissipation is helpful also because of heat generated by gigabit modulation and switching.

The advantages in thermal conductivity come at the cost of reduced vertical index contrast. When the material above or below the free standing membrane is no longer air, the index of the top or bottom cladding layer increases. This increase in the cladding index reduces the vertical index contrast and increases the critical angle for total internal reflection. The result is more internal wavevectors that can radiate away vertically from the photonic crystal device. As mentioned earlier, even in air clad geometries, the vertical loss is often dominant. Including a thermal heat sink can greatly deteriorate the optical properties of the device.

In the remainder of this chapter, we present numerical results quantifying the degree of optical loss induced by vertical radiation for different substrate indices. We will discuss novel waveguide and cavity designs that incorporate non-symmorphic space group symmetry in the form of a glide plane. Bandstructure properties will be investigated using space group theory, and modal properties of the Fourier space distributions will be derived. These designs are predicted to have increased immunity to vertical radiation loss [34, 48], enabling the use of heat-sinking lower substrates whose refractive index is larger than that of air. These lower substrates have the potential for enabling continuous wave laser operation and improving mechanical stability. Both of these improvements will help make two dimensional photonic crystals a more promising candidate for integrated photonics applications.

2. Photonic Crystal Waveguides

2.1. Numerical Analysis

Photonic crystal geometries are a complicated electromagnetic problem and require numerical methods for their analysis. Many methods exist that are appropriate for various aspects of this analysis. In this work we will be using the finite difference time domain (FDTD) method [68]. FDTD is favorable due to its simplicity, generality and favorable scaling with problem size. FDTD is an explicit discretization of Maxwell's curl equations in space and time. The output are the three dimensional vector electromagnetic fields in some region of space solved for some time interval. Typically either an initial condition is specified or a time-dependent source is added to provide input energy to the system. The fields can be monitored as a function of time, and signal processing techniques can be used on the ensuing time sequences to extract information about the geometry under investigation. The geometry interacts with the FDTD method by discretizing the electric and magnetic permittivities in space and storing them in computer memory. In what follows we discretize the photonic crystal geometries using at least 20 points per lattice constant. We employ perfectly matched layer absorbing boundary conditions to prevent reflection from the termination of the computational domain. Details on the application of FDTD to specific devices will be described in subsequent sections.

2.2. Photonic Crystal Waveguide Bandstructure

In Figure 3(a) we display a photonic crystal waveguide band structure diagram. The horizontal axis is propagation constant, and the vertical axis is normalized frequency. Both axes are normalized with respect to the photonic crystal lattice constant (a). The black lines indicate the photonic crystal line defect waveguide modes. The blue shaded regions denote the one dimensional projection of the two dimensional photonic crystal cladding modes onto the propagation direction. The line cutting diagonally across the diagram is the light line, and the region to the left of the light line is the light cone projection onto the waveguide dispersion diagram.

The dispersion diagram shown in Figure 3(a) corresponds to a triangular lattice with a single missing row line defect waveguide formed along the $\Gamma - K$ direction. The $\Gamma - K$ direction corresponds to the defect orientation shown in Figure 2(a). The geometry is a suspended air-clad membrane with a thickness to lattice constant ratio of $d/a = 0.6$. The holes have a radius to lattice constant ratio of $r/a = 0.30$. The membrane index is set to $n = 3.5$. Because the structure possesses mirror symmetry along the vertical direction at

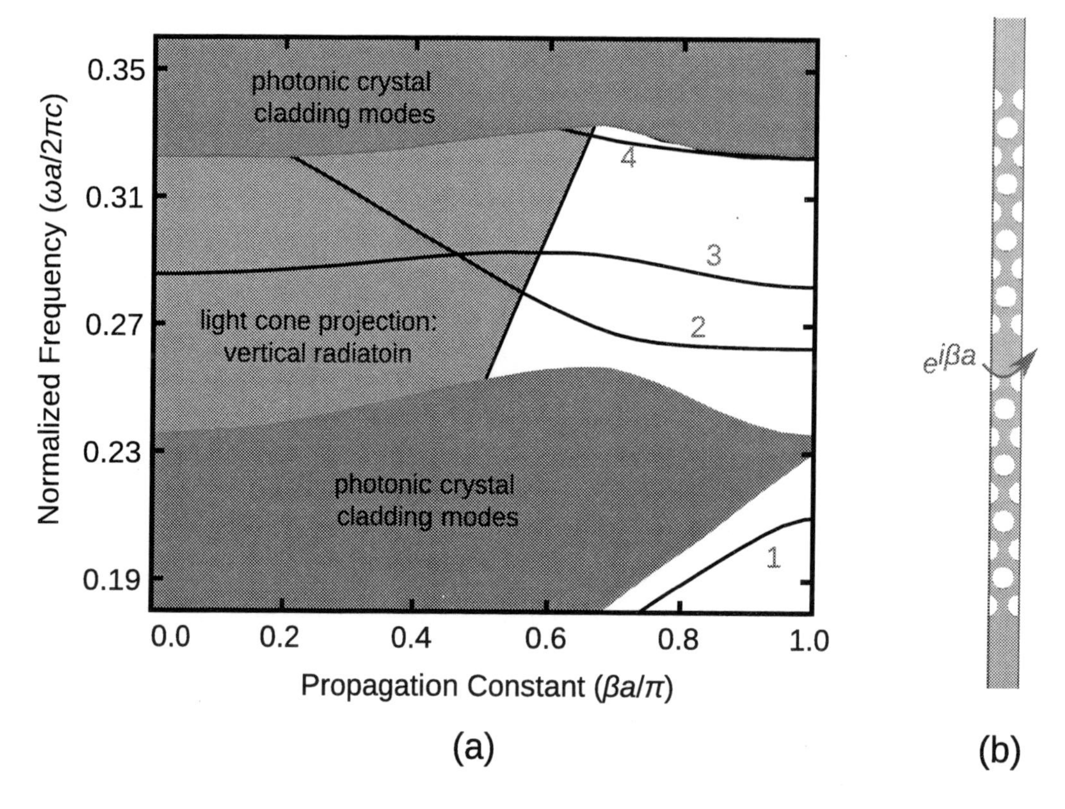

Figure 3. (a) Dispersion diagram for the low frequency TE-like bands in a two dimensional photonic crystal line defect waveguide. The blue shaded regions correspond to the photonic crystal cladding modes. The gray shaded region is the light cone projection. The dispersion was calculated using the three dimensional finite difference time domain method with spatially looped boundary conditions. (b) The unit cell used for the numerical calculation.

the midplane of the membrane ($z = 0$), the guided slab modes are categorized into TE-like or TM-like according to the polarization and parity of the associated fields. Thin slabs are adopted so that only the lowest order TE-like and TM-like modes are guided in the frequency range of interest. It turns out the in-plane electrical field is even for the lowest order TE-like mode and odd for the lowest order TM-like mode. These two modes are orthogonal to each other. The bandstructure in Figure 3(a) corresponds to the lowest order TE-like mode of the slab.

The waveguide dispersion diagram was calculated using the FDTD method. Because of the periodicity of the waveguide structure, only a single unit cell needs to be analyzed. The unit cell is shown in Figure 3(b). The boundaries along the propagation direction are related to each other by a phase factor $\exp(i\beta a)$ resulting from Bloch's theorem [31, 32]. The user specifies the propagation constant β, and a time sequence is obtained containing on the order of 10^5 time steps. A discrete Fourier transform is applied to the time sequence which reveals the propagating modes of the structure corresponding to the user specified β value. This is done for discrete β values spanning the range 0 to π/a.

The various shaded regions on the dispersion diagram provide a wealth of information

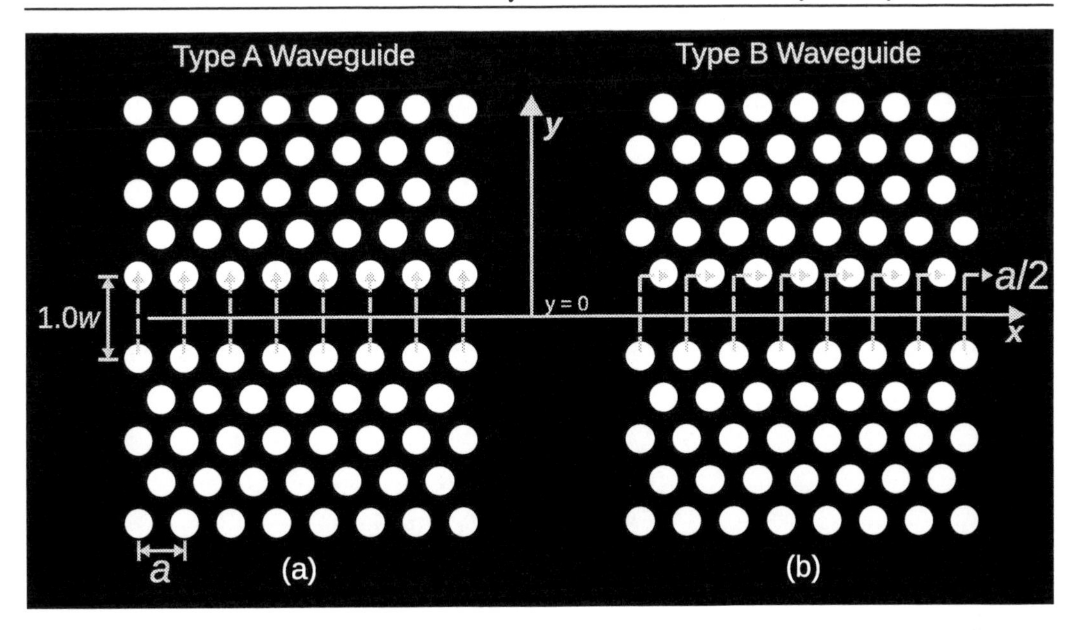

Figure 4. (a) Schematic diagram of a type A line defect photonic crystal waveguide $w = \sqrt{3}a$. (b) Schematic diagram of a type B line defect photonic crystal waveguide. The half lattice constant ($a/2$) shift is indicated.

about the loss mechansims of the waveguide modes and their available low-loss bandwidths. The region between the two shaded regions labeled "photonic crystal cladding modes" is known as the photonic crystal bandgap. The modes of the uniform lattice were also obtained using the FDTD method with Bloch boundary conditions [33]. This photonic crystal lattice possesses a TE-like bandgap in the normalized frequency range 0.256-0.323. For structures fabricated using a lattice constant of around 400 nm, the low loss (1550 nm) and low dispersion (1300 nm) wavelengths important for telecommunication fall into the bangap of this lattice.

In order for the waveguide mode to experience Bragg reflection and maintain in-plane confinement, its frequency must not overlap with the photonic crystal cladding modes. For the waveguide bands in the bandgap in Figure 3(a), only the band labeled '3' spans the entire first Brillouin zone without intersecting the photonic crystal modes. The bandwidth of the waveguide bands labeled '2' and '4' is limited due to their overlap with the cladding modes.

The shaded region labeled "light cone projection: vertical radiation" corresponds to those wavevectors in the slab that have a small in-plane component and whose out-of-plane component is large enough so that they are not totally internally reflected. Waveguide modes whose frequencies fall in the "light cone projection" region have large out-of-plane radiation loss [31,32,41,55]. Because of the large radiation losses suffered by modes whose frequencies fall in the "light cone projection" region, typically photonic crystal waveguides are considered unusable at these frequencies. Therefore, the light line represents an upper bound on the usable frequency bandwidth of these waveguides.

2.3. Waveguides With Non-symmorphic Space Group Symmetry

2.3.1. Qualitative Aspects

Figure 4 depicts two photonic crystal waveguide geometries. In Figure 4(a) the single line defect waveguide that has been discussed in the previous section is shown. In Figure 4(b) a similar geometry is shown, but one side of the photonic crystal lattice is shifted by one half lattice constant ($a/2$) along the propagation direction. This half lattice constant shift is known as a "glide," and the $x-z$ plane is a glide-plane. In previous work [4, 34], line defect waveguides without a glid-plane have been called "type A," whereas those with the half lattice constant shift have been termed "type B."

Numerical studies of type B photonic crystal waveguides have shown that the vertical radiation of waveguide bands that fall in the light cone projection region of the band diagram have significantly reduced vertical radiation loss [34]. The reduced vertical loss is related to the symmetry properties of the lattice. Qualitatively speaking, the half lattice constant shift introduces a phase shift between the field profiles on either side of the line defect waveguide core. When the two fields meet on the waveguide axis, they destructively interfere reducing the energy radiated vertically from the center of the waveguide. Kuang has shown in reference [34] that the out-of-plane radiation loss of the photonic crystal slab waveguide on semicondutor substrates can be reduced by an order of magnitude going from the type A structure to type B.

The significantly reduced radiation loss for waveguide bands above the light line associated with the type B waveguide is useful for expanding the low loss bandwidth of these waveguides. This is particularly important when a heat sinking substrate is introduced as discussed in Section 1. As the index of the lower substrate is increased, the slope of the light line decreases and overlaps a larger portion of the waveguide band reducing the usable bandwidth. The type B waveguide design could overcome this issue by reducing the out-of-plane radiative losses.

Later in this chapter, we will present photonic crystal resonant cavity designs that utilize a type B waveguide double heterostructure [48]. The resonant modes of double heterostructure cavities are closely related to the properties of the corresponding waveguide from which the cavity is made [45]. The loss and modal properties of the heterostructure bound state resonances can be understood by indentifying their frequencies and dominant spatial wavevector components and then using this information to find their location on the corresponding waveguide dispersion diagram. These concepts will be discussed in greater detail in a later section. For now this provides further motivation for obtaining a detailed understanding of photonic crystal waveguide dispersion diagrams.

Figure 5 displays a waveguide dispersion diagram with dispersion bands corresponding to six different waveguides. The black curves correspond to the dispersion of a type A waveguide. The magenta curves correspond to the dispersion of a type B waveguide. The remaining curves illustrate the waveguide dispersion for waveguides in which the two sides of the photonic crystal lattice have been shifted by some intermediate value between $0.0a$ and $0.5a$. Figure 5 illustrates how the waveguide dispersion of a type A waveguide evolves into that of a type B waveguide.

One notable feature of Figure 5 is the transition from a band crossing near $\beta a/\pi = 0.44$ for the type A waveguide to a band anticrossing for all nonzero shifts in the photonic crys-

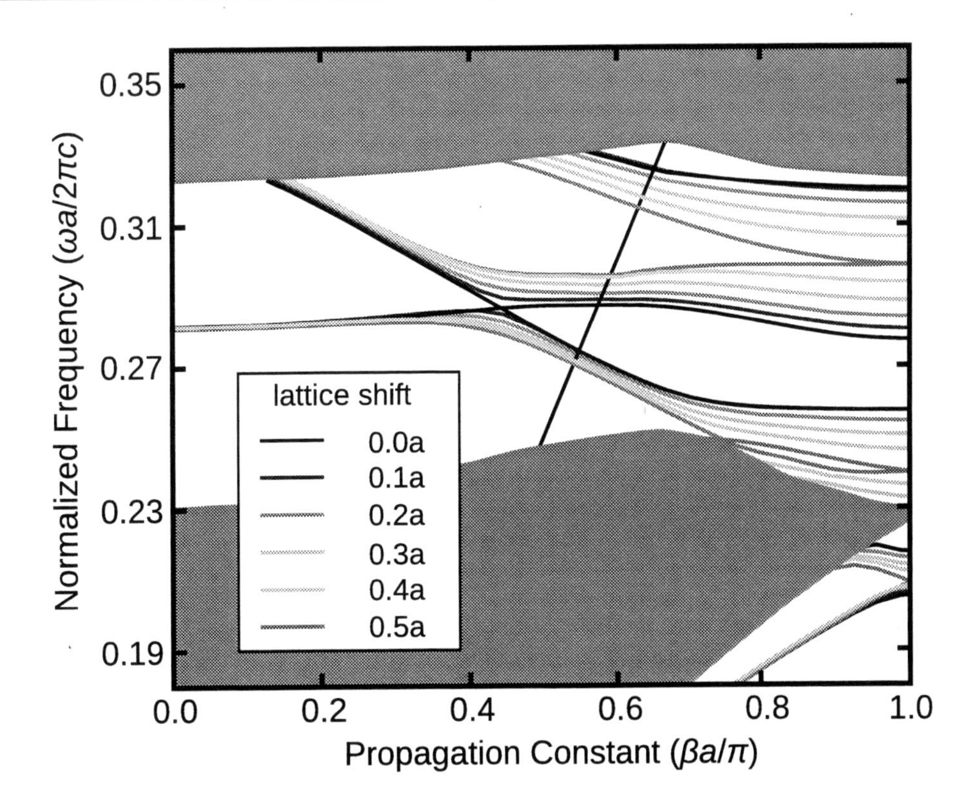

Figure 5. Photonic crystal waveguide dispersion diagram showing the dispersion bands for six different waveguide configurations. A lattice shift of $0.0a$ corresponds to the nominal type A cavity in which the photonic crystal lattices on either side of the waveguide core are aligned. A lattice shift of $0.5a$ corresponds to the type B cavity in which the photonic crystal lattices on either side of the waveguide core are offset by $a/2$ as shown in Figure 4(b). The other dispersion curves correspond to waveguides with intermediate shifts from $0.1a$ to $0.4a$.

tal lattice. This phenomenon can be qualitatively understood from a coupled mode theory perspective [75]. In general, modes associated with different bands (labeled by n_i) in a photonic crystal waveguide are orthogonal, and modes associated with the same band but at different propagation constants are orthogonal: $\int \varepsilon(\vec{r})\vec{E}_{n_1}^{\beta_1}(\vec{r}) \cdot \vec{E}_{n_2}^{\beta_2 *}(\vec{r}) d^3\vec{r} = \delta(\beta_1 - \beta_2)\delta_{n_1,n_2}$. The two bands that either cross or anti-cross at $\beta a/\pi = 0.44$ have three-dimensional field distributions whose overlap integrals equate to zero: $\int \varepsilon(\vec{r})\vec{E}_{n_1}^{\beta}(\vec{r}) \cdot \vec{E}_{n_2}^{\beta *}(\vec{r}) d^3\vec{r} = 0$. If we consider the shifting of one side of the photonic crystal lattice to be a perturbation to the unshifted type A waveguide, coupled mode theory tells us that the coupling constants characterizing the interaction between waveguide modes involve spatial overlap integrals of the form

$$\kappa \propto \int \vec{E}_{n_1}^{\beta}(\vec{r}) \cdot \vec{E}_{n_2}^{\beta *}(\vec{r})(\varepsilon_A(x,y,z) + \Delta\varepsilon(x,y,z))d^3\vec{r} \tag{1}$$

where $\vec{E}_{n_i}^{\beta}(\vec{r}) = \vec{E}_{n_i}^{\beta}(x,y,z)$ is the electric field of a type A waveguide mode labeled by

$i = \{1,2\}$, $\varepsilon_A(x,y,z)$ represents the dielectric distribution of the type A waveguide and $\Delta\varepsilon(x,y,z) = \varepsilon(x,y,z) - \varepsilon_A(x,y,z)$ represents deviation from the type A waveguide structure induced by the photonic crystal lattice shift. Note that the x and y coordinates are labeled in Figure 4, and the z coordinate axis is labeled in Figure 2(c). For the type A waveguide, $\Delta\varepsilon(x,y,z) = 0$ and $\varepsilon_A(x,y,z) = \varepsilon_A(x,-y,z)$ which suggests that the $\vec{E}^{\beta}_{n_i}(x,y,z)$ are either even or odd along the y-direction. If $\vec{E}^{\beta}_{n_1}(x,y,z)$ is odd and $\vec{E}_{n_2}(x,y,z)$ is even along the y-direciton (or vice versa) and $\varepsilon(x,y,z)$ is even, κ vanishes, and there is no coupling between the bands associated with $\vec{E}^{\beta}_{n_1}$ and $\vec{E}^{\beta}_{n_2}$. Previous studies of the spatial modes associated with the two bands that cross for the type A waveguide have shown that one mode is odd and one mode is even along the y-direction [55]. This results in $\kappa = 0$ for the two type A waveguide modes at $\beta a/\pi = 0.44$, and crossing is allowed. When one side of the photonic crystal lattice is shifted, $\Delta\varepsilon(x,y,z) \neq 0$ and $\Delta\varepsilon(x,y,z) \neq \Delta\varepsilon(x,-y,z)$. If κ is evaluated using the unperturbed even and odd modes of the type A waveguide, then $\kappa \neq 0$, which causes the modes to anti-cross. This argument provides a qualitative explanation of why the type A waveguide bands cross at $\beta a/\pi = 0.44$, and bands associated with lattice shifted geometries anticross.

Another interesting feature displayed in Figure 5 is the evolution toward pairwise degeneracy of the type B waveguide bands at the Brillouin boundary ($\beta = \pi/a$). In addition to its theoretical interest, understanding this feature is particularly important for the design of heterostructure cavities whose bound states form near extrema in the dispersion bands [44].

Extrema in photonic crystal waveguide dispersion bands often form near the Brillouin zone boundary. For the type B waveguide which has pairwise degeneracy at the Brillouin zone boundary, one expects that corresponding heterostructure cavities will have bound state resonances with closely spaced frequencies as a result. Understanding the nature of the degeneracy is important for future design requirements that may include improved free spectral range or mode selectivity [46].

The pairwise degeneracy is also interesting for slow light applications. As the waveguide dispersion bands flatten, the group velocity approaches zero. As mentioned earlier, bands tend to flatten near the Brillouin zone boundary making this part of the dispersion diagram of particular interest for slow light. As a consequence of band flattening, the slow light bandwidth is usually small. For bands that become simultaneously pairwise degenerate and flat, the slow light bandwidth could potentially be doubled. This represents another useful feature of the type B waveguide geometry.

In the following subsection we discuss the modal properties of the type B waveguide from a spatial Fourier transform perspective. This is followed by a group theory analysis which predicts the pairwise degeneracy of the waveguide bands at $\beta = \pi/a$.

Before proceding with the analysis, we present a comparison between the type A and the type B waveguide dispersion bands in Figure 6(a). The same waveguide dispersion bands from Figure 5 are repeated but without the bands for intermediate shifts. One sees that parts of the type B waveguide dispersion bands corresponding to the lower lying bands in the bandgap are partially obstructed by the photonic crystal cladding modes. In Figure 6(b) we display the waveguide dispersion corresponding to a type B waveguide with a reduced core width of $0.8w$ where w is labeled in Figure 4. Reducing the waveguide core width lowers the effective index of the waveguide causing its bands to increase in frequency. One can tune the waveguide bands this way [55]. Here we have shifted the lowest bands in the bandgap to higher frequencies, so they can span the entire Brillouin zone without

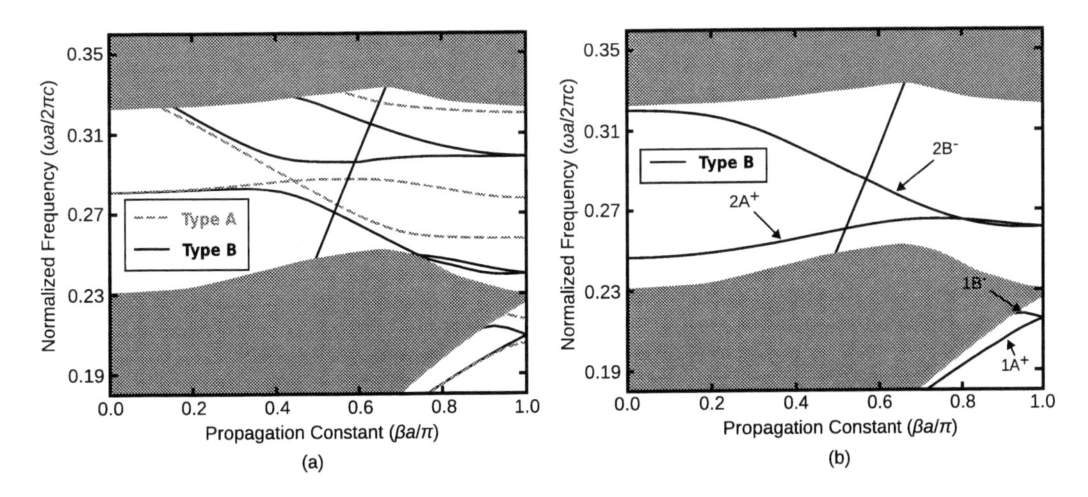

Figure 6. (a) Photonic crystal waveguide dispersion diagram showing dispersion bands for type A and type B waveguides. (b) Photonic crystal waveguide dispersion diagram for a type B waveguide with $0.8w$ core thickness.

intersecting the photonic crystal modes. It also improves the frequency overlap between the lowest frequency type A band in the bandgap and the lowest frequency type B bands in the bandgap making comparison between the two more straight forward.

It is also intersting to note that this reduced width tuning is beneficial for device applications. The reduced width type B waveguide has a larger bandwidth (beneficial for wavelength division multiplexed telecommunication applications) and smaller mode volume (beneficial for lowering power and nonlinear applications). We note also that this tuning does not effect any of the space group symmetries associated with the glide operation. The only significant difference is that the core width is no longer associated with a single line defect, but this does not have any consequences for the present discussion.

2.3.2. Spatial Fourier Space Analysis

In this section we discuss the modal properties of type A and type B waveguides. We use the Bloch properties of the waveguide modes to derive symmetry relations governing the transverse y-dependence of the mode profiles. For the type B waveguide, the y-direction symmetry is neither even nor odd, and we show that the modal properties of this waveguide are more clearly seen in the Fourier space decomposition of the spatial mode profile.

Due to the discrete translational invariance of the refractive index of a photonic crystal waveguide, the ith component of the electric or magnetic field (F_i) obeys Bloch's theorem [26]. Mathematically, this implies

$$F_i(x,y,z) = u_{i,\beta}(x,y,z)e^{-i\beta x} \tag{2}$$

where $u_{i,\beta}(x+a,y,z) = u_{i,\beta}(x,y,z)$ is a periodic function of x with period a. The transverse field properties of F_i are contained in $u_{i,\beta}(x,y,z)$. The longitudinal properties of F_i consist of the periodic function $u_{i,\beta}$ multiplied by the phase term $\exp(-i\beta x)$ where β is the propagation constant.

Consider a type A waveguide. The refractive index profile contains mirror symmetry about the x axis. Mathematically this implies $n_A(x,y,z) = n_A(x,-y,z)$. Because of this mirror symmetry, the field solutions will be eigenfunctions of the mirror symmetry operator denoted by σ_y. Therefore $\sigma_y F_A = \alpha F_A$ where α is the eigenvalue of the operator σ_y acting on the type A waveguide field represented by F_A. Note that we have dropped the subscript i and replaced it with A. It should be assumed that F represents a component of the electric or magnetic field. The label A refers to the type A waveguide.

Applying the operator σ_y to $F_A(x,y,z)$ yields

$$\sigma_y F_A(x,y,z) = F_A(x,-y,z) = \alpha F_A(x,y,z). \tag{3}$$

Applying the operator twice returns the field to its original configuration

$$\sigma_y \sigma_y F_A(x,y,z) = F_A(x,y,z) = \alpha^2 F_A(x,y,z). \tag{4}$$

The latter equality implies $\alpha^2 = 1$ or $\alpha = \pm 1$. From Equation 3 one has

$$\sigma_y F_A(x,-y,z) = \alpha F_A(x,y,z) = \pm F_A(x,y,z) \tag{5}$$

which shows that the field profiles of the type A waveguide are either even or odd about the center of the waveguide core. This result is intuitive and has been known for some time [55].

Now we apply these results to the spatial Fourier decomposition of the type A waveguide modes. Because the function u_β appearing in Equation 2 (the subscript i is dropped for conciseness) is periodic, it can be written as a Fourier series

$$u_\beta(x,y,z) = \sum_{m=-\infty}^{m=+\infty} f_m^\beta(y,z) e^{-ix\frac{2\pi}{a}m} \tag{6}$$

where $f_m^\beta(y,z)$ is the mth spatial Fourier series component of the mode with propagation constant β. So the photonic crystal waveguide mode can be expressed as

$$F_A(x,y,z) = \sum_{m=-\infty}^{m=+\infty} f_m^\beta(y,z) e^{-ix(\beta+\frac{2\pi}{a}m)}. \tag{7}$$

Using Equation 7 in Equation 5 yields

$$\begin{aligned}
\sigma_y F_A(x,y,z) &= F_A(x,-y,z) \\
&= \sum_{m=-\infty}^{n=+\infty} f_m^\beta(-y,z) e^{-ix(\beta+\frac{2\pi}{a}m)} \\
&= \pm F_A(x,y,z) \\
&= \pm \sum_{m=-\infty}^{m=+\infty} f_m^\beta(y,z) e^{-ix(\beta+\frac{2\pi}{a}m)}.
\end{aligned} \tag{8}$$

Comparing the second and fourth lines shows that $f_m^\beta(-y,z) = \pm f_m^\beta(y,z)$. This means that for a mode with even (odd) symmetry along the y-direction, every spatial Fourier series component of that mode will also have even (odd) symmetry along the y-direction. Again, this is an intuitive and expected result.

Now consider the type B waveguide. Because of the half lattice constant shift between the photonic crystal lattices on either side of the waveguide core, the structure no longer possesses mirror symmetry about the x axis: $n_B(x,y,z) \neq n_B(x,-y,z)$ where B denotes type B waveguide. The type B waveguide does possess a glide plane. The glide operation consists of a mirror operation about the x axis (σ_y) followed by a half lattice constant translation which will be denoted by $T_x(a/2)$. The glide operator g is given by $g = \sigma_y T_x(a/2)$.

We will follow the same procedure performed for the type A waveguide. Because the refractive index of the type B waveguide is invariant under application of g, the type B waveguide modes will be eigenfunctions of the glide operator.

Applying the operator g to $F_B(x,y,z)$ yields

$$gF_B(x,y,z) = F_B(x+a/2,-y,z) = \gamma F_B(x,y,z). \tag{9}$$

where γ is the eigenvalue of F_B under g. Applying the operator twice results in a one half lattice constant shift

$$ggF_B(x,y,z) = gF_B(x+a/2,-y,z) = F_B(x+a,y,z) = \gamma^2 F_B(x,y,z). \tag{10}$$

From Equation 2, $F_B(x+a,y,z) = F_B(x,y,z)\exp(i\beta a)$. Therefore $\gamma = \pm\exp(i\beta a/2)$. This result implies

$$gF_B(x,y,z) = F_B(x+a/2,-y,z) = \pm e^{i\beta a/2} F_B(x,y,z). \tag{11}$$

In the type A wavguide, we found that the operator σ_y gave rise to modes with even and odd symmetry along the y-direction as shown in Equation 5. From the analogous result shown in Equation 11 for the type B waveguide, a concise conclusion about the symmetry of the modes cannot yet be made.

In order to investigate the consequenses of Equation 11, we turn to the spatial Fourier decomposition of the waveguide modes as in Equation 7. Applying the g operator to F_B yields

$$\begin{aligned} gF_B(x,y,z) &= \sum_{m=-\infty}^{m=+\infty} f_m^{\beta}(-y,z) e^{-i(x+a/2)(\beta+\frac{2\pi}{a}m)} \\ &= e^{-i\beta a/2} \sum_{m=-\infty}^{m=+\infty} e^{-i\pi m} f_m^{\beta}(-y,z) e^{-ix(\beta+\frac{2\pi}{a}m)} \\ &= \pm e^{-i\beta a/2} F_B(x,y,z) \\ &= \pm e^{-i\beta a/2} \sum_{m=-\infty}^{m=+\infty} f_m^{\beta}(y,z) e^{-ix(\beta+\frac{2\pi}{a}m)}. \end{aligned} \tag{12}$$

Comparing the second and fourth lines shows that the Fourier components of the type B waveguide modes have the following property

$$e^{-i\pi m} f_m^{\beta}(-y,z) = \pm f_m^{\beta}(y,z). \tag{13}$$

Table 1. Symmetry of spatial Fourier series components according to Eq. 13.

m	+ ($1A^+$, $2A^+$)	− ($1B^-$, $2B^-$)
−2	$f^{\beta}_{-2}(-y,z) = f^{\beta}_{-2}(y,z)$ even	$f^{\beta}_{-2}(-y,z) = -f^{\beta}_{-2}(y,z)$ odd
−1	$-f^{\beta}_{-1}(-y,z) = f^{\beta}_{1}(y,z)$ odd	$-f^{\beta}_{-1}(-y,z) = -f^{\beta}_{-1}(y,z)$ even
0	$f^{\beta}_{0}(-y,z) = f^{\beta}_{0}(y,z)$ even	$f^{\beta}_{0}(-y,z) = -f^{\beta}_{0}(y,z)$ odd
1	$-f^{\beta}_{1}(-y,z) = f^{\beta}_{1}(y,z)$ odd	$-f^{\beta}_{1}(-y,z) = -f^{\beta}_{1}(y,z)$ even
2	$f^{\beta}_{2}(-y,z) = f^{\beta}_{2}(y,z)$ even	$f^{\beta}_{2}(-y,z) = -f^{\beta}_{2}(y,z)$ odd

Equation 13 is a relationship describing the even and odd symmetry of the individual spatial Fourier components of the type B waveguide mode. For the type A waveguide, we found that $f^{\beta}_{m}(-y,z) = \pm f^{\beta}_{m}(y,z)$ as a result of the even and odd symmetry of the modes along the y-direction. Equation 13 states that the spatial Fourier components of a single waveguide mode *alternate* between even and odd symmetry along the y-direction for sequential increments of m. Recall that a type B waveguide mode is characterized by an eigenvalue γ under g where γ can be either $+\exp(i\beta a/2)$ or $-\exp(i\beta a/2)$. The $\pm$ in Equation 13 refers to which eigenvalue is associated with the mode under consideration. Consider a waveguide mode characterized by $+\exp(i\beta a/2)$. From Equation 12, the spatial Fourier components will satisfy $\exp(-i\pi m) f^{\beta}_{m}(-y,z) = +f^{\beta}_{m}(y,z)$. The component f_0 corresponding to $m = 0$ is the component in the first reciprocal space unit cell (the first reciprocal space unit cell refers to β values between $-\pi/a$ and π/a). Equation 13 says that $f_0(-y,z) = f_0(y,z)$, or $f_0(y,z)$ has even symmetry along y. The component f_1 corresponding to $m = 1$ is the component in the second reciprocal space unit cell (the second reciprocal space unit cell refers to β values between π/a and $3\pi/a$). Equation 13 says that $-f_1(-y,z) = f_1(y,z)$, or $f_1(y,z)$ has odd symmetry along y. The component f_2 corresponding to $m = 2$ is the component in the third reciprocal space unit cell (the third reciprocal space unit cell refers to β values between $3\pi/a$ and $5\pi/a$). Equation 13 says that $f_2(-y,z) = f_2(y,z)$, or $f_2(y,z)$ has even symmetry along y. The same idea holds for negative values of m as well.

For modes characterized by the eigenvalue $\gamma = -\exp(i\beta a/2)$, a similar analysis can be done. The results will be the same except the even and odd components will be exchanged. The results of Equation 13 are illustrated in Table 1 for both eigenvalues and for m values in the range $m = \{\ldots -2, -1, 0, 1, 2, \ldots\}$.

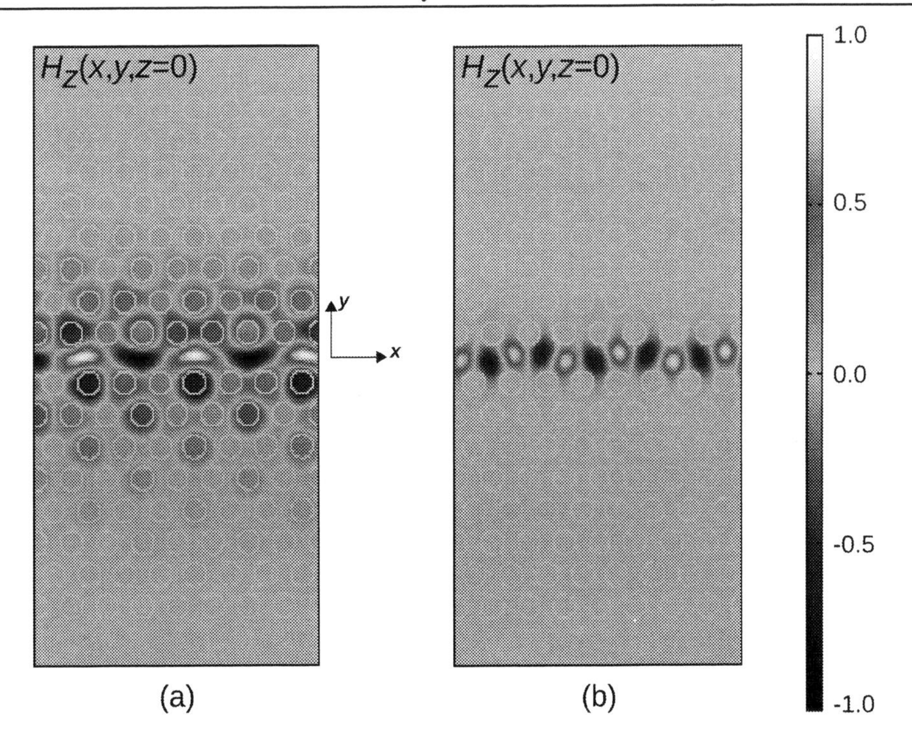

Figure 7. Magnetic field ($H_z(x,y,z=0)$) distribution in a type B photonic crystal waveguide with a core thickness of $0.8w$ corresponding to $\beta = 2.0/a$. (a) Field associated with band labeled $2A^+$ in Figure 6(b) and $\omega a/(2\pi c) = 0.2637$. (b) Field associated with band labeled $2B^-$ in Figure 6(b) and $\omega a/(2\pi c) = 0.2776$.

2.3.3. Spatial Mode Profiles

In this section we present numerically calculated spatial mode profiles for the type B waveguide. Figure 7 illustrates the $H_z(x,y,z=0)$ field profiles associated with the modes labeled $2A^+$ and $2B^-$ in Figure 6(b) for $\beta a = 2.0 = 0.634\pi$. $z = 0$ corresponds to the mid-plane of the slab where the H_z component is completely scalar for the fundamental TE-like slab mode. From Figure 7 it is clear that these modes are neither even nor odd about y. Qualitatively speaking, the modes appear to "zig-zag" along the waveguide core as a result of the photonic crystal lattice being a half lattice period out of phase on either side of the core. It is also interesting to note that the mode associated with the band labeled $2A^+$ in Figure 6(b) has weaker confinement along the y-direction than that of mode $2B^-$. This is due to the small frequency spacing between the $2A^+$ waveguide mode and the low frequency band associated with the photonic crystal cladding modes. That is, mode $2A^+$ is weakly confined by the photonic crystal lattice due to its shallow placement in the bandgap.

In Figures 8(a) and 8(d) we display the spatial Fourier transform of the modes depicted in Figure 7(a) and (b), respectively. These were calculated via two-dimensional discrete Fourier transform of the data depicted in Figure 7. The horizontal axes in Figure 8 are β, and the vertical axes are k_y. The coloring represents Fourier space magnitude. The spatial Fourier decompositions shown in Figure 8 are made up of discrete vertical "stripes" equally spaced along β. These "stripes" represent the discrete spatial Fourier series components

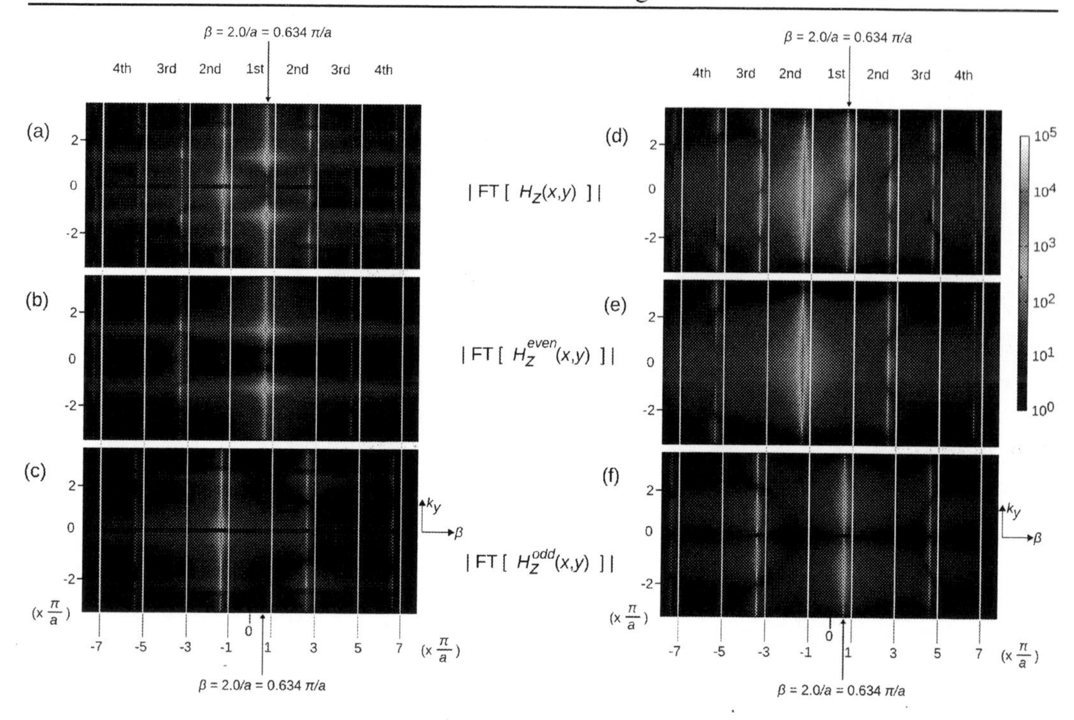

Figure 8. Two dimensional spatial Fourier transforms of the fields shown in Figure 7. (a) Corresponds to Figure 7(a). (d) Corresponds to Figure 7(b). (b) and (e) are the spatial Fourier transforms of the even components of the fields. (c) and (f) are the spatial Fourier transforms of the odd components of the fields. The user specified β value is labeled in (a) and (c).

making up the series in Equation 7. In principle, these "stripes" should have an infinitesimally small width in the β direction. Their finite width in Figure 7 is due to the finite width filter window used to isolate the spatial mode energy of interest. They are spaced every $2\pi/a$, and their location along the β axis is given by $\beta = 2.0/a + m2\pi/a$. The term $2.0/a$ corresponds to the user defined β value in the first Brillouin zone and can take any value between 0 and π/a. In Figures 8(a) and 8(d), the spatial Fourier peak in the first reciprocal lattice unit cell is identified at $\beta = 2.0/a$. It is clear that this component has the largest amplitude compared to the Fourier components in other reciprocal lattice unit cells.

Just below the spatial Fourier transforms in the top portion of Figure 8, we have plotted the spatial Fourier transforms of $H_z^{even}(x,y,z=0)$ and $H_z^{odd}(x,y,z=0)$. $H_z^{even}(x,y,z=0) = \frac{1}{2}(H_z(x,y,z=0)+H_z(x,-y,z=0))$ and $H_z^{odd}(x,y,z=0) = \frac{1}{2}(H_z(x,y,z=0)-H_z(x,-y,z=0))$. Note that the Fourier space distribution for $H_z^{odd}(x,y,z=0)$ passes through zero at $k_y = 0$. This results in the dark line along $k_y = 0$ in Figire 8(c) and (f) (its thickness and abruptness is due to the limited resolution in the k_y-direction).

H_z^{even} and H_z^{odd} represent the decomposition of H_z into its even and odd (along y) components. From Figures 8(b) and (c) we see that the Fourier series components making up the even component of the mode are located in the first and third reciprocal space unit cells, and the Fourier series components making up the odd component of the mode are located in the second and fourth reciprocal space unit cells. Comparison to the results of Table 1, shows that the mode labeled $2A^+$ in Figure 6(b) has the characteristics of the eigenfunction associated with the '+' sign (hence the '+' label already in place) implying its eigenvalue under g is $\gamma = +e^{-i\beta a/2}$.

In Figures 8(d)-(f), we plot the spatial Fourier transform for the mode labeled $2B^-$ in Figure 6(b). In this case, we see that the Fourier series components making up the even component of the mode are located in the second and fourth reciprocal space unit cells, and the Fourier series components making up the odd component of the mode are located in the first and third reciprocal space unit cells. Comparison to the results of Table 1, shows that the mode labeled $2B^-$ in Figure 6(b) has the characteristics of the eigenfunction associated with the '-' sign (hence the '-' label already in place) implying its eigenvalue under g is $\gamma = -e^{-i\beta a/2}$.

We have also analyzed the symmetry properties of the modes labeled $1A^+$ and $1B^-$ in Figure 6(b). The mode labeled $1A^+$ transforms according to the '+' sign in Eq. 13 and the mode labeled $1B^-$ transforms according to the '-' sign. For the two sets of bands shown in Figure 6(b), one sees that for the two bands that become degenerate at the Brillouin zone boundary, one will have y-direction symmetry defined by the '+' sign in Eq. 13, and the other will have y-direction symmetry defined by the '-' sign.

3. Group Theory Analysis of Two-Dimensional Photonic Crystal Waveguides

In this section two dimensional photonic crystal waveguides are characterized using group theory [70]. Several group theory analyses have been done of two dimensional photonic crystal geometries based on point group symmetry [27, 30, 57, 60]. But point groups are subgroups of space groups, and the theory of space groups allows for the classification of a broader class of photonic crystal lattice geometries. In this section, we examine the symmetry properties of two-dimensional photonic crystal waveguides in the context of space groups and focus on the non-symmorphic space group of the type B waveguide. The pairwise degeneracy at the Brillouin zone boundary for the type B waveguide can be predicted based on the properties of non-symmorphic space groups.

Space groups [10] consist of point group operations and translation group operations. They can be described by the Seitz operator $\{R|t\}$ which is defined by a point operation R followed by a translation t. The Seitz operator operates on an arbitrary position vector r as $\{R|t\}r = Rr + t$. The space group of a Bravais lattice involves all the translations of the lattice vectors that are linear combinations of the primitive lattice vectors.

The space group of a Bravais lattice may also contain operations involving a translation that is smaller than a primitive lattice translation combined with a rotation or reflection. Examples of such operations include "screw rotations" and "glide reflections". A screw rotation, whose translation is parallel to the rotation axis, does not exist when the space-dimension is lower than three. Glide reflection exists in two or three dimensions. A space group is "non-symmorphic" if it contains screw or glide operations. Otherwise it is "symmorphic". Furthermore, all the elements of a symmorphic space group can be written as the direct-product (denoted by $\otimes$) of translation groups and point groups. This is not the case for non-symmorphic space groups. Glide reflection is the only "non-symmorphic op-

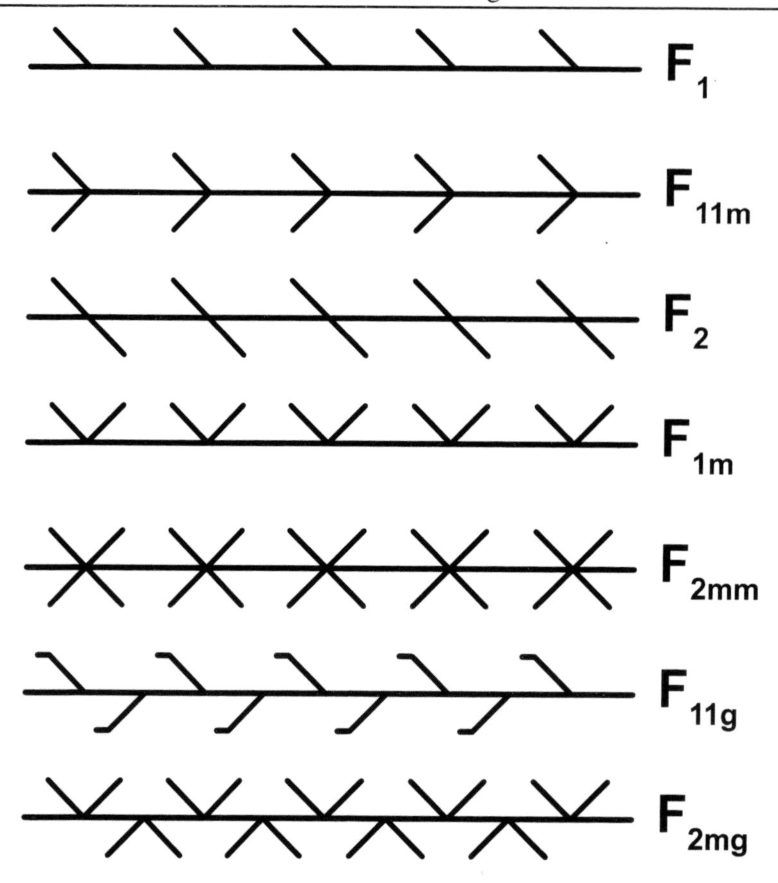

Figure 9. The seven Frieze groups [42]. F_{11g} and F_{2mg} are non-symmorphic groups containing glide reflections. The rest are symmorphic.

eration" in the two dimensional space group.

Among the crystallographic space groups of three dimensional lattices, there are 73 symmorphic groups and 157 non-symmorphic groups. In the case of two dimensional lattices, there are 13 symmorphic groups and 4 non-symmorphic groups. These groups are often refered to as wallpaper groups, due to their likeness to a repeating wallpaper pattern on a two dimensional wall or plane. The space group of two dimensional patterns containing only one dimensional translation is called the Frieze group. This is the space group of a two dimensional photonic crystal waveguide. It has 5 symmorphic groups and 2 non-symmorphic groups.

Figure 9 illustrates the basic symmetry structure of the seven Frieze groups [42]. The symbols for the group starts with "F" as in Frieze. The number "1" and "2" refers to the rotation operator C_1 and C_2. "m" indicates the existence of a mirror operation and "g" means a glide operation. The type A waveguide structure has F_{2mm} (symmorphic group) symmetry, and the type B waveguide structure has F_{2mg} (non-symmorphic group) symmetry.

3.1. Type A Photonic Crystal Waveguides: Symmorphic Space Group Analysis

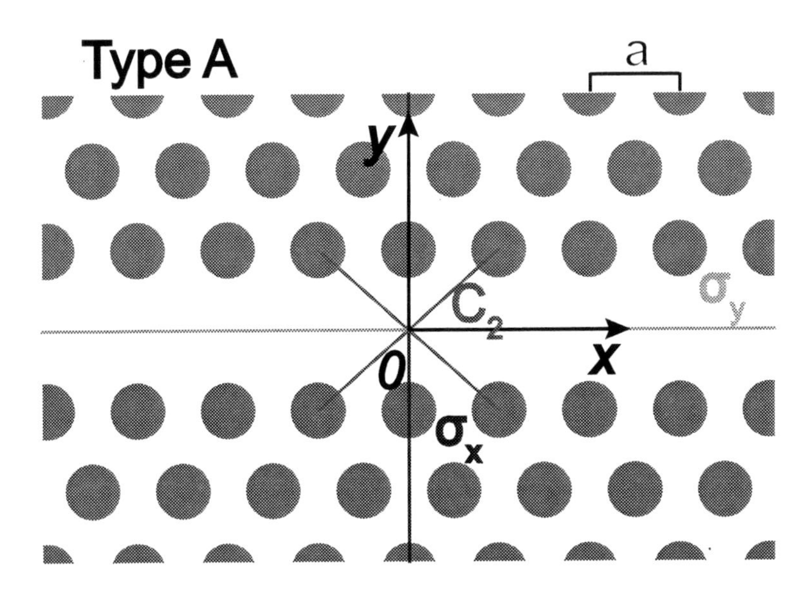

Figure 10. Illustration of a type A photonic crystal waveguide and its symmetry operations.

In the following two subsections, we will analyze the type A and type B photonic crystal waveguides using the Frieze groups F_{2mm} and F_{2mg}. These two groups are particularly interesting because they are the symmorphic and non-symmorphic Frieze groups of the highest symmetry. The other five Frieze groups can be constructed by removing one or two symmetry operations from F_{2mm} or F_{2mg}.

As mentioned previously the dielectric structure of a type A photonic crystal waveguide possesses the symmetry of the symmorphic space group F_{2mm}. F_{2mm} is isomorphic to the direct-product of a translation group and a point group ($T_{na} \otimes C_{2v}$). The translation group $T_{na} = \{E|na\}$ is Abelian and has only one-dimensional representations which are complex numbers. In the Seitz operator, E is a unity operator, n is an arbitrary integer and a is the lattice constant introduced in Figure 4 and illustated in Figure 10. The point group is $C_{2v} = (E, C_2, \sigma_x, \sigma_y)$ as illustrated in Figure 10, where σ is the reflection operator. Because of the simplicity associated with the translation group, analysis is typically focused on point groups alone for symmorphic space groups.

Consider a Bloch wave propagating in a Bravais lattice with a wavevector $\vec{k}$. The symmetry properties of propagating waves in Bravais lattices are characterized via the point group of wavevectors. Also known as the "little group," the point group of wavevectors consists of point group elements that transform $\vec{k}$ into itself or an equivalent point that is connected by reciprocal lattice vectors. It is denoted as $G^{\vec{k}}$ and is a subgroup of the point group. The label $\vec{k}$ here implies that the point group characterizing the transformation properties of a wavevector $\vec{k}$ in a Bravais lattice is a function of the direction and magnitude of $\vec{k}$.

In the case of the type A waveguide, the structure is periodic along the x-direction only, so we can represent the wavevector $\vec{k}$ using a scalar k. For this waveguide G^k is a subgroup of C_{2v}. At the Brillouin zone center and edge, $G^{k=0} = G^{k=\pi/a} = C_{2v}$. For "general" k points inside the Brillouin zone, the point group is $G^{0<k<\pi/a} = C_{1h} = (E, \sigma_y)$. The character tables of the point groups C_{2v} and C_{1h} are shown in Table 2 and Table 3, respectively. Their compatibility relation is evident from the eigenvalues of the σ_y operation. The dispersion curves of the type A waveguide are assigned to the representations in the character tables in Figure 11 according to the H_z field components of the waveguide modes. These eigenmodes are irreducible representations of the group.

Table 2. Character table of C_{2v} point group.

$C_{2v}(2mm)$	E	C_2	σ_x	σ_y
A_1	1	1	1	1
A_2	1	1	-1	-1
B_1	1	-1	1	-1
B_2	1	-1	-1	1

Table 3. Character table of C_{1h} point group.

$C_{1h}(m)$	E	σ_y
A	1	1
B	1	-1

Table 2 shows that all the representations of C_{2v} are one-dimensional. Therefore, there is no degeneracy imposed by symmetry. Any crossing of waveguide bands in the dispersion diagram is accidental (modes "A" and "B" in the middle of the TE bandgap shown in Figure 11 for example).

3.2. Type B Photonic Crystal Waveguides: Non-symmorphic Space Group Analysis

The symmetry operations of the type B waveguide are illustrated in Figure 12. They include e, c, m and g, where

$$\begin{aligned} e &= \{E|0\}, \\ c &= \{C_2|0\}, \\ m &= \{\sigma_x|a/2\}, \\ g &= \{\sigma_y|a/2\}. \end{aligned} \tag{14}$$

The coordinate origin is choosen at the center of the C_2 rotation, so it is quarter-lattice-constant shifted away from a lattice point along the x-direction at the center of the defect

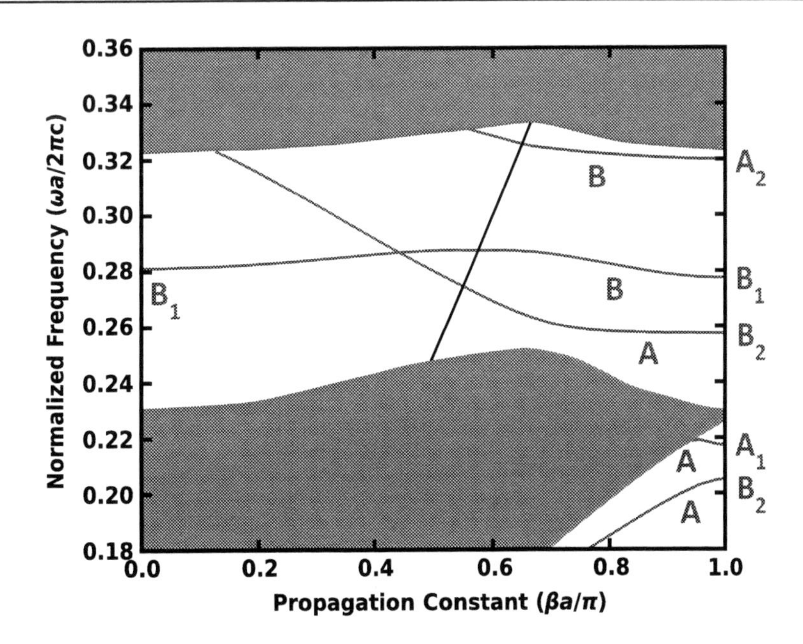

Figure 11. The dispersion diagram of the type A waveguide. The representations are assigned according to the H_z field component of the waveguide mode profiles.

region. The mirror operation σ_x is thus not with respect to the vertical line through the origin. The g operator is the glide reflection.

In contrast to the type A case, these non-translation operations (e,c,m,g) do not form a group. This is due to the existence of the glide operation. Consider the effect of applying g twice: $g^2 = \{E|a\}$ becomes a pure translation which is not a member of the non-translation elements just listed. Therefore, it is not possible to express the non-symmorphic group as the direct-product of a translation group and a point group. An additional complication arises because the translation group, $T_{na} = \{E|na\}$, has an infinite number of elements. Fortunately, it is an invariant subgroup (normal divisor) and can be divided out. The resulting factor group can be analyzed with a finite number of elements and is homomorphic to the original space group [22].

At $k = 0$, the whole translation group $T_{na} = \{E|na\}$ is the normal divisor, since $e^{k\times na} = e^{0\times na} = 1$. The factor group $F_{2mg}^{k=0}/T_{na} = (eT_{na}, cT_{na}, mT_{na}, gT_{na}) = (\{E|na\}, \{C_2|na\}, \{\sigma_x|a/2+na\}, \{\sigma_y|a/2+na\})$ where the elements of the factor group are cosets of the normal divisor. This factor group is isomorphic to the point group C_{2v}, and the character table is listed in Table 4. The representations in Table 4 are all one-dimensional. Therefore one expects the type B waveguide bands to be nondegenerate at $k = 0$.

At $k = \pi/a$, the translation group $T_{2na} = \{E|2na\}$ is the normal divisor, since $e^{k\times 2na} = e^{\pi/a\times 2na} = 1$. Then the factor group $F_{2mg}/T_{2na} = (e', \overline{e}', c', \overline{c}', m', \overline{m}', g', \overline{g}')$ [19, 21]. It has eight elements, and they are

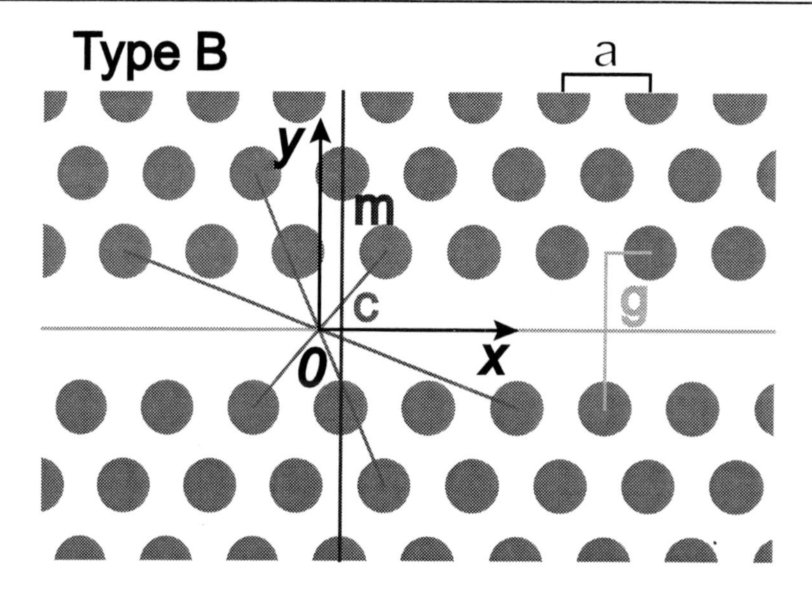

Figure 12. Illustration of a type B waveguide and its symmetry operators.

Table 4. Character table of the factor group of $F_{2mg}^{k=0}/T_{na}$.

$F_{2mg}^{k=0}/T_{na}$	eT_{na}	cT_{na}	mT_{na}	gT_{na}
Γ_1	1	1	1	1
Γ_2	1	1	-1	-1
Γ_3	1	-1	1	-1
Γ_4	1	-1	-1	1

$$
\begin{aligned}
e' &= \{E|2na\}, \\
\overline{e}' &= \{E|a+2na\}, \\
c' &= \{C_2|2na\}, \\
\overline{c}' &= \{C_2|a+2na\}, \\
m' &= \{\sigma_x|a/2+2na\}, \\
\overline{m}' &= \{\sigma_x|a/2+a+2na\}, \\
g' &= \{\sigma_y|a/2+2na\}, \\
\overline{g}' &= \{\sigma_y|a/2+a+2na\}.
\end{aligned} \tag{15}
$$

This factor group is isomorphic to the point group C_{4v}, and its character is shown in Table 5. There are four one-dimensional representations and one two-dimensional representation. For a Bloch wave representation at the Brillouin zone boundary ($k = \pi/a$), a one lattice vector translation $\{E|a\}$ changes the representation by a minus sign ($e^{k\times a} = e^{\pi/a\times a} = -1$). The only representation that changes the sign for the characters between $e' = \{E|2na\}$ and

$\overline{e}' = \{E|a+2na\}$ is K which is two-dimensional. K is then the only compatible representation at $k = \pi/a$.

Table 5. Character table of the factor group of F_{2mg}/T_{2na}. The two-dimensional irreducible representation (K) is also shown in the unity matrix form. [16]

F_{2mg}/T_{2na}	e'	$\overline{e}'$	$c', \overline{c}'$	$m', \overline{m}'$	$g', \overline{g}'$
	1	1	1	1	1
	1	1	1	-1	-1
	1	1	-1	1	-1
	1	1	-1	-1	1
K	2	-2	0	0	0
	$\begin{pmatrix} 1 & 0 \\ 0 & 1 \end{pmatrix}$	$\begin{pmatrix} -1 & 0 \\ 0 & -1 \end{pmatrix}$	$\begin{pmatrix} i & 0 \\ 0 & -i \end{pmatrix}$	$\begin{pmatrix} 0 & 1 \\ -1 & 0 \end{pmatrix}$	$\begin{pmatrix} 0 & i \\ i & 0 \end{pmatrix}$

Because K is a 2D representation, the waveguide modes of a type B waveguide should be two-fold degenerate at the Brillouin zone boundary. Thus the double-degeneracy of all the type B waveguide dispersion bands at the Brillouin zone boundary can be predicted based on the non-symmorphic space group symmetry of the waveguide. This "bands sticking together" effect is associated with the glide reflection symmetry that makes the space group non-symmorphic.

The dispersion curves of the type B waveguide are assigned with representations from the character tables in Figure 13, according to the H_z field components of the waveguide modes. For a general point in the Brillouin zone, none of the space symmetry operations leave the k vector invariant or shifted by a reciprocal lattice vector of the waveguide (other than translation by a). Therefore, the representation for a general point in the Brillouin zone is trivial (i.e. the identity representation) and not labeled.

The 2-by-2 unitary matrix irreducible representation of K is shown at the bottom of Table 5. It is helpful in revealing the relation between the real-space field distributions of the two degenerate modes at the Brillouin zone boundary. It is apparent that the operators (m, m') and (g, g') transform one mode to the other with only an extra phase factor (-1 or i) while the dielectric structure remains the same after these operations. This implies they share the same frequency at the Brillouin zone boundary and are, thus, degenerate there.

4. Photonic Crystal Double Heterostructure Cavities Utilizing Glide Plane Symmetry

Detailed understanding of the waveguide dispersion diagram is critical for proper design and utilization of photonic crystal waveguides. The previous section discussed several aspects of the band structure and modal properties of type B waveguides with a particular emphasis on the pairwise degeneracy at the Brillouin zone boundary. In this section photonic crystal double heterostructure cavities are discussed. Because the modal properties of heterostructure cavities are related to the modal properties of the straight waveguide

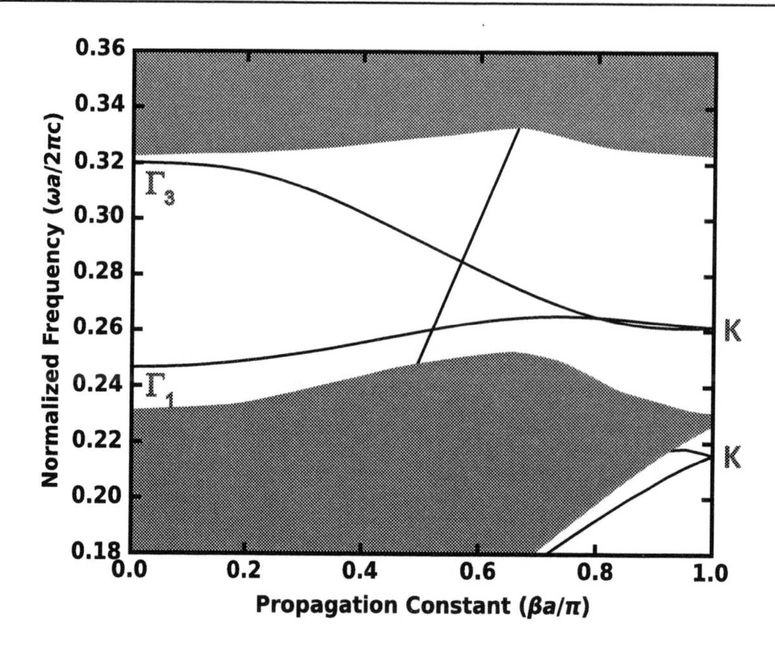

Figure 13. The dispersion diagram of the type B waveguide. The representations are assigned according to the H_z field component of the waveguide mode profiles.

from which they are formed [45], heterostructure cavity design is greatly facilitated by the bandstructure analysis provided in the previous sections.

Photonic crystal double heterostructure cavities are formed from otherwise straight waveguides by perturbing a few periods of that waveguide [1, 44, 65, 66, 69]. A schematic diagram of such cavities formed from type A and type B waveguides is shown in Figure 14. The photonic crystal double heterostructure cavity is interesting due to its large Q factor ($>10^6$), small mode volume (on the order of a cubic wavelength) and natural integration with output waveguides [37–39]. For the geometries shown in Figure 14, the lattice constant is enlarged along the x-direction. This causes a local decrease in frequency of the dispersion bands associated with the perturbed section of the waveguide. If the waveguide band possesses a local minimum as a function of β, a segment of the frequency band associated with the perturbed region will be shifted into the mode gap of the underlying straight waveguide. Because the perturbed section supports propagating modes at frequencies in the bandgap of the neighboring unperturbed sections, light is confined to the defect region. The connection between dispersion band minima and bound state formation allows one to predict many of the modal properties of heterostructure cavities based on the location of these local minima in the dispersion of the underlying straight waveguide.

4.1. Toward Continuous Wave Photonic Crystal Lasers

As discussed in the Introduction, use of two dimensional photonic crystal devices utilizing glide plane symmetry is motivated in large part by the desire to grow or bond quantum well active material to heat sinking substrates such as silicon dioxide or sapphire. Dissipating heat from free standing semiconductor membranes perforated with air holes is difficult

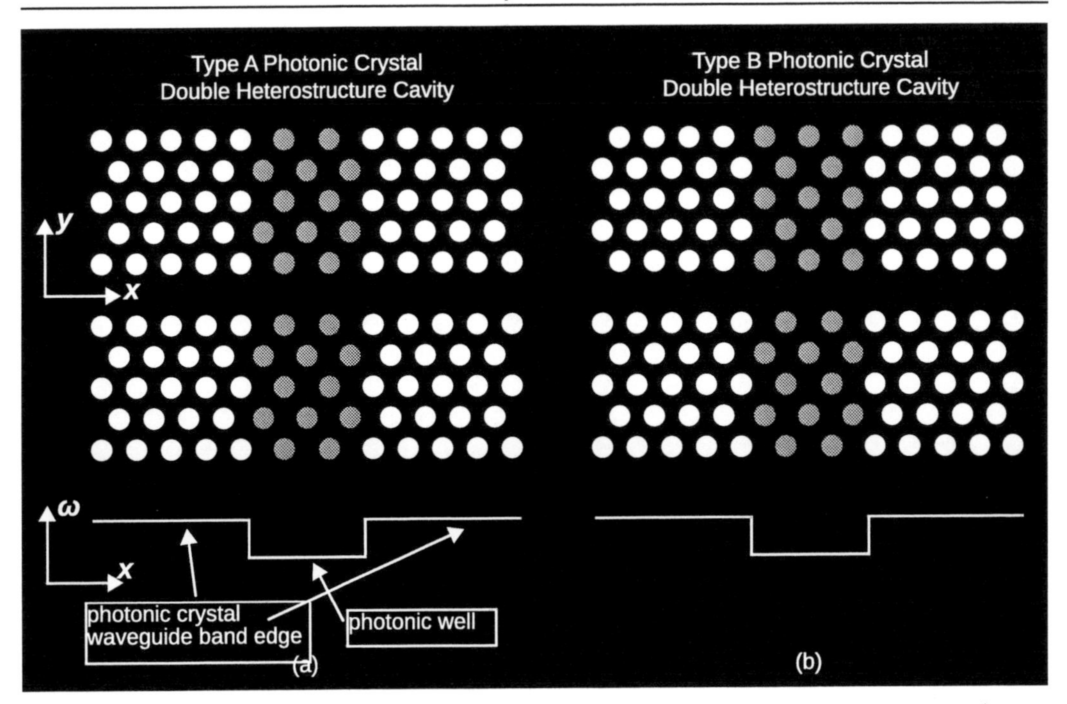

Figure 14. Schematic diagrams depicting the structure of photonic crystal double heterstructure cavities. The lattice constant of the colored holes has been increased along the x-direction by a few percent. (a.) The cavity is formed from a type A waveguide. (b.) The cavity is formed from a type B waveguide.

due to the low thermal conductivity of air. Thermal resistances of 1000 K/mW have been measured in suspended membrane photonic crystal defect lasers [24, 56]. Because of the problem of heating in photonic crystal membrane lasers, there have been relatively few reports of CW lasing in photonic defect cavities [11, 23, 40, 53, 56, 62]. Although CW lasing was achieved in the L3 [53] and point shift [56] cavities owing to the reduced mode volumes and high Q factors, using a thermally conductive lower substrate (such as silicon dioxide or sapphire) will allow the laser to be driven with higher incident pump power and therefore higher output power. A thermal resistance of 43 K/mW has been measured for a sapphire bonded photonic crystal defect cavity which shows that a sapphire heat sink can improve the heat dissipation by a factor of 20 [40]. Other advantages of a dielectric substrate include better thermal conduction for the entire chip, better CMOS compatibility and improved mechanical stability [11, 23, 62]. In addition, CW lasing with dielectric lower substrates represents a first step toward demonstrating electrically injected photonic crystal lasers that utilize a semiconductor vertical slab structure with low refractive index contrast. Candidate substrates such as silicon dioxide and sapphire have thermal conductivities two orders of magnitude larger than air as shown in Table 6.

A disadvantage of employing a dielectric heat sink is the reduced vertical index contrast. Total internal reflection is responsible for optical confinement in the vertical direction. However, as the index contrast between the semiconductor slab and the substrate is reduced, fewer wavevector components will be reflected and lost to free space. As the optical losses

Table 6. Candidate materials for heat sinking lower substrates and their refractive indices and thermal conductivities.

Material	Refractive Index	Thermal Conductivity (W/cm-K)
Air	1.003	0.00024
Silicon Dioxide	1.46	0.014
Sapphire	1.75	0.34

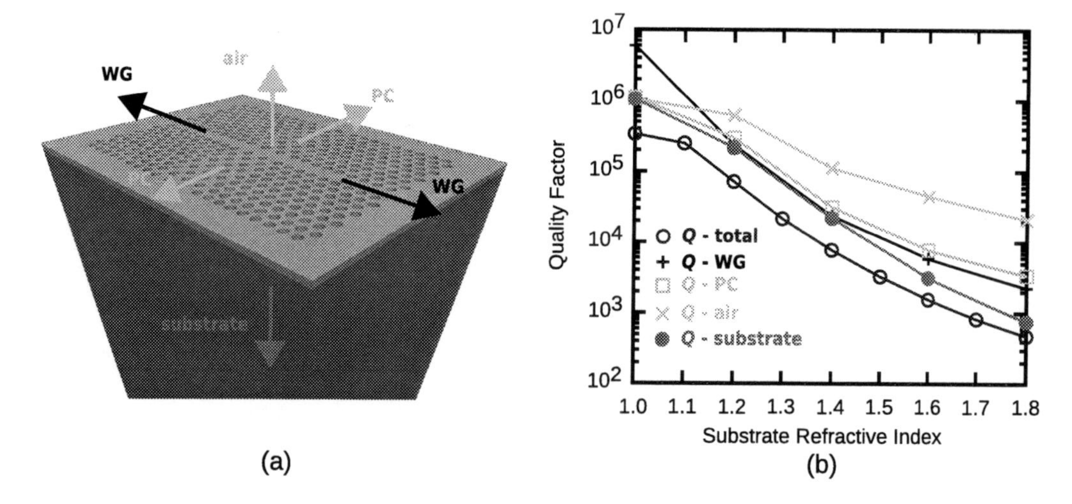

Figure 15. (a) Directional radiative loss channels in a photonic crystal double heterostructure cavity with a heat-sinking lower substrate. (b) Directional Q factors characterizing loss along the directions shown in (a.) as a function of substrate index.

increase, the laser threshold increases accordingly. Previous measurements and analysis have suggested that high output power photonic crystal membrane lasers operate at a passive quality (Q) factor of around a few thousand [38, 39]. It was also determined that a minimum Q factor of 1000 is required for continuous wave lasing for cavities bonded to a sapphire substrate [11, 62].

Figure 15(b) shows the Q factor of a photonic crystal double heterostructure cavity as a function of the index of the lower substrate. The results shown in Figure 15(b) were obtained from three dimensional finite-difference time-domain calculations employing perfectly matched layer absorbing boundary conditions. The Q factor was calculated by estimating the full width at half maxium of the numerically calculated resonance spectrum. Padé interpolation was used to overcome limitations in the discrete Fourier transform [17, 47]. The index of the semiconductor slab was set to $n = 3.4$, the hole radius was set to $r = 0.29a$ and the slab thickness was set to $d = 0.6a$. The Q factor of the nominally high Q type A cavity drops nearly exponentially as a function of lower substrate index. Also shown in Figure 15(b) are Q factors characterizing the directional radiation channels associated with the cavity. The directional Q factors were obtained by calculating the time-averaged Poynting vector and performing a two-dimensional spatial integration of the

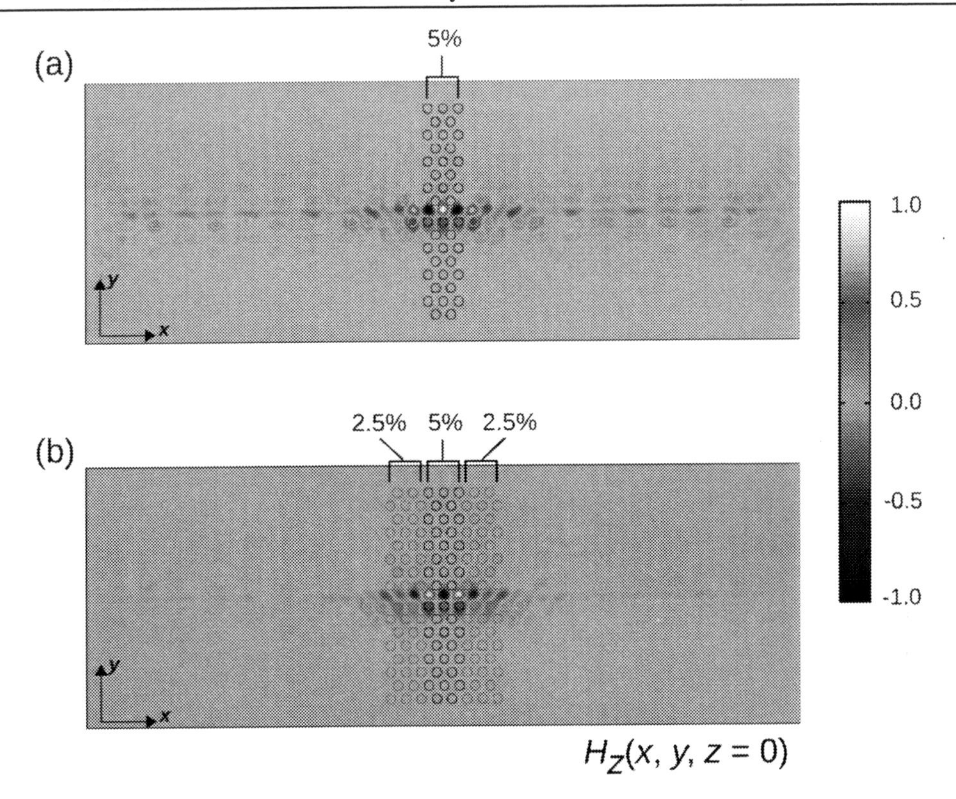

Figure 16. z-component of the magnetic field $H_z(x, y, z = 0)$ at the miplane of the slab for type B heterostructure cavities calculated using the three dimensional finite-difference time-domain method. (a.) The heterostructure cavity is formed by perturbing a few waveguide periods by 5%. (b.) The heterostructure cavity is formed by sandwiching a 5% perturbation between two 2.5% perturbations.

Poynting vector over a plane perpendicular to the corresponding loss direction. The various directional Q factors add as inverses to give the total Q factor as shown in Equation 16.

$$1/Q = 1/Q_{WG} + 1/Q_{PC} + 1/Q_{sub} + 1/Q_{air} \tag{16}$$

These directional loss calculations confirm that the primary loss mechanism is radiation into the substrate. As mentioned earlier, studies of type B waveguides have predicted significanly reduced out-of-plane radiation in the presence of high index substrates [34]. Because the modes of photonic crystal heterostructure cavities have properties closely related to the underlying straight waveguide from which they are formed, one would reasonably expect that the substrate radiation in these cavities could be reduced by forming the heterostructure from a type B waveguide [48].

A schematic of a type B heterostructure cavity is shown in Figure 14(b). The spatial profile of the z-component of the magnetic field $H_z(x, y, z = 0)$ is illustrated in Figure 16(a). The cavity in Figure 16(a) has a hole radius of $r = 0.29a$ and slab thickness of $d = 0.6a$. The index is $n = 3.4$. The width of the waveguide has been reduced to $0.8w$ for reasons discussed in Section 2.3.1.

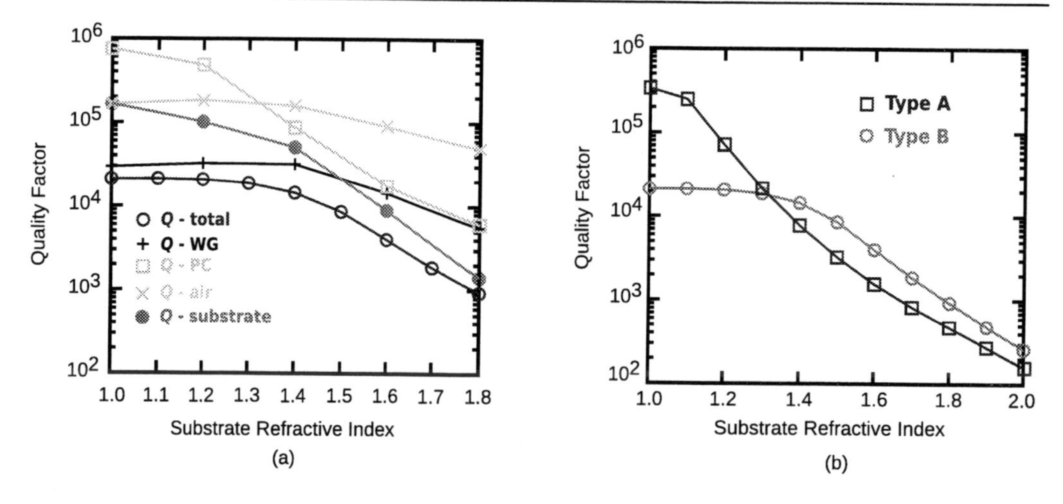

Figure 17. (a) Directional loss analysis for the type B heterostructure cavity shown in Figure 16(b). The loss directions are the same as those illustrated in Figure 15(a). (b) Comparison between the passive Q factors of type A and type B heterostructure cavities as a function of substrate index.

Figure 16(b) shows a type B heterostructure cavity with an additional 2.5% tapering layer between the 5% perturbed section and the straight waveguide. Previous work has used this design to lower the out-of-plane radiation by reducing the number of high frequency spatial Fourier components [67, 69, 73]. Here we use the tapering to soften the glide plane symmetry breaking. When the lattice is perturbed to form the heterostructure cavity, the geometry no longer has rigorous glide plane symmetry. As this symmetry is broken, the two bands that become pairwise degenerate at the Brillouin zone boundary can couple to each other. In the field profile shown in Figure 16(a), significant field amplitude can be seen in the straight waveguide sections of the cavity. This suggests that in-plane radiation along the waveguide sections represents a significant loss channel in these cavities. In Figure 16(b), the field amplitude in the straight waveguide sections is reduced. The Q factor of the cavity in Figure 16(a) is 1500. The Q factor of the cavity shown in Figure 16(b) is 21000 showing the significance of this loss direction and the role of the tapering section in reducing it.

Three dimensional finite-difference time-domain analysis of the loss in type B heterostructure cavities is shown in Figure 17(a). Even with the tapered perturbation introduced to soften the glide-plane symmetry breaking, the radiation loss is still dominated by leakage along the straight waveguide sections. It is significant that in the type B heterostructure cavity radiation loss into the substrate is not the dominant loss mechanism, so one is free to insert heat sinking substrates with refractive indices larger than air without significantly decreasing the passive Q factor. Figure 17(a) shows that this is true so long as the substrate refractive index is less than 1.4. For substrate indices larger than 1.4, radiation into the substrate becomes dominant and one is faced with the same substrate loss dominated situation as the type A cavity.

Figure 17(b) illustrates a comparison between the passive Q factors of the type A and type B cavities. Although the type A cavity has a significantly larger nominal Q factor for a suspended membrane, the Q factor of the type B cavity is relatively constant for substrate

Figure 18. A schematic diagram depicting a vertical slab structure defined using semiconductor alloys for electrically addressed photonic crystal lasers. The lavender layer is the high index active region. The green layers have a lower index. The gold strip on top is an electrical contact. A section is cut away to see the vertical layer structure and the hole etch depth.

refractive indices between 1.0 and 1.4. It is clear that the type B cavity has a larger passive Q factor for substrate refactive indices greater than 1.5. It is also notable that the type B cavity has a passive Q factor larger than 1000 when the substrate refractive index is consistent with sapphire ($n_{\text{sap}} = 1.75$, see Table 6) making the type B cavity a promising geometry for continuous wave photonic crystal heterostructure edge emitting lasers.

4.2. Toward Electrically Addressed Photonic Crystal Lasers

To conclude the discussion on photonic crystal heterostructure cavities, we note that another advantage of maintaining a sufficiently large passive Q factor for cavities with reduced vertical index contrast is the possibility of defining the vertical slab structure using different alloys of semiconductor material [13, 49, 59, 74]. If the entire laser stucture is made from semiconductor materials, then a vertical p-n junction can be defined, and electrical current can be injected from the top. One can also deeply etch the photonic crystal holes into the materials all the while maintaining a thermal conductivity similar to that of the materials shown in Table 6.

Laser operation under electrical current injection is an important step in demonstrating the practicality of photonic crystal lasers, as it removes the requirement of having a secondary external pump laser.

Figure 18 shows a schematic diagram of a photonic crystal heterostructure cavity with a vertical slab structure that includes both a bottom substrate as well as a thin top cladding layer. The purpose of the top cladding layer is to reduce the vertical symmetry breaking

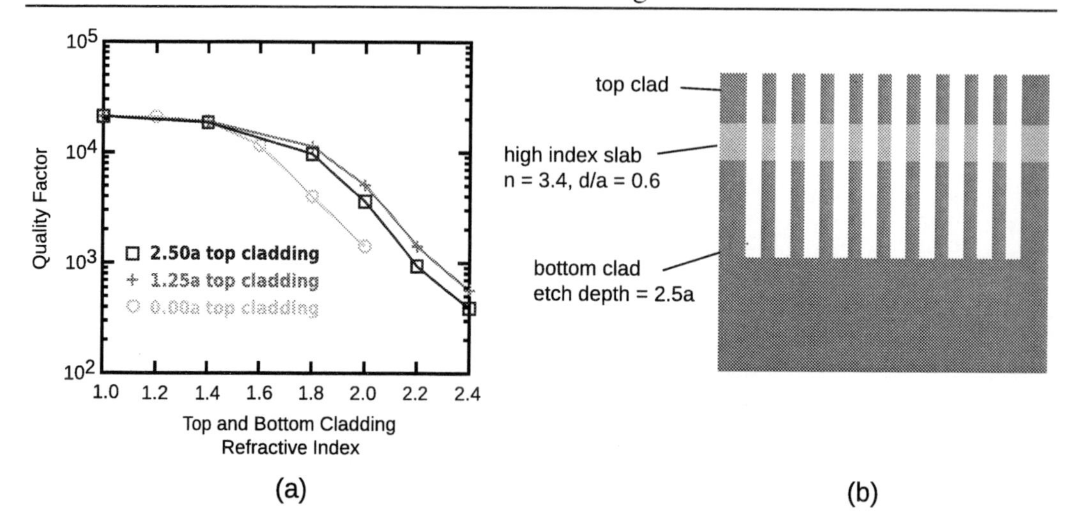

Figure 19. (a) Passive Q factor of a type B heterostructure cavity with a layered vertical slab structure illustrated in Figure 18 and 19(b). (b) Side view schematic diagram of a type B heterostructure cavity with a layered vertical slab structure.

which lowers coupling between the TE-like and TM-like modes of the slab. Figure 19(a) shows the passive Q factor of the cavity shown in Figure 18 and 19(b). One sees that with a top cladding layer of $1.25a$, the passive Q factor of the type B cavity remains above 1000 so long as the index of the cladding layers remains below 2.3. The lowest index achievable using InGaAsP materials at near infrared wavelengths is around 3.2 [7]. Therefore, cladding layers with indices around 2.3 are not feasible using InGaAsP. On the other hand, AlGaN and SiC materials do have indices in the range 2.0-2.6 [8, 20], and these materials could be used as substrates on which InGaAsP active materials could be grown or bonded for initial demonstrations of electrically injected photonic crystal lasers.

5. Conclusion

In this Chapter we have discussed two dimensional photonic crystal devices utilizing glide-plane symmetry. Previous work has predicted that photonic crystal devices utilizing glide plane symmetry have reduced out-of-plane radiation when the vertical index contrast is lowered. Here we investigated the band structure and modal properties of waveguides employing glide plane symmetry. We then analyzed the passive Q factors of photonic crystal double heterostructure cavities formed from type B waveguides. It was shown that a type B photonic crystal heterostructure has a passive Q greater than 1000 for a substrate refractive index consistent with sapphire making this geometry a good candidate for continuous wave laser operation.

References

[1] T. Asano, B.-S. Song, and S. Noda. Analysis of the experimental Q factors (~ 1 million) of photonic crystal nanocavities. *Optics Express*, **14**(5):1996–2006, 2006.

[2] Toshihiko Baba. Slow light in photonic crystals. *Nature Photonics*, **2**:465–473, 2008.

[3] Toshihiko Baba and Daisuke Mori. Slow light engineering in photonic crystals. *Journal of Physics D*, **40**:2659–2665, 2007.

[4] H. Benisty. Modal analysis of optical guides with two-dimensional photonic band-gap boundaries. *Journal of Applied Physics*, **79**(10):7483–7492, 1996.

[5] Romeo Bernini, Stefania Campopiano, Luigi Zeni, and Pasqualina M. Sarro. Arrow optical waveguides based sensors. *Sensors and Actuators B*, **100**:143–146, 2004.

[6] W. Bogaerts, D. Taillaert, B. Luyssaert, P. Dumon, J. Van Campenhout, P. Bienstman, D. Van Thourhout, and R. Baets. Basic structures for photonic integrated circuits in silicon-on-insulator. *Optics Express*, **12**(8):1583–1591, 2004.

[7] B. Broberg and S. Lindgren. Refractive index of InGaAsP layers and inp in the transparent wavelength region. *Journal of Applied Physics*, **55**(9):3376–3381, 1984.

[8] D. Brunner, H. Angerer, E. Bustarret, F. Freudenberg, R. Hopler, R. Dimitrov, O. Ambacher, and M. Stutzmann. Optical constants of epitaxial AlGaN films and their temperature dependence. *Journal of Applied Physics*, **82**(10):5090–5096, 1997.

[9] Emily F. Burmeister, John P. Mack, Henrik N. Poulsen, Milan L. Mašanović, Biljana Stamenić, Daniel J. Blumenthal, and John E. Bowers. Photonic integrated circuit optical buffer for packet-switched networks. *Optics Express*, **17**(8):6629–6635, 2009.

[10] Gerald Burns and A.M. Glazer. *Space Groups for Solid State Scientists*. Academic Press, Inc., 1990.

[11] J. R. Cao, Wan Kuang, Zhi-Jian Wei, Sang-Jun Choi, Haixia Yu, Mahmood Bagheri, John D. O'Brien, and P. Daniel Dapkus. Sapphire-bonded photonic crystal microcavity lasers and their far-field radiation patterns. *Photonics Technology Letters*, **17**(1):4–6, 2005.

[12] Luca P. Carloni, Partha Pande, and Yuan Xie. Networks-on-chip in emerging interconnect paradigms: Advantages and challenges. In *Proceedings of the ACM/IEEE International Symposium on Networks-on-Chip*, 2009.

[13] Liang Chen and Elias Towe. Design of high-Q microcavities for proposed two-dimensional electrically pumped photonic crystal lasers. *IEEE Journal of Selected Topics in Quantum Electronics*, **12**(1):117–123, 2006.

[14] Sang-Yeon Cho and Nan Marie Jokerst. A polymer microdisk photonic sensor integrated onto silicon. *IEEE Photonics Technology Letters*, **18**(20):2096–2098, 2006.

[15] E. Chow, A. Grot, L. W. Mirkarimi, M. Sigalas, and G. Girolami. Ultracompact biochemical sensor built with two-dimensional photonic crystal microcavity. *Optics Letters*, **29**(10):1093–1095, 2004.

[16] A.P. Cracknell. Tables of the irreducible representations of the 17 two-dimensional space groups and their relevance to quantum mechanical eigenstates for surfaces and thin films. *Thin Solid Films*, **21**(1):107 – 127, 1974.

[17] Supriyo Dey and Raj Mittra. Efficient computation of resonant frequencies and quality factors of cavities via a combination of the finite-difference time-domain technique and the padé approximation. *Microwave and Guided Wave Letters*, **8**(12):415–417, 1998.

[18] Dirk Englund, David Fattal, Edo Waks, Glenn Solomon, Bingyang Zhang, Toshihiro Nakaoka, Yasuhiko Arakawa, Yoshihisa Yamamoto, and Jelena Vučković. Controlling the spontaneous emission rate of single quantum dots in a two-dimensional photonic crystal. *Physical Review Letters*, **95**:013904 1–4, 2005.

[19] L. M. Falicov. *Group Theory and Its Physical Applications*. The University of Chicago Press, 1966.

[20] Gary L. Harris. *Properties of Silicon Carbide*. Institute of Engineering and Technology, London, UK, 1995.

[21] Volker Heine. *Group Theory in Quantum Mechanics*. Pergamon Press, New York, 1960.

[22] Conyers Herring. Character tables for two space groups. *J. Franklin Inst.*, **233**:525, 1942.

[23] J. K. Hwang, H. Y. Ryu, D. S. Song, I. Y. Han, H. K. Park, D. H. Jang, and Y. H. Lee. Continuous room-temerature operation of optically pumped two-dimensional photonic crystal lasers at 1.6 μm. *Photonics Technology Letters*, **12**(10):1295–1297, 2000.

[24] Kyoji Inoshita and Toshihiko Baba. Fabrication of gainasp/inp photonic crystal lasers by icp etching and control of resonant mode in point and line composite defects. *IEEE Journal on Special Topics in Quantum Electronics*, **9**(5):1347–1354, 2003.

[25] Yongqiang Jiang, Wei Jiang, Lanlan Gu, Xiaonan Chen, and Ray T. Chen. 80-micron interaction length silicon photonic crystal waveguide modulator. *Applied Physics Letters*, **87**:221105, 2005.

[26] John D. Joannopoulos, Robert D. Meade, and Joshua N. Winn. *Photonic Crystals*. Princeton University Press, Princeton, NJ, 1995.

[27] Se-Heon Kim and Yong-Hee Lee. Symmetry relations of two-dimensional photonic crystal cavity modes. *Quantum Electronics, IEEE Journal of*, **39**(9):1081–1085, Sept. 2003.

[28] H Kogelnik and C. V. Shank. Stimulated emission in a periodic structure. *Applied Physics Letters*, **18**(4):152–154, 1971.

[29] T. F. Krauss. Slow light in photonic crystal waveguides. *Journal of Physics D*, **40**:2666–2670, 2007.

[30] Wan Kuang, Jiang R. Cao, Tian Yang, Sang-Jun Choi, Po-Tsung Lee, John D. O'Brien, and P. Daniel Dapkus. Classification of modes in suspended-membrane, 19-missing-hole photonic-crystal microcavities. *J. Opt. Soc. Am. B*, **22**(5):1092–1099, 2005.

[31] Wan Kuang, Cheolwoo Kim, Andrew Stapleton, Woo Jun Kim, and John D. O'Brien. Calculated out-of-plane transmission loss for photonic-crystal slab waveguides. *Optics Letters*, **28**(19):1781–1783, 2003.

[32] Wan Kuang, Woo Jun Kim, Adam Mock, and John D. O'Brien. Propagation loss of line-defect photonic crystal slab waveguides. *IEEE Journal of Selected Topics in Quantum Electronics*, **12**(6):1183–1195, 2006.

[33] Wan Kuang, Woo Jun Kim, and John D. O'Brien. Finite-difference time domain method for nonorthogonal unit-cell two-dimensional photonic crystals. *Journal of Lightwave Technology*, **25**(9):2612–2617, 2007.

[34] Wan Kuang and John O'Brien. Reducing the out-of-plane radiation loss of photonic crystal waveguides on high-index substrates. *Optics Letters*, **29**(8):860–862, 2004.

[35] Peter Lodahl, A. Floris van Driel, Ivan S. Nikolaev, Arie Irman, Karin Overgaag, Daniel Vanmaekelbergh, and Willem L. Vos. Controlling the dynamics of spontaneous emission from quantum dots by photonic crystals. *Nature*, **430**:654–657, 2004.

[36] Marko Lončar, Axel Scherer, and Yueming Qiu. Photonic crystal laser sources for chemical detection. *Applied Physics Letters*, **82**(26):4648–4650, 2003.

[37] L. Lu, A. Mock, M. Bagheri, E. H. Hwang, J. D. O'Brien, and P. D. Dapkus. Double-heterostructure photonic crystal lasers with reduced threshold pump power and increased slope efficiency obtained by quantum well intermixing. *Optics Express*, **16**(22):17342–17347, 2008.

[38] L. Lu, A. Mock, E. H. Hwang, J. O'Brien, and P. D. Dapkus. High-peak-power efficient edge-emitting photonic crystal nanocavity lasers. *Optics Letters*, **34**(17):2346–2648, 2009.

[39] L. Lu, A. Mock, T. Yang, M. H. Shih, E. H. Hwang, M. Bagheri, A. Stapleton, S. Farrell, J. D. O'Brien, and P. D. Dapkus. 120 μW peak output power from edge-emitting photonic crystal double heterostructure nanocavity lasers. *Applied Physics Letters*, **94**(13):111101, 2009.

[40] Ling Lu, Adam Mock, Mahmood Bagheri, Jiang-Rong Cao, Sanj-Jun Choi, John D. O'Brien, and P. Daniel Dapkus. Gain compression and thermal analysis of a sapphire-bonded photonic crystal microcavity laser. *IEEE Photonics Technology Letters*, **21**(17):1166–1168, 2009.

[41] Sharee J. McNab, Nikolaj Moll, and Yurii A. Vlasov. Ultra-low loss photonic integrated circuit with membrane-type photonic crystal waveguides. *Optics Express*, **11**(22):2927–2939, 2003.

[42] R. Mirman. *Point Groups, Space Groups, Crystals, Molecules*. World Scientific, 1999.

[43] Toishi Mitsuru, Dirk Englund, Andrei Faraon, and Jelena Vučković. High-brightness single photon source from a quantum dot in a directional-emission nanocavity. *Optics Express*, **17**(17):14618–14626, 2009.

[44] A. Mock, L. Lu, and J. D. O'Brien. Spectral properties of photonic crystal double heterostructure resonant cavities. *Optics Express*, **16**(13):9391–9397, 2008.

[45] Adam Mock, Wan Kuang, M. H. Shih, E. H. Hwang, J. D. O'Brien, and P. D. Dapkus. Spectral properties of photonic crystal double-heterostructure resonant cavities. In *Laser and Electro-Optics Society Annual Meeting Technical Digest*, page ML4, Montreal, Cananda, October 30-November 2 2006.

[46] Adam Mock, Ling Lu, Eui Hyun Hwang, John O'Brien, and P. Dan Dapkus. Modal analysis of photonic crystal double-heterostructure laser cavities. *Journal of Selected Topics in Quantum Electronics*, **15**(3):892–900, 2009.

[47] Adam Mock and John D. O'Brien. Direct extraction of large quality factors and resonant frequencies from padé interpolated resonance spectra. *Optical and Quantum Electronics*, **40**(14):1187–1192, 2009.

[48] Adam Mock and John D. O'Brien. Strategies for reducing the out-of-plane radiation in photonic crystal heterostructure microcavities for continuous wave laser applications. *IEEE Journal of Lightwave Technology*, **28**(7):1042–1050, 2010.

[49] Masato Morifuji, Yousuke Nakaya, Takashi Mitamura, and Masahiko Kondow. Novel design of current driven photonic crystal laser diode. *IEEE Photonics Technology Letters*, **41**(9):1131–1141, 2009.

[50] M. Nakamura, A. Yariv, H. W. Yen, S. Somekh, and H. L. Garvin. Optically pumped gaas surface laser with corregation feedback. *Applied Physics Letters*, **22**(10):515–516, 1973.

[51] Susumu Noda, Masayuki Fujita, and Takashi Asano. Spontaneous-emission control by photonic crystals and nanocavities. *Nature Photonics*, **1**:449–458, 2007.

[52] Susumu Noda, Katsuhiro Tomoda, Noritsugu Yamamoto, and Alongkarn Chutinan. Full three-dimensional photonic bandgap crystals at near-infrared wavelengths. *Science*, **289**(5479):604–606, 2000.

[53] Masahiro Nomura, Satoshi Iwamoto, Katsuyuki Watanabe, Naoto Kumagai, Yoshiaki Nakata, Satomi Ishida, and Yasuhiko Arakawa. Room temperature continuous-wave lasing in photonic crystal nanocavity. *Optics Express*, **14**(13):6308–6315, 2006.

[54] M. Notomi, A. Shinya, S. Mitsugi, E. Kuramochi, and H.-Y. Ryu. Waveguides, resonators and their coupled elements in photonic crystal slabs. *Optics Express*, **12**(8):1551–1561, 2004.

[55] Masaya Notomi, Akihiko Shinya, Koji Yamada, Jun-ichi Takahashi, Chiharu Takahashi, and Itaru Yokohama. Structural tuning of guiding modes of line-defect waveguides of silicon-on-insulator photonic crystal slabs. *Journal of Quantum Electronics*, **38**(7):736–742, 2002.

[56] Kengo Nozaki, Shota Kita, and Toshihiko Baba. Room temperature continuous wave operation and controlled spontaneous emission in ultrasmall photonic crystal nanolaser. *Optics Express*, **15**(12):7506–7514, 2007.

[57] Makoto Okano and Susumu Noda. Analysis of multimode point-defect cavities in three-dimensional photonic crystals using group theory in frequency and time domains. *Phys. Rev. B*, **70**(12):125105, Sep 2004.

[58] Yan Pan, Prabhat Kumar, John Kim, Gokhan Memik, Yu Zhang, and Alok Choudhary. Firefly: Illuminating future network-on-chip with nanophotonics. In *International Symposium on Computer Architecture*, pages 429–440, 2009.

[59] Hong-Gyu Park, Se-Heon Kim, Min-Kyo Seo, Young-Gu Ju, Sung-Bock Kim, and Yong-Hee Lee. Characteristics of electrically driven two-dimensional photonic crystal lasers. *IEEE Journal of Quantum Electronics*, **41**(9):1131–1141, 2005.

[60] Kazuaki Sakoda. *Optical Properties of Photonic Crystals*. Springer, Germany, 2001.

[61] M. H. Shih, W. J. Kim, Wan Wuang, J. R. Cao, H. Yukawa, S. J. Choi, J. D. O'Brien, and W. K. Marshall. Two-dimensional photonic crystal mach–zehnder interferometers. *Applied Physics Letters*, **84**(4):460–462, 2004.

[62] M. H. Shih, Wan Kuang, Tian Yang, Mahmood Bagheri, Zhi-Jian Wei, Sang-Jun Choi, Ling Lu, John D. O'Brien, and P. Daniel Dapkus. Experimental characterization of the optical loss of sapphire-bonded photonic crystal laser cavities. *Photonics Technology Letters*, **18**(3):535–537, 2006.

[63] C. L. C. Smith, D. K. C. Wu, M. W. Lee, C. Monat, S. Tomljenovic-Hanic, C Grillet, B. J. Eggleton, D. Freeman, Y. Ruan, S. Madden, B. Luther-Davies, H. Giessen, and Y.-H. Lee. Microfluidic photonic crystal double heterostructures. *Applied Physics Letters*, **91**:121103–1–121103–3, 2007.

[64] H. Soda, K. Iga, C. Kitahara, and Y. Suematsu. Gainasp/inp surface emitting injection lasers. *Japan Journal of Applied Physics*, **18**(12):2329–2330, 1979.

[65] Bong-Shik Song, Takashi Asano, and Susumu Noda. Heterostructures in two-dimensional photonic-crystal slabs and their application to nanocavities. *Journal of Physics D*, **40**:2629–2634, 2007.

[66] Bong-Shik Song, Susumu Noda, Takashi Asano, and Yoshihiro Akahane. Ultra-high-Q photonic double-heterostructure nanocavity. *Nature Materials*, **4**:207–210, 2005.

[67] K. Srinivasan and O. Painter. Momentum space design of high-Q photonic crystal optical cavities. *Optics Express*, **10**(15):670–684, 2002.

[68] Allen Taflove and Susan C. Hagness. *Computational electrodynamics*. Artech House, Massachusetts, 2000.

[69] Y. Tanaka, T. Asano, and S. Noda. Design of photonic crystal nanocavity with Q-factor of $\sim 10^9$. *Journal of Lightwave Technology*, **26**(11):1532–1539, 2008.

[70] Michael Tinkham. *Group Theory and Quantum Mechanics*. Dover Publications, Inc., New York, 1964.

[71] Rodney S. Tucker. The role of optics and electronics in high-capacity routers. *Journal of Lightwave Technology*, **24**(12):4655–4673, 2009.

[72] Yurii A. Vlasov, Martin O'Boyle, Hendrik F. Hamann, and Sharee J. McNab. Active control of slow light on a chip with photonic crystal waveguides. *Nature*, **438**:65–69, 2005.

[73] J. Vučković, M. Lončar, H. Mabuchi, and A. Scherer. Optimization of the Q factor in photonic crystal microcavities. *Journal of Quantum Electronics*, **38**(7):850–856, 2002.

[74] Y. Wakayama, A. Tandaechanurat, S. Iwamoto, and Y. Arakawa. Design of high-Q photonic crystal microcavities with a graded square lattice for application to quantum cascade lasers. *Optics Express*, **16**(26):21321–21332, 2008.

[75] Amnon Yariv and Pochi Yeh. *Optical Waves in Crystals*. Wiley, New Jersey, USA, 2003.

[76] T. Yoshie, A. Scherer, J. Hendrickson, G. Khitrova, H. M. Gibbs, G. Rupper, C. Ell, O. B. Shchekin, and D. G. Deppe. Vacuum rabi splitting with a single quantum dot in a photonic crystal nanocavity. *Nature*, **432**:200–203, 2004.

In: Photonic Crystals
Editor: Venla E. Laine, pp. 245-275

ISBN: 978-1-61668-953-7

Chapter 11

NONLINEAR PHOTONIC CRYSTALS OF STRONTIUM TETRABORATE

A.S. Aleksandrovsky[1,2], A.M. Vyunishev[1,2] and A.I. Zaitsev[1,2]

[1] L.V. Kirensky Institute of Physics SB RAS, Krasnoyarsk, Russia.
[2] Siberian Federal University, Krasnoyarsk, Russia.

Abstract

Structural, optical and nonlinear optical properties of strontium tetraborate (SBO) single crystals are summarized. SBO presently is not considered to be a ferroelectric, however, domain structures consisting of alternating oppositely poled domains, strongly resembling those present in such ferroelectrics as potassium titanyl phosphate and lithium niobate, were recently discovered in this crystal. Such geometrical properties as orientation, size, and degree of randomization, of these domain structures are described. The problem of origin of domain structures in SBO is considered, as well as the possibility of their characteristics control. From the point of view of optical properties, domain-structured samples of strontium tetraborate are classified as randomized nonlinear photonic crystals. Comparative study of nonlinear diffraction and random quasi-phase-matching of nanosecond and femtosecond laser pulses in these nonlinear photonic crystals is presented. Prospects of creation of nonlinear optical converters of laser radiation into VUV spectral region based on domain structured strontium tetraborate are discussed.

Introduction

The media with homogeneous linear optical properties and spatially modulated nonlinear optical ones are classified as nonlinear photonic crystals (NPC) [1]. The structures of this kind can be produced, for instance, in ferroelectrics due to spontaneous or artificial reversal of the polar axis of the crystal structure, leading to formation of ferroelectric domains, while the linear optical properties remain unchanged except for narrow areas in the vicinity of the domain walls. 1D structures in ferroelectrics were extensively studied as nonlinear optical converters of laser radiation [2]. Recently 1D NPC were detected in strontium tetraborate that is not known to be a ferroelectric, but is of great interest due to its possible nonlinear optical applications in the vacuum ultraviolet (VUV) region of spectrum.

In present days, nonlinear optics is a well-developed area of science with many achievements being introduced into industrial products. Existing crystalline nonlinear media allow efficient conversion of near infrared and visible laser radiation into shorter wavelengths, including near ultraviolet (UV) spectral range [3]. Most widely used nonlinear media are potassium titanyl phosphate (KTP) and lithium niobate. These crystals possess both high nonlinear coefficients and large enough birefringence necessary for achieving angular phase matching. Their transparency in the UV is limited to wavelengths above 0.33 – 0.35 μm, and for conversion to shorter wavelengths such crystals as those of KDP and borates families are used, the latter being preferable due to higher nonlinear susceptibility. Most widely investigated borates, in their turn, possess some limitations connected with angular phase matching requirements. For instance, direct generation of the second harmonic in beta barium borate (β-BBO) is possible down to wavelengths of order 210 nm, with effective nonlinear coefficient approaching zero. Alteration of phase matching conditions is possible with the help of quasi-phase-matching [2] that is widely used with KTP and lithium niobate. This method of phase matching is attractive because it allows employing the highest components of nonlinear susceptibility tensor for a given crystal. Quasi-phase-matching requires the presence of alternating domains with opposite directions of static polarization. Such domain structures can be most easily obtained in ferroelectrics. The only known ferroelectric among crystals transparent in the VUV region is $BaMgF_4$, but its nonlinear susceptibility is very low [4].

Strontium tetraborate (SBO) is not presently known to be a ferroelectric. Evident proofs for existence of ferroelectric domains in any other borates were not reported, too. Surprisingly, our recent studies revealed the existence of domain structures in as-grown crystals of SBO [5]. These structures fairly satisfy the definition of NPC stated above. In the present chapter we discuss the properties of SBO domains and the results of nonlinear optical studies of NPC structures in SBO.

Overview of SBO Properties

Polycrystalline SBO was synthesized for the first time in 1964 [6] via solid-state reaction. Its crystalline structure belongs to orthorhombic symmetry class *mm2*. Space group most widely used in recent papers concerning this crystal is $Pnm2_1$, though it is not a standard one [7]. For the sake of consistency we will use this non-standard group, too. Unit cell parameters for SBO reported by various researchers slightly differ. Our measurements were carried out on Bruker SMART APEX II single crystal diffractometer using Mo K_α radiation. The unit cell parameters were found to be: a = 4.4145(4) Å, b = 10.6827(10) Å, c = 4.2234(4) Å. The symmetry of the unit cell assumes that SBO will possess pyroelectric and nonlinear optical properties. The direction of static polarization of the crystal in the chosen space group coincides with c crystallographic axis. Single crystals of SBO were obtained in [8], and their electro-optical coefficients were measured. Maximum value of electro-optical coefficient was found to be 1.61 pm/V, which is noticeable but rather impractical one. Typically the single crystals of SBO could be obtained by Czochralski method [8, 9, 5] due congruent character of melting of starting materials. Kyropoulos method was also employed for SBO growth in [10].

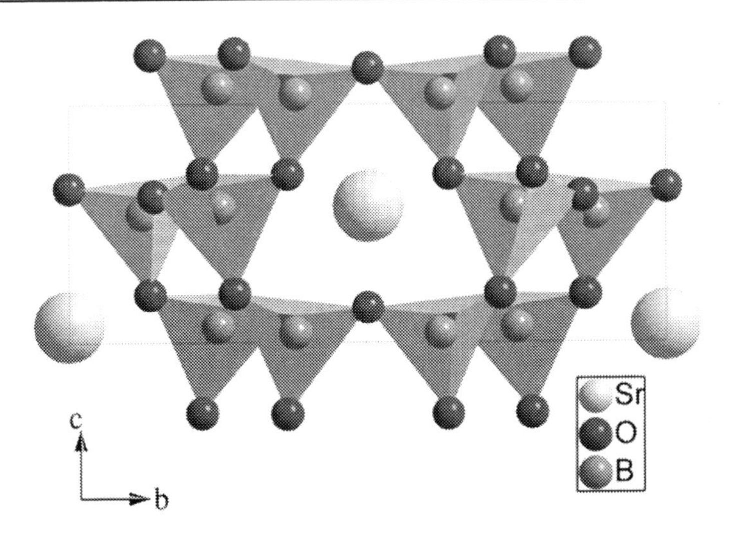

Figure 1. The crystal structure of SBO viewed in the direction of *a* axis. Notation of the axes corresponds to $Pnm2_1$ space group.

The feature of SBO crystal structure (figure 1) is the presence of exclusively tetrahedrally coordinated boron ions forming a continuous B – O network. The crystalline structure of the most known borates contains whether both BO_3 and BO_4 groups or exclusively BO_3 groups. Among crystals containing exclusively BO_4 groups, such as $R^{5+}BO_4$ (R^{5+} denotes pentavalent cation), continuous polyanion network is rarely present. Only several crystals like lead tetraborate (PBO) and europium tetraborate, both belonging to the same structural type as SBO, copper metaborate (CuB_2O_4), γ-HBO_2 and recently discovered orthorhombic bismuth triborate (δ-BiB_3O_6) [11, 12] contain exclusively networked BO_4 groups. In borate glasses commonly both BO_3 and BO_4 groups are present, too, the ratio of them being dependent on additional components of the glass. In case of SBO we can see some consequences of structural feature mentioned above. In particular, SBO is characterized by unusually small volume per ion, or, in other words, demonstrates very dense packing of the constituents. The velocities of bulk acoustic waves in SBO crystal differ markedly from those calculated by the well-known empirical formula [13] relating the BAW velocity in oxides to the ratio of the melting point to the average atomic mass of the material. SBO is non-hygroscopic, in contrast to such widely used borate as lithium triborate (LBO). The microhardness of SBO is noticeably higher than in other borates. SBO demonstrates stability against color center formation under X-ray irradiation. We observed that exposure of SBO crystals to X-rays with energy up to 40 keV in Bruker S4 Pioneer fluorescent spectrometer during the measurement times up to several tens of minutes does not lead to an observable coloration, while the same conditions in the case of huntite-structured borates produce very intense coloration. Presence of BO_4 groups influences even the luminescent properties of SBO doped with rare earths [14, 15], reducing concentration quenching and enhancing the quantum yield. The commercial dark (ultraviolet) phosphor Eu:SBO owes its good operational properties largely due to this effect.

The features of SBO that make it attractive for nonlinear optics are the transparency at the wavelengths below 200 nm [9, 16], large values of second order nonlinear coefficients [16, 17] and high radiation damage threshold. According to measurements made for 0.1 mm thick samples [16], the fundamental absorption edge in SBO lies at 130 nm. However, absorption at 125 nm is not large enough to make observation of nonlinearly generated signal at this wavelength completely impossible. A relatively weakly pronounced transparency window could be found even at wavelengths as short as 120 nm. The components of SBO second order susceptibility tensor were measured in [16, 17] (see table 1). For clarity, we use direct notation of components through the crystallographic axes, as it was used in [18], and notation of [16] is given in brackets. The difference in values measured with femtosecond and nanosecond pulses can be attributed to the influence of group velocity dispersion that may affect measurements with short pulses. According to our measurements, nonlinear coefficient d_{ccc} of SBO is the highest among borates. However, the drawback that hinders the application of SBO in nonlinear optics is its small birefringence that leads to the absence of angular phase matching. The search for the possibility of creation of domains with different value or sign of nonlinear susceptibility is necessary to overcome this drawback.

Domains in SBO

The possibility of a regular domain structures creation in the crystals used in nonlinear optics is based, particularly, on the existence of two stable positions of some ions inside the unit cell of a crystal. The switching between these positions can be performed at the elevated enough temperatures under whether a static or a pulsed electric field. Formation of the domain structures, commonly with randomness in the domain size, is also possible in crystals as a result of phase transition from a higher symmetric (commonly non-ferroelectric) to a lower-symmetric (ferroelectric) phase, and is influenced by inhomogeneities of chemical content formed during the crystal growth. The latter case is extensively studied for lithium niobate (see, e.g. [19] and references therein). It was shown that during the growth of lithium niobate crystal in the presence of impurities the concentration of the latter in the crystal becomes spatially modulated. The scale of this modulation can be controlled by the growth conditions. After the end of the growth procedure the crystal is in a paraelectric phase and, consecutively, with no domains. The cooling of the as-grown crystal to the room temperature makes it to undergo the phase transition into the ferroelectric phase. The orientation of static polarization after this phase transition appears to be dependent on the concentration of impurity, so that domains are formed with domain walls coinciding with the maxima and minima of impurity distribution. In case when concentration of intended impurity in lithium niobate is negligible, domains can nevertheless be formed from non-congruent melt of stoichiometric content, due to small inhomogeneity of relative concentration of the main constituents [20]. In both cases domain walls are obligatorily perpendicular to the surface between the crystal and the melt.

SBO is not presently considered to be a ferroelectric. However, the earlier study [10] mentioned that angular dependence of second harmonic in SBO demonstrated some unspecified nontrivial behavior. One of the possible explanations for such behavior could be, to our opinion, the existence of some superstructure inside the crystal, which, as it was established for huntites, may affect angular behavior of second harmonic output [21]. In order to investigate this possibility, we performed the study of SBO growth and search for the domains [5].

Table 1. Nonlinear coefficients of strontium tetraborate (SBO).

Nonlinea coefficient	Value of nonlinear coefficient (pm/V)	
	Petrov et al [15] (800 nm, 50 fs)	Zaitsev et al [16] (1.064 μm, 15 ns)
$d_{caa}(d_{31})$	0.8	1.7
$d_{ccc}(d_{33})$	1.5	3.5
$d_{aac}(d_{15})$		0.9
$d_{bbc}(d_{24})$		0.7
$d_{cbb}(d_{32})$	1.1	2.0

The single crystals of strontium tetraborate were grown by Czochralski method. The starting materials used were $SrCO_3$ (>99.5%) and H_3BO_3 (>99.8%). We have chosen *b* axis as the vertical growth direction. The starting temperature on the melt surface was 1000°C, the seed-rotation rate was 10 rotations per minute, the pulling rate was 2.4 mm/day, and the cooling rate was 1° C/day. After five days a transparent colorless SBO single crystal was obtained with an approximate size of 30 x 20 x 30 mm³ along *a*, *b* and *c* axes. This crystal had several rather well developed facets; the facets corresponding to the planes (100), (010) and (011) were especially pronounced.

It is well known that chemical etching reveals the domain structure of the crystals. The maximum selectivity for SBO is expected on (001) plane of the crystal. As-grown crystal was lacking this facet, and we made it by polishing. Then the crystal was etched in 5 wt. % water solution of nitric acid at 95°C during 10 minutes. A part of this crystal after etching is depicted in figure 2.

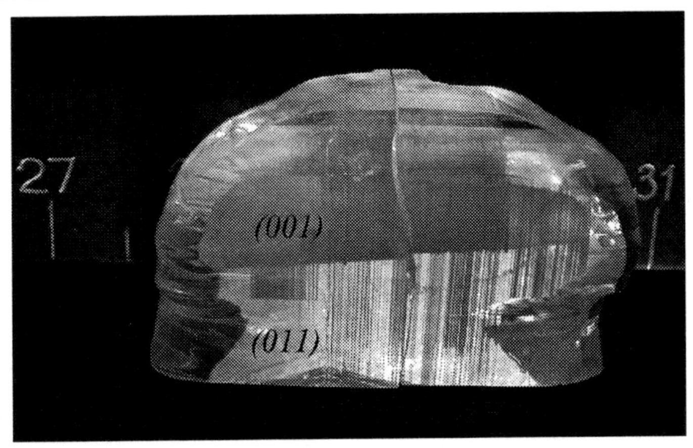

Figure 2.[i] A part of SBO crystal after the etching.

The front facet corresponds to (011) plane, and the facet above it corresponds to (001). The etching stripes are present on both surfaces. Generally these stripes protrude across the entire surface of (011) plane along [010] direction, while in the left part of the facet the abrupt disappearance of the stripes is seen. In order to enable further microscopic investigations, we have cut the sample oriented along the main crystallographic axes from the crystal described above. The dimensions of the sample were 5 mm along *a* axis, 11 mm along *b* axis and 9 mm

[i] Reprinted from Ref. 5. Copyright (2008), with permission from Elsevier.

along *c* axis. After optical polishing the sample was investigated under polarizing microscope. No any evident inhomogeneity was observed, admitting that refraction index is constant in the whole sample. After the repeated etching of the sample the etching stripes appeared again.

We examined the etching patterns of (100) and (010) planes of the SBO sample (these planes are free from etching stripes). These patterns are the same for both opposite faces perpendicular to *a* and *b* axes. However, they quite differ from etching patterns on the faces perpendicular to *c* axis. The etching patterns on two opposite faces perpendicular to *c* axis of SBO sample containing the etching stripes are shown in figure 3. One of the faces exhibit strong etching and must be attributed to *c*+ direction, while opposite face (*c*-) remains almost unetched. The latter can be seen from a comparison of figures 3a and 3b. The right part of the (001) face in figure 2b is weakly etched, and the scratches after incomplete polishing are well seen. The right part of the same facet in figure 3a is strongly etched, so that no scratches have been remained. Note that these facets that are separated by 9 mm along *c* axis are very similar but their etching patterns are inverted with respect to the etching behavior.

The repeated growth experiments show that the formation of domain structures similar to that described above is reproducible and happens every time when the growth's conditions specified above are fulfilled. The comparison of the properties of the domain structures obtained in a separate growth experiments will be presented below in this chapter, in the section devoted to the nonlinear optical studies of these structures.

The stability of the stripe pattern along *b* axis was investigated at larger magnifications of an optical microscope. A milder etching regime was used for these studies; namely, the etching was performed at 30°C during 1 minute. The etching depth was determined by Linnik interferometer and was found to be equal to 80 nanometer. Figure 4 depicts the etching patterns separated by 250 micrometer along *b* axis. Both patterns almost ideally coincide with each other. As far as the optical microscopy resolution allows, we could find on the (001) facet the areas of only two types, namely, almost unetched areas and the areas with the fast etching rate.

The etching behavior observed in SBO is typical for 180° domain structures in ferroelectric crystals like KTP and lithium niobate. On the other hand, the observed etching patterns can not be ascribed to any other planar structural defects such as stacking faults, because figures 3 and 4 evidently show three-dimensional geometry of neighboring blocks inside the crystal. Comparison of figure 3a and figure 3b reveals that the structure in a crystal protrudes inside the crystal to the depths of order of at least a centimeter. The existence of only two types of etched areas on the surfaces perpendicular to polar axis *c* implies that these areas correspond to the opposite directions of the static polarization of the crystal. We deduce, then, that our sample of strontium tetraborate contains domains with opposite orientation of polar axis. Domains have the shape of sheets with the domain walls perpendicular to the *a* axis. The shape and orientation of domains in SBO is identical to those in KTP, which belongs to the same point symmetry class. The domain structure in SBO can spread to the depths of order of centimeter along both *b* and *c* axes. The widths of domains can vary from several tens of micrometers down to several tenths of micrometer. Thinner domains are suspected to be present inside the domain structure, but they cannot be surely observed and measured via optical microscopy. Evidently, the domain structures, being well-ordered as concerned domain orientation and preservation of thickness of certain domain along coordinates in *bc* plane, is at the same time strongly randomized as concerned the variation of thickness between neighboring domains in the direction of *a* axis. These conclusions are supported by the results of nonlinear optical studies of the crystal containing domain structure that will be described below in this chapter.

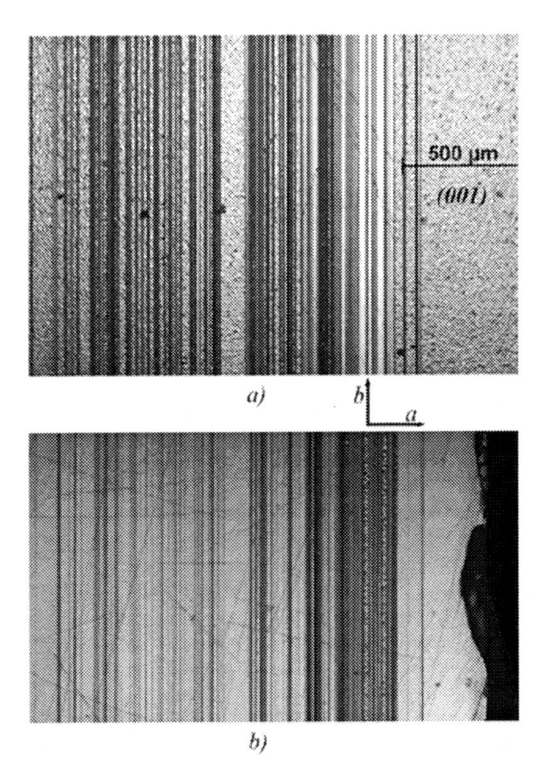

Figure 3.[i] The etching patterns of 9 mm thick SBO sample for two opposite planes perpendicular to *c* axis. The coordinate system shows the direction of crystallographic axes.

One of important characteristics of domain structures is the domain wall thickness. The transition from one domain to another can be considered as a kind of crystal structure defect. When domains are formed via electric field switching, the static polarization does not change its direction suddenly, at the thickness of one atomic layer. The same takes place in case of growth domains in lithium niobate [20]. Moreover, the quality of domain walls in pure lithium niobate is rather poor, with strong deviations from planar shape. The figure 5 presents the electron microscopic image of a part of etched domain structure in SBO taken using JEOL JSM7001F with 30000 magnification. To our opinion, the inhomogeneity of the etched surface of a domain must be attributed to the character of etching process and not to the inhomogeneity of the domain itself. The planarity of domain walls in SBO is much better than in lithium niobate, and no signs of finite thickness of the domain walls could be found at the magnification used.

The origin of the domain structure found in SBO is of a special interest. The stable formation of regular domain structure in ferroelectric lithium niobate was observed in [20] while growing the crystals from the incongruent melt of stoichiometric composition. The domain structures obtained in these conditions were definitely connected with the growth striations induced by intentional asymmetry of temperature field with respect to the seed rotation axis. Varying the rotation rate during the process of growth could control the period of the regular domain structure in lithium niobate. The domain walls are of cylindrical shape as determined by the rotational geometry. The formation of domain structure is assumed to happen during post-growth cooling from the growth temperature of lithium niobate (1260°C) experiencing phase transition to the lower-symmetry ferroelectric phase at 1210°C.

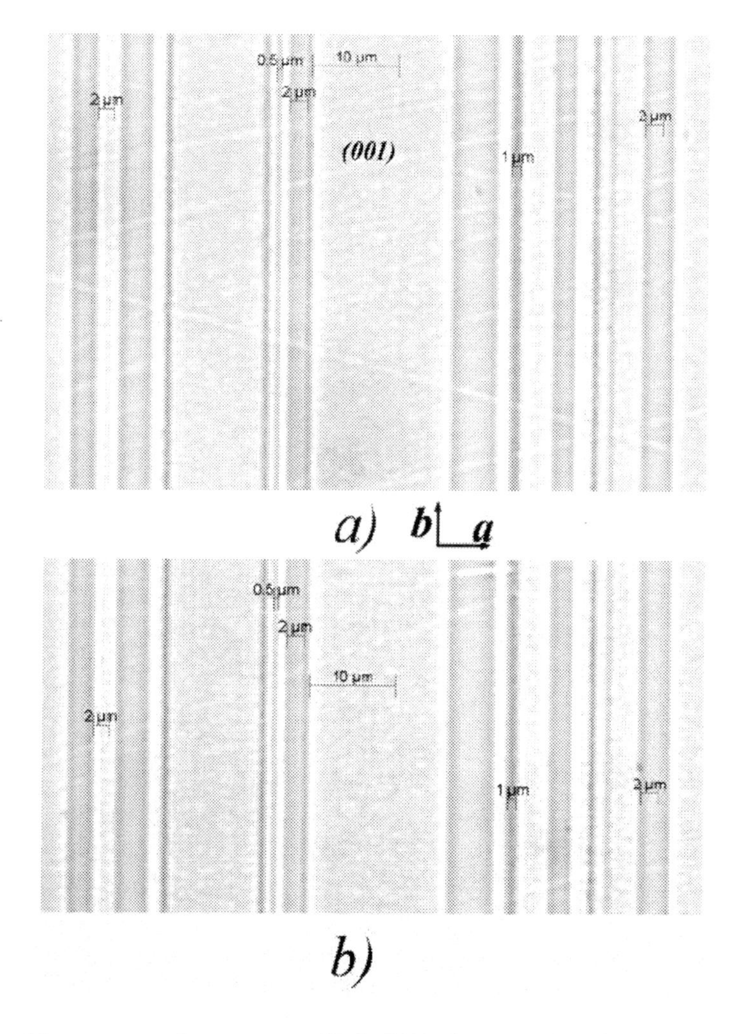

Figure 4.[i] The etching patterns of two areas at SBO (001) plane separated by 250 micrometers along *b* axis. Coordinate system shows the direction of crystallographic axes.

In case of strontium tetraborate surface growth striations can be observed, however, their observability depends on the conditions of growth. We observed surface growth striations in another growth experiment with SBO. The obtained sample showed weak surface growth striations looking like concentric rectangles around the seed, with the sides along *a* and *c* axes. In any case, domain walls are oriented orthogonally to the possible growth striations characteristic for SBO. Thus, no correlation can be found between the growth striations in SBO and the orientation of domain walls. We can conclude that the domains in strontium tetraborate are formed via the mechanism quite different from that in lithium niobate. From another side, strontium tetraborate is not known to be a ferroelectric, and no any phase transitions are known for this crystal. Hence, domains cannot form in the course of any known phase transition from one solid-state phase into another one. However, the initial formation of domain seeds is possible in the course of phase transition from isotropic melt to lower-symmetry crystalline phase. On the surface of growth facet that is not perpendicular to the static polarization direction, the positions and the order of atoms adjoining from the melt into the crystal structure is governed not only by the potential minima in the nearest-neighbors crystal field, but also by the projection of the static polarization on the surface, i.e. by the contribution of the crystal field from other atoms in a unit cell. Generally, one of these factors may dominate, but if both of them are of comparable influence, than certain instability of adjoining atoms positions will exist. More stable geometry of growing crystal in these conditions will be domain structure consisting of alternating oppositely poled domains. This structure will possess reduced value of static polarization projection onto the growth facet, and hence, the source of instability will be eliminated. Note that the mechanism under consideration leads to the formation of domain walls that are obligatorily not parallel to the growth facet, in the contrast to the case of lithium niobate. This conclusion fairly agrees with results presented above.

Figure 5. The electron microscopic image of etched domain in SBO (JEOL JSM7001F, at 30000 magnification).

As one can deduce from figure 2, the range of conditions for instability leading to formation of domain structure is very narrow. The complete elimination of the source of instability would take place if oppositely poled domains would have the thickness equal to the unit cell dimension along the growth facet. The existence of much more thick domains means that inhomogeneity of some parameters on the crystal surface and in the melt near it may lead whether to the formation in a chosen part of the growing surface a single crystalline part of the growing crystal or to the formation of inverted domain. At the present state of our technology conditions for the formation of narrowest possible domains are not fulfilled at the whole surface of growth facet. The physical quantities that locally favor or suppress formation of inverted domains cannot be definitely specified at the present stage of our investigations. The most probably these are value of temperature near the growth facet or value of temperature gradient. The time interval during the crystal growth when instability most efficiently plays its role for the domain structure formation is the time when the growth facet, for instance, (011) facet that initially was absent at the seed crystal, is formed.

Partial proof for this picture of domains formation is the result of the following growth experiment. We have obtained single crystal of $(Sr,Ca)B_4O_7$ from the melt with addition of 20% of calcium. XRD study has established the same symmetry group of this crystal as for SBO. Domain structure could be found in the grown crystal; however, this structure was observable even without etching, due to evident inhomogeneity of the refraction index, in contrast to pure SBO with no signs of this inhomogeneity. After etching, the refraction index variation in $(Sr,Ca)B_4O_7$ was found to closely reproduce the etching pattern revealing domain structure. The evident explanation for this result is the gradient of Sr/Ca ratio inside the domains. This result admits, first of all, that the formation of a domain structure takes place at the moment of the crystallization from the melt, but not in the moment of some undetected phase transition below the temperature of crystallization. Next to that, it evidences the presence of static polarization field on the surface of the growing crystal and its influence on the distribution of Sr/Ca ratio along the growth facet, due to different effective charge of these ions. The spatial scale of this influence is determined by the size of domains in the grown crystal.

Another important feature of SBO domain structures is their reproducibility during the repeated growth. Our experiments show that if the properly cut seed initially containing domain structure is used, than the crystal grown on it will contain the domain structure identical to that of the seed. This means that both the monodomain growth and the instability leading to the formation of new domain structures can be avoided by the choice of orientation of the seed and by the proper growth conditions.

Nonlinear optical applications of SBO would require the formation of regular domain structures or the structures with a controllable degree of irregularity. The simplest way to create such structures could be the polarization switching in SBO. Examination of SBO crystal structure (see figure 1) makes us to think that this switching will require complete reconstruction of B-O network around strontium ions that could be possible only at temperatures close to the melting point being approximately 1000 C. In other words, instability, which can lead to the formation of inverted domains during the crystal growth process, is hardly preserved, when the densely packed crystal structure is completely formed. The practical implementation of the domain inversion technique after the end of SBO growth process may meet severe experimental difficulties.

Topology of Nonlinear Optical Processes in 1D Nonlinear Photonic Crystals

The main features of nonlinear photonic crystals is the presence of rectangular spatial modulation of nonlinear susceptibility and the absence of any modulation of linear dielectric permeability, the latter being the essential property of common (linear) photonic crystal. The regular 1D modulation allows the description of the structure by introducing a single reciprocal superlattice vector (RSV); the Fourier spectrum of nonlinear susceptibility function on a coordinate in this case contains only main value being inversely proportional to the period of modulation, and its odd overtones. The direction of RSV is perpendicular to the domain wall plane. In a regular nonlinear photonic crystal one can distinguish between two limiting cases of plane wave phase matching. When fundamental plane wave propagates collinearly to RSV, scalar quasi-phase matching will be exclusively observed, and the generated wave is collinear to the fundamental one. Oppositely, when the fundamental wave propagates in the plane of domain walls, a kind of non-collinear phase matching is exclusively observed, which often is called nonlinear diffraction [22]. However, maximal output for these processes for fixed fundamental wavelength is attained at different values of RSV modulus. Between these two cases, in general, both QPM and nonlinear diffraction must be observable, with efficiency swiftly dropping from maximal values with angle detuning.

A randomized nonlinear photonic crystal structure with irregular domain thickness, like that described above for SBO, is characterized by a spectrum of RSV that can contain values necessary for both efficient QPM and nonlinear diffraction. Figure 6 schematically represents the topology of fundamental wave directions where whether RQPM or nonlinear diffraction will dominate. The former dominates inside the cone with the axis coinciding the normal to domain walls, and the latter dominates in the rest of the sphere. The value of the cone angle discriminating between these two areas is dependent on the form of an RSV spectrum.

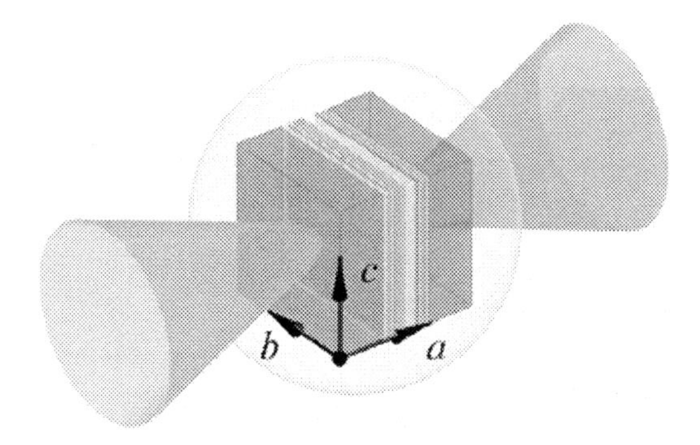

Figure 6. Topology of predominant directions for nonlinear processes in randomized 1D nonlinear photonic crystals. Cone with the axis coinciding the normal to domain walls is the area of the RQPM domination, while the rest of sphere is the area of the nonlinear diffraction domination

Nonlinear Diffraction of 1.064 μm Radiation in SBO

For the nonlinear optical characterization of nonlinear photonic crystal structures in SBO we used the phenomenon of nonlinear diffraction. Two samples obtained in separate growth experiments were studied. One of the samples under study, further referred as the Sample 1, had the following dimensions: 6 mm in the direction of *a* axis, 8 mm in the direction of *b* axis and 4.6 mm in the direction of *c* axis. Linear optical homogeneity of the crystal was verified with the help of a He-Ne laser at 633 nm. No sign of diffraction of this radiation due to the inhomogeneity of the refractive index was observed. Another sample referred as the Sample 2 had the dimensions 20x5x5 mm in the *a, b,* and *c* directions. The etching indicated that continuous area containing a nonlinear photonic crystal structure does not spread over all the volume of both samples, but has the thickness of order of 2 mm in the *a* axis direction. The dimensions of the nonlinear photonic crystal structure in *b* and *c* axes directions coincide with the dimensions of the samples.

Q-switched Nd:YAG laser radiation (several mJ per pulse, 15 ns duration) was used as the source of fundamental wave. This radiation was focused into the Sample 1 through the facet perpendicular to *b* axis by a 10 cm focal length lens to the spot of order of 50 μm that is much smaller than the size of our structure in the *a* axis direction. The polarization of the fundamental beam contained projections along both crystallographic axes *a* and *c*. A non-phase-matched second harmonic beam collinear to the fundamental beam was observed behind the crystal. However, when the propagation path of the fundamental beam falls within nonlinear photonic crystal structure, additional second harmonic beams much more intense than the non-phase-matched SH were observed. Under normal incidence of the fundamental, two groups of beams at equal angles were observed on the left and on the right of the fundamental beam. Each beam group consisted of three beams, two of them with the polarization along the vertical axis (corresponding to *c* axis of the crystal), and the third beam was horizontally polarized (along *a* axis of the crystal). In view of linear homogeneity of the crystal, the observed beams were attributed to the nonlinear diffraction due to inhomogeneity of the second order nonlinear susceptibility of the crystal. This is the additional proof that the crystal contains domains with identical linear optical properties but with different nonlinear optical properties. The absence of beams diffracted outside the horizontal plane (i.e., *ac* plane of the crystal) means that the domains have the form of sheets perpendicular to the *a* axis. However, in the case of periodic inhomogeneity, the RSV spectrum is very narrow, and nonlinear diffraction must be observable in a narrow range of angles near the non-collinear phase matching direction that corresponds to the central value of this spectrum.

On the contrary, rotation of our crystal around *c* axis does not lead to disappearance of the diffracted beams but causes their angular displacement. If the crystal is rotated clockwise when viewed from the above, then left group of beams (when viewed in the direction of propagation of the laser beam) slowly approaches the position of the fundamental beam, while the right group of beams rotates clockwise at a faster angular rate, approaching 90° when crystal is rotated by 45°. The intensity of the beams during rotation varies less than it would be expected for nonlinear diffraction on a periodic structure.

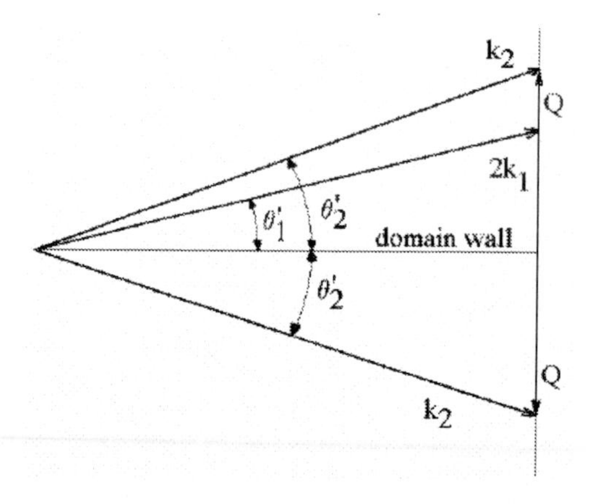

Figure 7. Phase matching diagram for nonlinear diffraction of second harmonic.

The expected angular behavior of the diffracted beams can be easily calculated from the vector condition of the phase matching for nonlinear diffraction (figure 7):

$$\vec{k}_2 = 2\vec{k}_1 \pm \vec{Q} \tag{1}$$

where $\vec{k}_2$ and $\vec{k}_1$ are the wave vectors of the diffracted second harmonic and of the fundamental, respectively, and $\vec{Q}$ is the main value of RSV. In the case of a rectangular dependence of the nonlinear susceptibility on the coordinate with oppositely poled domains of equal thickness d $|\vec{Q}| = \frac{\pi}{d}a,$ $a = 1,3....$. Consider the fundamental beam propagating inside the crystal at the angle $\theta_1^{'}$ to the domain wall (figure 7). The angles of propagation of the beams diffracted on the right and on the left of the fundamental beam, $\theta_2^{R,L}$, measured outside the crystal, must be:

$$\theta_2^{R,L} = \arcsin\sqrt{n_2^2 - n_1^2 + \sin^2\theta_1} \pm \theta_1 , \tag{2}$$

where θ_1 is the angle of the fundamental beam incidence to the crystal facet measured outside the crystal, and $n_{1,2}$ are the refractive indices of the fundamental and the harmonic, respectively. If the fundamental wave contains different polarizations, then the vector $2\vec{k}_1$ must be substituted by $\vec{k}_{1i} + \vec{k}_{1j}$, where *i* and *j* denote different polarizations, and n_1 must be substituted by $(n_{1i}+n_{1j})/2$. Generally, a total of six diffracted beams must exist on each side of the fundamental beam, corresponding to the conversion schemes *aaa, aac, acc, caa, cac,* and *ccc*. Due to the symmetry of the crystal studied, some nonlinear coefficients are zero (namely, *aaa*, *acc* and *cac*), and the number of beams is reduced to three on each side of the fundamental beam. The angles of nonlinear diffraction calculated for $\theta_1 = 0$ using refractive indices from [8], and those measured experimentally, are presented in table 2 and demonstrate good agreement.

The brightest second harmonic spot corresponded to the nonlinear diffraction process using d_{caa}, because polarization of fundamental wave was directed closely to *a* axis. The dependence of the diffraction angle on the angle of incidence of the fundamental for the brightest spot is shown in figure 8. The agreement between the calculations and the experiment is fairly satisfactory again. This proves that the observed phenomenon has a nonlinear diffraction nature and that nonlinear-optically active domain orientations do exist in the crystal under study. However, the diffracted beams observed in a wide range of the angles indicate that the values of $\vec{Q}$ form a much richer spectrum than in the case of a regular domain structure with a constant domain thickness. It is clear that domains have random thickness, so that the Fourier spectrum of the dependence of the second order nonlinear susceptibility $\chi^{(2)}$ on the coordinate in the direction perpendicular to the domain walls is a continuum of values rather than discrete harmonics. The boundaries of this spectrum of $Q^{L,R}$

can be estimated from the smallest and largest observed angular positions of the left and the right diffracted beams using expression:

$$Q^{L,R} = \frac{4\pi}{\lambda}\left[\sqrt{n_2^2 - n_1^2 + \sin^2\theta_1} \mp \sin\theta_1\right] \tag{3}$$

These values are found to be equal to approximately π/5.6 (for diffraction of the left beam at 45 deg rotation angle of the sample) and π/0.18 μm^{-1} (for diffraction of the right beam at 45 deg rotation angle of the sample) respectively, i. e. the effective domain widths contributing to the nonlinear diffraction to the right beam lie in the range from 180 nm to 5.6 μm.

On the base on the measured dependence of the diffraction angles on the angle of incidence, one can expect that the average thickness of domains in the studied sample would be of order of several microns. This allows one to predict that some kind of quasi-phase-matching in randomized structure is expected to be found when the average value of domain thickness is larger than the coherence length. The typical values of coherence length for the second harmonic generation in SBO are above 10 μm for the fundamental wavelengths above 1 μm, but they become shorter than 3 μm for the fundamental wavelengths less than 0.6 μm. The investigation of QPM in our samples will be discussed later in another subsection of this chapter.

Table 2. Calculated and measured nonlinear diffraction angles for 1.064 μm fundamental wavelength

Conversion scheme and nonlinear coefficient	aac, 15	ccc, d33	caa, d31
Calculated nonlinear diffraction angle	13.4°	14.5°	15.3°
Measured nonlinear diffraction angle	13.4°	14.2°	15.4°

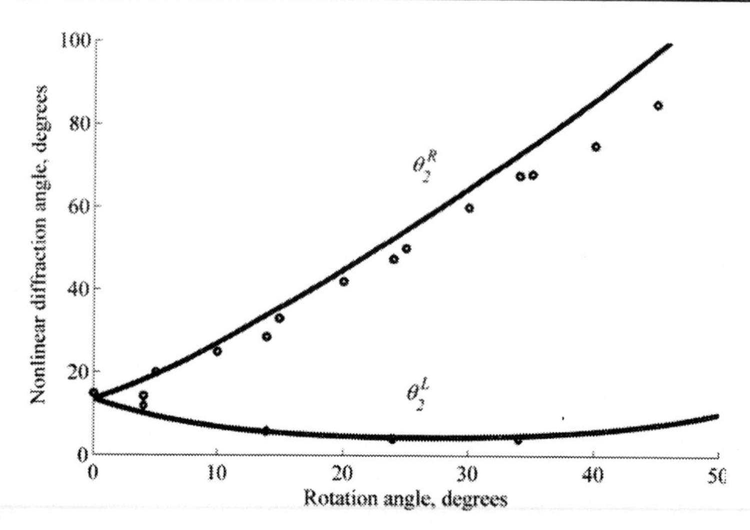

Figure 8. The dependence of nonlinear diffraction angles upon the crystal rotation angle for Sample 1. Solid lines are the calculations; circles and diamonds are the experimental data obtained for the right-hand beam and for the left-hand beam, respectively.

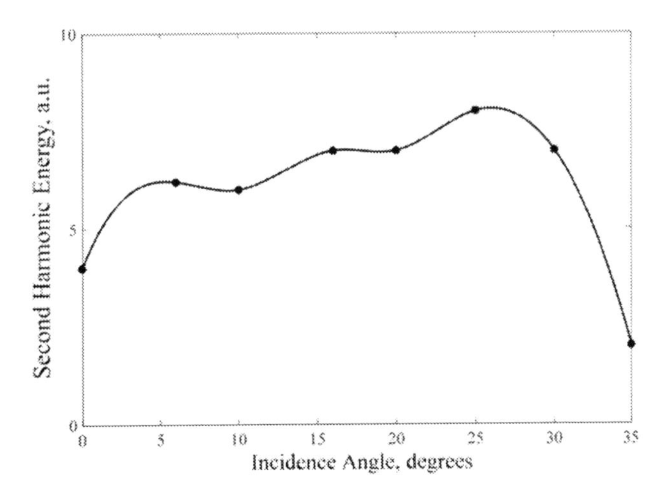

Figure 9. The dependence of diffracted second harmonic energy upon the angle of incidence onto the input facet of the Sample 1 (*ac* plane of the crystal). $\theta_1 = 0$ corresponds to $Q^{L,R} = \pi$ μm^{-1}, and $\theta_1 = 35^o$ corresponds to $Q^R = \pi / 0.22$ μm^{-1}.

The dependence of the diffracted second harmonic intensity on the rotation angle is shown in figure 9. This dependence reflects the behavior of the function $\chi^{(2)}(Q)$, being the Fourier-transform of the $\chi^{(2)}(x)$ function. This dependence slowly varies in the region from zero to 30°, after which it begins to drop noticeably. The presence of the wide spectrum of Q indicates that nonlinear diffraction at frequencies other than second harmonic can also be observed. Indeed, when the horizontally polarized fundamental and the vertically polarized second harmonic were simultaneously focused into the crystal, both the nonlinearly diffracted second and third harmonic beams were observed. For $\theta_1 = 0$, the experimental external diffraction angle for the third harmonic is equal to 16°. The expected value for this angle can be calculated using the relation obtainable from the vector phase matching diagram similar to that in figure 7:

$$\theta_3^{R,L} = \arcsin\sqrt{n_3^2 - \left[\frac{\sqrt{n_1^2 - \sin^2\theta_1} + 2\sqrt{n_2^2 - \sin^2\theta_1}}{3}\right]^2} \mp \theta_1 \tag{4}$$

The calculated value is equal to 16.5° and fairly agrees with experimental one. Obviously, this result makes us to suggest that randomized nonlinear photonic crystal structures can be more tolerable to the wavelength tuning than strictly periodic structures, not only in the nonlinear diffraction but also in the case of an ordinary nonlinear conversion.

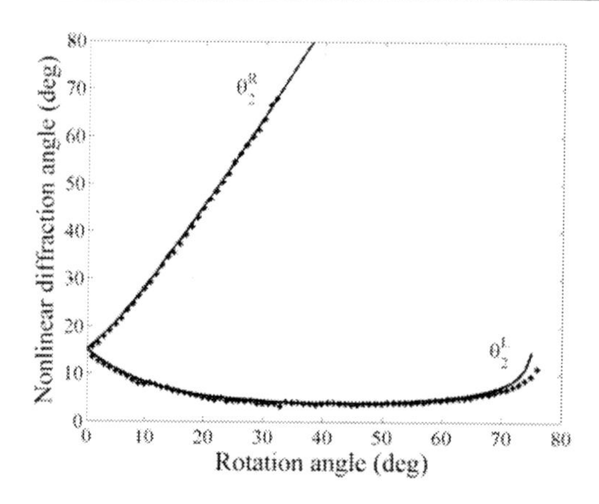

Figure 10. Dependence of nonlinear diffraction angle upon rotation angle of Sample 2 for the right and the left beams (crosses). Solid lines are the calculations

In order to investigate the stability of the nonlinear photonic crystal structure against uncontrollable deviations during the growth, we performed similar study for the Sample 2. The fundamental radiation with divergence 3 mrad and larger pulse energy than in the study of Sample 1 was not focused into the Sample 2, in a contrast to the case of Sample 1. The geometry of propagation was identical to the case of the Sample 1. Fundamental radiation spot covered all the domains in the Sample 2, so that the properties of the whole structure were tested. Polarization of the fundamental strictly coincided to the a axis, so only one spot corresponding to the nonlinear diffraction using d_{caa} was observed. Figure 10 presents the dependence of diffraction angle on the angle of Sample 2 rotation. The agreement with calculations is now much better than in the case of focused beam. The figure 11 presents angular dependences of diffracted energy for the Sample 2. Diffracted signal can be tracked for the values of effective domain widths in the range 0.2 to 7.6 μm.

Figure 12 presents the comparison of diffracted energy for Samples 1 and 2. We see that the spread of corresponding RSV spectrum in both samples is generally the same, while the details of the spectrum can noticeably vary from one growth experiment to another.

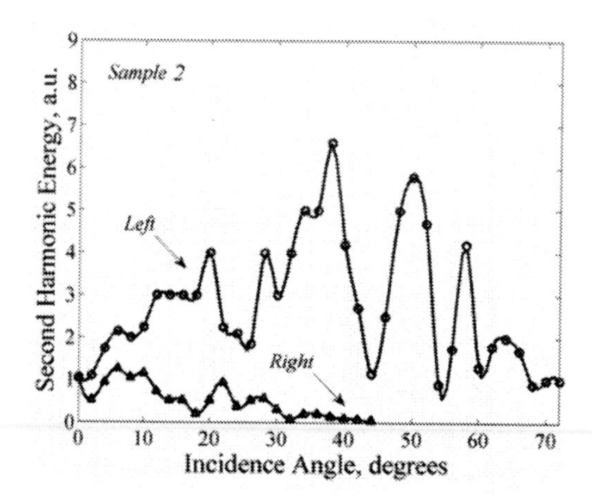

Figure 11. The dependence of nonlinearly diffracted energy for left (circles) and right (triangles) beams in Sample 2.

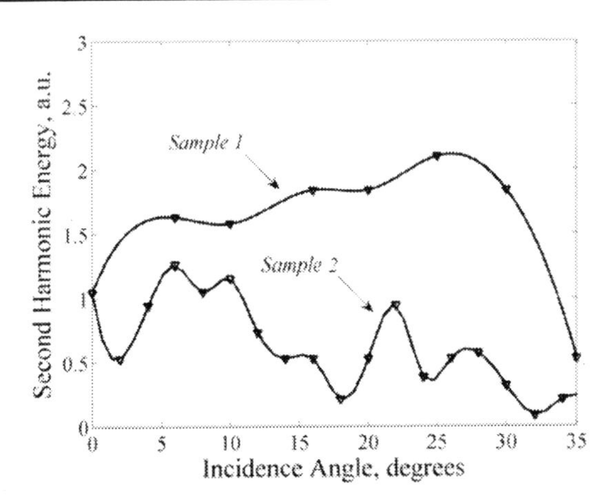

Figure 12. The dependence of nonlinearly diffracted energy for the right beams in the Samples 1 and 2.

Random QPM in SBO

As mentioned in the previous subsection of this chapter, wide RSV spectrum of nonlinear photonic crystal structures in SBO implies that some kind of QPM can be found in them. The peculiarities of quasi-phase-matched nonlinear conversion in randomized domain structures were studied in [23]. It was shown that as the degree of randomization increases, the generated radiation dependence on the length of a randomized medium changes from square law to linear one. The experimental investigation of this effect was done in [24] for difference frequency mixing in polycrystalline ZnSe. It was shown that when the average size of grains becomes close to the coherence length of the nonlinear process, the generated power experiences maximum. The effect is suggested to be called random quasi-phase-matching (RQPM). Further theoretical treatment of RQPM is reported in [25]. Modeling showed that while in any fixed computer generated nonlinear photonic crystal structure the generated power may vary in a random way, the averaging over sufficient number of random structures gives nearly linear dependence of generated power along its length. Of course, if any existing structure is properly geometrically characterized, one can calculate its output following approach used in [24,25] for the plane wave approximation. In this approach the field generated by a stack of domains with random thickness is calculated as the sum of fields generated by every selected domain in the stack, with the account for the phase acquired during the propagation along all the domains between selected one and the detection plane,

$$E_{2\omega} = \sum_{n=1}^{N} \left[\frac{2\omega^2 \chi_n^{(2)}}{k_{2\omega}\Delta k} E_\omega^{\;2} \cdot \left(\exp(i\Delta k d_n) - 1 \right) \cdot \exp\left(i\Delta k \sum_{r=n+1}^{N} d_n \right) \right], \quad (5)$$

where E_ω, and $E_{2\omega}$ are the electric field strengths of the fundamental and generated waves, ω is the fundamental frequency, d_n is the thickness of a selected domain, $\chi_n^{(2)} = (-1)^n \left|\chi^{(2)}\right|$ is its second order nonlinear susceptibility, N is the full number of domains, and Δk is the wave vector mismatch. The sample prepared for the characterization, further referred as the Sample 3, had the dimensions 5 mm along *a* axis direction, 11 mm along *b* axis and 9 mm along *c* axis. Domain structure was revealed by etching, as described above in this chapter, and geometrically characterized via optical microscopy. A part of the domain structure is presented in figure 13. This is the only part of the structure where part of domains disappears on the (001) facet. All the rest domains protrude over all the cross section of the crystal. Overall thickness of domain structure is 2 mm. The structure contains 262 domains with thickness from tens to tenths of μm. Mean value of domain thickness is 8 μm, while overall standard deviation is 4 μm. The thickness of domains in accordance with their position in the structure is presented in the inset of figure 14. Domains with thickness smaller than 0.5 μm could not be definitely identified and measured via optical microscopy. The domain structure observed in the sample for investigation is rather typical for crystal growth of SBO. Our structure is randomized much stronger than structures generated in numerical calculations [25], as can be seen from figure 14.

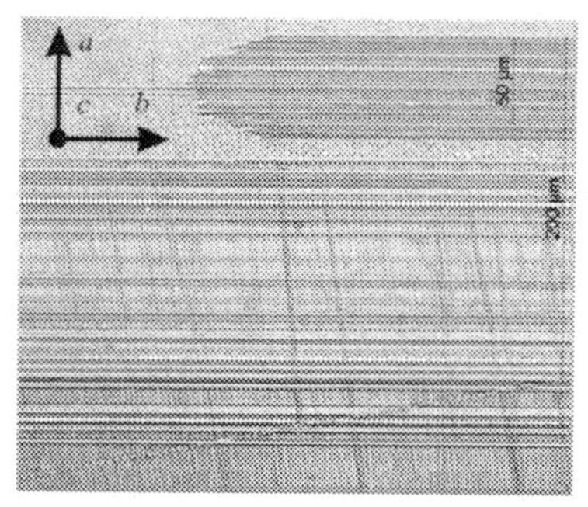

Figure 13. Part of the domain structure in SBO viewed from (001) facet. (Reprinted with permission from Phys. Rev. A, 78, 031802 [2008]. Copyright 2008 by the American Physical Society.)

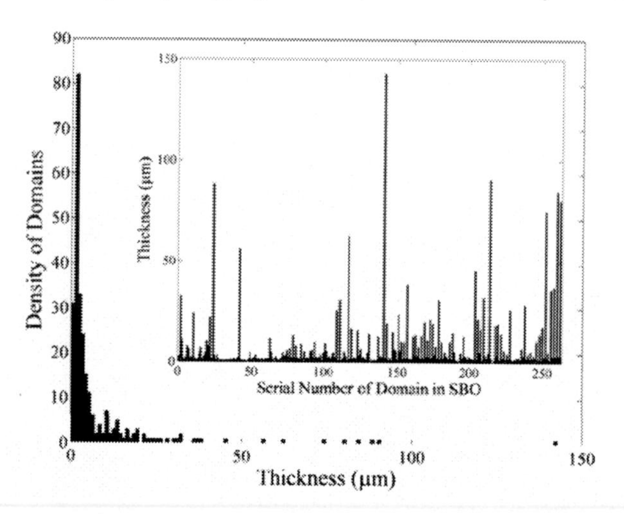

Figure 14. The density of domains as a function of domain size in Sample 3 of SBO crystal. Inset: The thickness of domains as the function of domain number. (Reprinted with permission from Phys. Rev. A, 78, 031802 [2008]. Copyright 2008 by the American Physical Society)

First of all, we used formula (5) to calculate the dependence of second harmonic efficiency in the Sample 3 on the fundamental wavelength. The data of refraction indices dispersion were taken from [8]. The resulting spectral SHG efficiency dependence is depicted in figure 15, together with the dependence of non-phase-matched generation, the latter being multiplied by 10. It consists of a number of relatively narrow peaks, where interference of the fields generated by all domains is somewhat constructive. These peaks are divided by holes where this interference is almost completely destructive. For instance, we have found no noticeable enhancement due to RQPM at the wavelengths of Nd:YAG laser and its second harmonic, which could be used for test of the calculations. However, the possibility exists to change this situation by rotation of the domain structure in a way when the fundamental radiation will propagate at certain angle to the domain walls. This rotation will act like simultaneous scaling of thickness of all domains. The formula (5) can be modified to calculate the SHG efficiency for this case:

$$E_{2\omega}=\sum_{n=1}^{N}\left[\frac{2\omega^2\chi_n^{(2)}}{k_{2\omega}(\theta_{\rm int})\Delta k(\theta_{\rm int})}E_\omega^{\,2}\cdot\left(\exp(i\Delta k(\theta_{\rm int})\frac{d_n}{\cos(\theta_{\rm int})})-1\right)\cdot\exp\left(i\Delta k(\theta_{\rm int})\sum_{r=n+1}^{N}\frac{d_r}{\cos(\theta_{\rm int})}\right)\right] \quad (6)$$

where $\theta_{\rm int}$ is internal angle of propagation, $\Delta k(\theta_{\rm int})$ is the wave vector mismatch with account for anisotropy of refraction indices.

The results of calculation via formula (6) for arbitrarily chosen part of the spectrum are presented in figure 16. The rotation of the crystal by 9.6 degrees results in the shift of spectral dependence and the change of destructive interference to constructive one at the wavelength 791.5 nm.

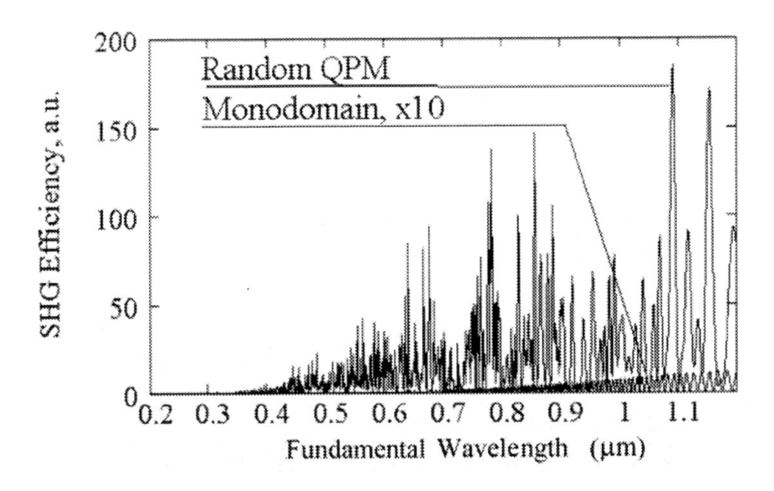

Figure 15. Calculated dependence of SHG efficiency due to RQPM on the fundamental wavelength for the Sample 3. (Reprinted with permission from Phys. Rev. A, **78**, 031802 [2008]. Copyright 2008 by the American Physical Society).

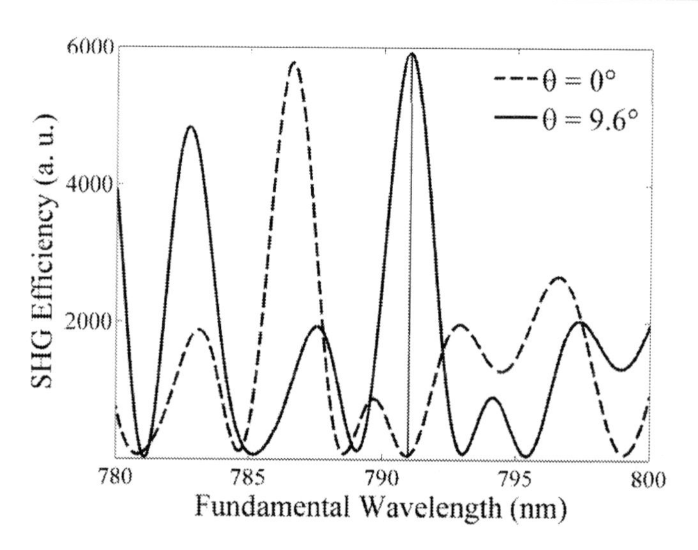

Figure 16. Calculated SHG efficiency in case of RQPM for Sample 3 near the maximum of tuning curve of Ti:sapphire laser. Dashed curve corresponds to the propagation along crystallographic axis *a* ($\theta = 0°$). Solid line corresponds to the crystal rotated by external angle $\theta = 9.6°$.

The evident test for the checking of the considerations stated above is the study of angular dependence of RQPM [26]. For this experiment we have chosen the process of doubling of the second harmonic of Nd:YAG laser, resulting in generation of 266 nm radiation. The pulse energy at 532 nm was up to 3 mJ with pulse duration 15 ns. Divergence of the beam was 3 mrad. Radiation propagated in *ab* plane of the crystal so that zero angles corresponded to the direction of *a* axis. The polarization of a 532 nm fundamental wave was along *c* axis, so that maximal component of the nonlinear susceptibility d_{ccc} could be employed. The rotation axis coincided with *c* axis of crystal. Generated second harmonic at 266 nm was filtered with a monochromator and detected with Hamamatsu H5773 photomultiplier module. The polarization of radiation at 266 nm was verified to coincide with *c* axis. Figure 17 depicts angular dependence of RQPM. For better fitting with the experiment we have modified the formula (6) to the form:

$$E_{2\omega} = \sum_{n=1}^{N}\left[\frac{2\omega^2\chi_n^{(2)}}{k_{2\omega}(\theta)\Delta k(\theta)}E_{\omega}^{2}\cdot\left(\exp\left(i\Delta k(\theta)\frac{d_n}{\cos(\theta)}\right)-1\right)\exp\left(i\Delta k(\theta)\sum_{r=n+1}^{N}\frac{d_r}{\cos(\theta)}\right)\right]\cdot\sqrt{F(\theta_{ext})} \tag{7}$$

where $F(\theta_{ext})$ is the envelope factor for Maker fringes in monodomain crystal that contains, in general case, the account for angular dependence of refraction indices, effective nonlinear susceptibility and multiple reflections. The possibility to introduce this factor is based on the fact that from the point of view of linear optical properties our sample is homogeneous and identical to monodomain crystal. The envelope factor for *mm2* symmetry group crystals was calculated in [27]:

$$F(\theta_{ext}) = \frac{\left|P_2^{NL}\right|^2 w^{(2\omega)}(w^{(\omega)}+\cos(\theta_{ext}))}{(w^{(2\omega)}-\omega^{(\omega)})^2(w^{(2\omega)}+\omega^{(\omega)})(w^{(2\omega)}+\cos(\theta_{ext}))^3}, \tag{8}$$

where

θ_{ext} is the external incidence angle, $P_2^{NL} = \cos^2(\theta_{ext}) / (w^{(\omega)} + \cos(\theta_{ext}))^2$, $w^{(2\omega)} = (n_2^{(2\omega)2} - \sin^2(\theta_{ext}))^{1/2}$, $w^{(\omega)} = (n_2^{(\omega)2} - \sin^2(\theta_{ext}))^{1/2}$.

Theoretical curve in figure 17 calculated using (7) was least square fitted to experimental points by adjusting the value of $\Delta k(\theta)$. Note that for the geometry used in the experiment this value becomes independent on the internal angle θ. The calculated curve is found to be very sensitive to the variation of Δk in the range of 0.1%. The best fit is obtained for Δk = 1.0035 Δk_{calc}). Coincidence between experiment and theory is rather satisfactory up to external incidence angles of order 20 degrees. Slight misfit in the angular position of the first maximum must be attributed to small deviation of observed etched domain pattern from real domain structure. The role of uncertainty of domain thickness measurement increases at larger angles and leads to larger discrepancy in the range above 20°.

Nonlinear diffraction, generally speaking, can have a certain contribution to the dependence depicted in figure 17. When fundamental wave propagates at small angle to the direction of RSV, non-collinear interaction that involves RSV must result in generation of additional second harmonic wave [2] that does not coincide with the direction of RQPM second harmonic wave. We checked the position of 266 nm spot with respect to the 532 nm radiation spot and found that the former is slightly shifted to the direction opposite to rotation direction. However, this shift is fairly explained by difference of refractive indices that must affect directions of harmonic and fundamental after refraction of two collinear waves on exit facet of the crystal. On the other hand, RSV values contributing to nonlinear diffraction are larger that the value Q_0 necessary for QPM in the direction of *a* axis, while in case of RQPM rotating the crystal off zero incidence angle leads to employing RSV values smaller than Q_0. In view of this, for domain structures with necessary statistics and most probable domain size larger than coherence length one must expect that contribution of RQPM will be larger than contribution from nonlinear diffraction. For our domain structure domain histogram is not smooth and statistics is not large enough, and these considerations cannot be straightly applied to it. Examination of calculated RSV spectrum of the domain structure indicates that there is a decrease of Fourier amplitude for RSV values corresponding to domain thickness lower than coherence length (in the range from 2.5 to 2.48 μm). However, angular dependence cannot be explained by the RSV spectrum in the area above the coherence length. So, the final evidence for domination of RQPM is fair agreement of experiment and calculation in figure 17. Moreover, angles of nonlinear diffraction at small rotations are of order of beam divergence. This means that the results obtained within more complicated theoretical framework including nonlinear diffraction will not differ from our present calculations to the large extent.

The enhancement factor due to RQPM was measured by comparing the intensity at 266 nm from domain-structured Sample 3 and from 432-µm thick (171 coherence lengths) reference single domain sample of the same orientation. The latter exhibited standard Maker fringes that were found to be in good agreement with the calculations using known values of refraction coefficients. The amplitude of central maximum of Maker fringes does not depend on the thickness of the reference sample but is determined by the intensity generated at one coherence length, so the reference sample thickness must not be equal to the thickness of nonlinear photonic crystal sample. The ratio of 266 nm power from domain structure in the second maximum at 10° to the 0° maximum of Maker fringes pattern from single domain sample corrected by ratio of envelope function values at corresponding angles gives RQPM enhancement factor $F_{RQPM} = 501$. Our calculations predict the value of $F_{RQPM} = 500$. Since the coherence length for our conversion scheme is 2.526 μm, and average value of domain thickness is 8 μm, it is clear that F_{RQPM} can be even more enhanced if the structure with smaller domains could be grown.

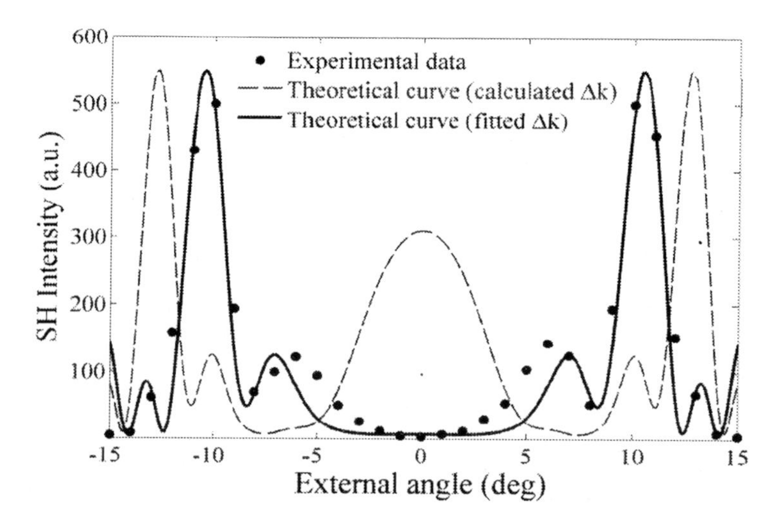

Figure 17. Angular dependence of SHG in case of RQPM in Sample 3 (dots). Dashed curve is the calculation according the formula (7) with calculated value of Δk. Solid line is least square fit at $\Delta k = 1.0035\Delta k_{calc}$.

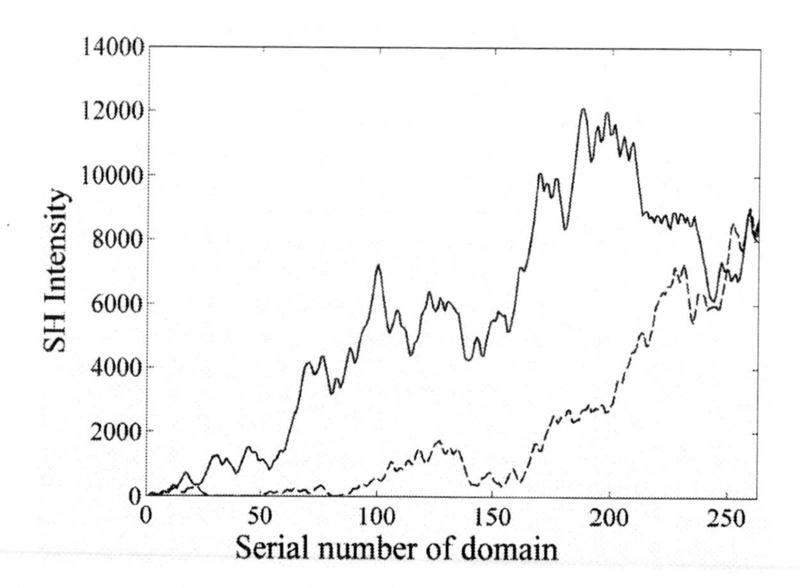

Figure 18. Calculated variation of second harmonic intensity (arb. units) along the length of structure in Sample 3. Fundamental wavelength is 532 nm.

Domain structure of the Sample 3 is rather typical for SBO at current state of our technology. Examination of spectral dependence of RQPM in figure 15 shows that the peaks with rather high F_{RQPM} protrude from near IR to near UV. Maximum second harmonic output due to RQPM is expected for the fundamental wave in the near infrared, while maximum enhancement factors are attained for fundamental wave in the near UV. For the doubling of radiation at 355 nm calculated enhancement factor for the Sample 3 is 2700, and the bandwidth of the peak at this wavelength is of order of 100 GHz. For doubling of 266 nm radiation the enhance factor is about 2400. However, if regular domain structure optimized for a given wavelength could be created, the larger enhancement factors can be attained. For instance, a regular structure with the thickness equal to the thickness of domain structure in the Sample 3, being optimized for doubling of the radiation at 355 nm, will enhance output by approximately 3500 with respect to the Sample 3.

Using formulas (5) – (7), one can calculate the behavior of second harmonic intensity along any geometrically characterized domain structure. Figure 18 presents the results of calculation using (5) for the Sample 3 in two opposite directions, along +a and –a axes. As one must expect, the intensity value generated at the exit of the structure for both cases is equal. This calculation shows, however, that if we would remove last part of domains then output will be increased by approximately 25%.

Band Structure of 1D NPC

The linear photonic crystal is characterized by its band structure that is defined as two-dimensional plot in coordinates like photon energy (commonly plotted along vertical axis) and tangential projection of wave vector (commonly plotted along horizontal axis). Certain ranges of these variables correspond to whether waves travelling inside the linear photonic crystal with real effective propagation constant or with complex one. The latter case corresponds to the band gap where the transmission of an infinite photonic crystal becomes zero. The rotation of 1D linear photonic crystal is known to shift the position of the bands to higher photon energy or shorter wavelengths (blue rotational shift), since the band structure of the latter originates from the interference of the waves reflected from layers' boundaries.

The concept of NPC band structure can be introduced in analogy with the linear photonic crystal [26]. The generated power for a certain nonlinear optical process in NPC depends on the range of values of fundamental (or generated) photon energy (or wavelength) and the wave vector direction according to Eq. 6. According to the equations of such kind one can calculate the band structure of any known NPC structure for any kind of nonlinear conversion process of interest. NPC band structure can be plotted as the contour plot in the coordinates, for instance, like fundamental wavelength and fundamental wave vector incidence angle, while generated power is assumed to be plotted along the third coordinate axis (normally to the two-dimensional band structure plot. The origin of the NPC band structure is the interference of the waves generated by all domains in NPC. According to (6), the NPC band stucture defined in the manner like this is expected to experience red rotational shift, contrary to the case of a linear PC. The band structure of regular 1D NPC is too simple and hardly can be useful for description of its properties and analysis of experimental results. The band structure of random 1D NPC is much more complicated and seems to be helpful both in description and in analysis of NPC properties. For instance, it helps in fitting of experimental spectral/angular dependencies of generated radiation with calculations in case of large number of closely positioned subbands.The example of NPC band structure calculation is presented in Fig.19.

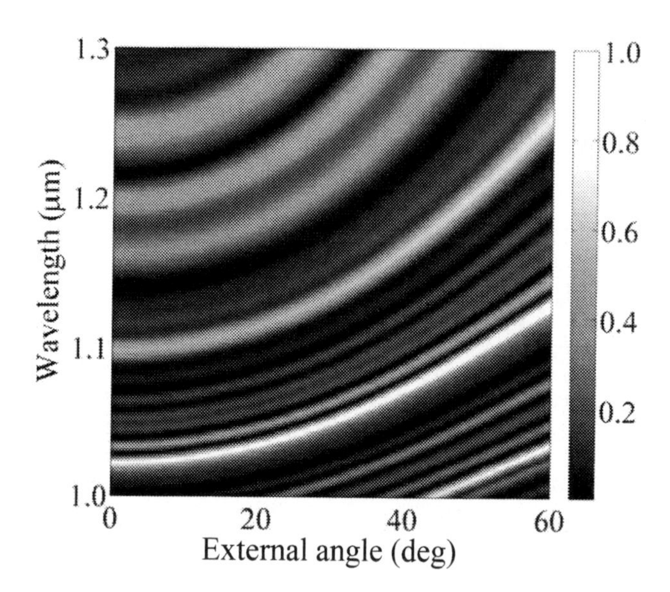

Figure 19. Example of calculated band structure of typical NPC in SBO for secong harmonic generation process in "natural" coordinates (fundamental wavelength and fundamental wave incidence angle onto the input facet of NPC).

Nonlinear Diffraction of Femtosecond Ti: Sapphire Laser Radiation in SBO

As one can see from figure 16, typical calculated spectral width of RQPM peaks in existing nonlinear photonic crystal structures of SBO is expected to be of order of 2-3 nm, when the fundamental radiation falls within the tuning curve of Ti:sapphire lasers. This value is noticeably smaller than typical bandwidths of femtosecond pulses generated by commercial laser systems. On the contrary, wide RSV spectra revealed in SBO via nonlinear diffraction of 1.064 μm radiation, as described above in this chapter, makes use of nonlinear diffraction promising for conversion of femtosecond pulses. For instance, estimations from the angular dependence of nonlinear diffraction at 1.064 μm fundamental wavelength for Sample 1 (figure 9) show that efficient nonlinear diffraction at zero incidence angle is expected down to fundamental wavelengths of order 450 – 500 nm.

Nonlinear diffraction of femtosecond pulses was investigated in [28] with the use of 2D nonlinear photonic crystal structure of strontium barium niobate (SBN). This structure produces whether conical or planar second harmonic emission, the latter being more efficient due to employing the maximal nonlinear coefficient of SBN. However, generated radiation is distributed over 180° inside the crystal, and only part of it leaves the crystal due to internal reflection. Estimated overall (internal) efficiency for the experiment [28] was 0.38%, the value that can be considered as high in view that fundamental wave source used was a simple Ti:sapphire oscillator, but not an oscillator/amplifier system. However, second harmonic scattered over large angle can be considered as a certain disadvantage inherent to 2D nonlinear photonic crystals. Another limitation of SBN is its transparency window ends at 400 nm. 1D nonlinear photonic crystals like SBO, as can be seen from our results reported above, generate unidirectional nonlinearly diffracted beams. Transparency of SBO allows nonlinear conversion to much shorter wavelengths including VUV.

Our present experiment, however, deals with conversion into blue and near UV spectral regions and is aimed onto the investigation of peculiarities of femtosecond pulse conversion in SBO. Nonlinear photonic crystal under study was the same Sample 1, that was characterized above in this chapter via nonlinear diffraction at 1.064 μm fundamental wavelength. In the present study we used a femtosecond Spectra Physics Tsunami Ti:sapphire oscillator pumped by 5 W MilleniaPro. Pulse width was 40 - 100 fs full width at half maximum (FWHM), with the average power up to 1 W, pulse energies up to 12.5 nJ, repetition rate of 80 MHz, and tunability in the range of 710–930 nm with the maximum of average power at 800 nm. The beam from the laser is focused inside the Sample 1 (with dimensions 6×6.5×5 mm^3) by a 10 cm focal length lens, resulting in a focal spot of 100 μm and peak intensity up to 4 GW/cm^2. The spectral properties of the fundamental and the generated second harmonic are measured by the Ocean Optics HR4000 spectrometer with 0.75 nm spectral resolution. Fundamental radiation propagated along *b* crystallographic axis, and its polarization coincided the *c* axis (figure 20a). Due to high degree of randomization, the RSV spectrum of the structure in the Sample 1 may vary over its volume. The fundamental beam waist is noticeably smaller than the thickness of the nonlinear photonic crystal (2 mm). For these reasons, the translation of the fundamental beam along *a* axis was employed in the course of measurements, together with small rotation tuning around *c* axis.

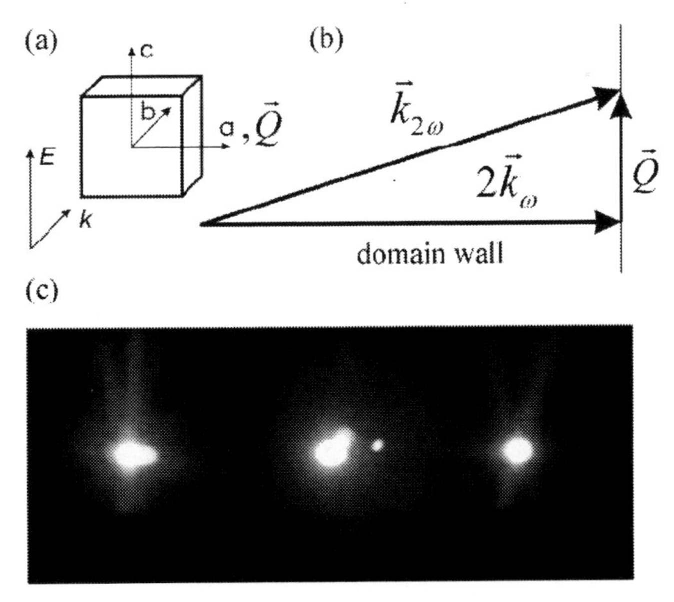

Figure 20.[ii] a) Geometry of fundamental wave propagation during the experiment on nonlinear diffraction of Ti:sapphire laser b) Phase matching diagram c) Nonlinear diffraction pattern of second harmonic radiation from SBO. The fundamental wavelength is 800 nm, and average power is 1W.

[ii] Reprinted from Ref. 30. Copyright (2009), with permission from Elsevier.

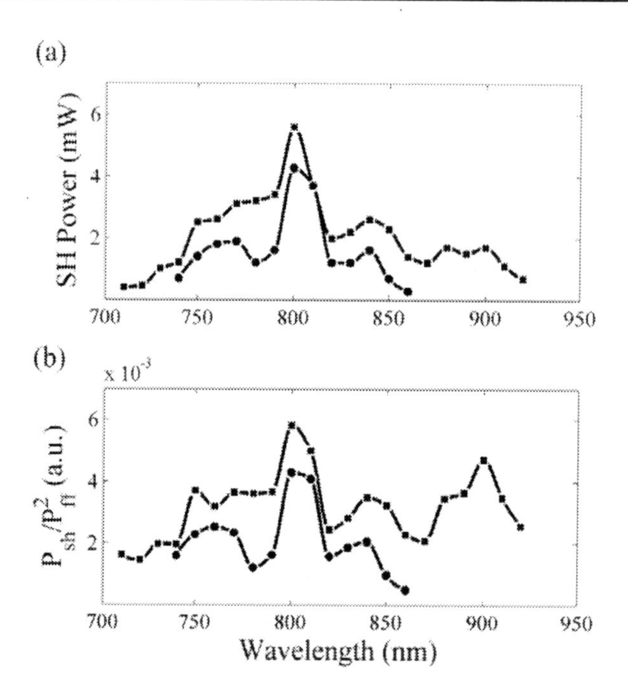

Figure 21.[ii] a) Spectral dependence of second harmonic power by tuning the femtosecond Ti:sapphire oscillator: b) RSV spectrum of SBO nonlinear photonic crystal; squares – with translational tuning of second harmonic, circles – for fixed position of the beam.

The observed pattern of nonlinear diffraction under fundamental beam propagation along *b* axis fairly agrees with the expected one (see figure 20c.). Polarization of the generated SH is found to coincide with both the polarization of the fundamental and with the *c* axis of the crystal. For a zero incidence angle of the fundamental beam the external nonlinear diffraction angle θ_2 equals to 16.6 deg. This value is in a fair agreement with the value of 16.9 deg calculated according to formula (2).

Under the central spot in figure 20c, which is produced by the fundamental beam, there is the spot of phase-mismatched collinearly generated radiation, with the intensity much smaller than the intensity of nonlinearly diffracted beams. Polarization of the generated radiation coincides with that of the fundamental, i.e. d_{ccc} component of second order susceptibility is employed. Removing the crystal from the beam and introducing it back, we did not found any observable changes in the shape of the fundamental beam. So we conclude that thermal focusing is absent for the regime used in our experiment, in contrast to SBN crystal [28]. The maximum average power of the second harmonic measured in our experiment is 5.6 mW per both beams at 400 nm, with the efficiency being 0.57%, the value being higher than that obtained for SBN, despite nonlinear susceptibility of the latter is much higher. We attribute this to the RSV spectrum of the sample under study, since the peak intensity in the SBN experiment was higher due to self-focusing [28]. The quality of the second harmonic beam in the near field is rather good.

Figure 21a (squares) shows the tuning curve of the sample under study, i.e. the dependence of second harmonic power on the central wavelength of the laser. The fundamental beam position was optimized to maximize harmonic output in every measurement point. Additionally, rotation of the sample by small angles of order of several degrees was used for optimization together with translation, in order to find the best part of the RSV spectrum. However, the effect of rotation was proven to be small. The tuning curve of the laser itself has not been removed from the dependence in figure 21a. As one can see, the tuning curve of the harmonic generally resembles that of the laser. However, the local peaks and minima of this dependence evidence that RSV spectrum is not smooth in the range of values involved in the nonlinear diffraction in the given wavelength region. Figure 21a (circles) presents the tuning curve for a stable position of the fundamental beam against the nonlinear photonic structure, initially tuned to the maximum SHG at 800 nm. It demonstrates that the selected part of the crystal contains most of the RSV spectrum necessary for efficient conversion, with exception for certain spectral regions where the RSV spectrum is more favorable in other parts of the domain structure. Figure 21b presents the second harmonic power normalized to the square of fundamental power. This dependence clearly represents the RSV spectrum of nonlinear photonic crystal averaged over the bandwidth of the laser. Figure 21b proves that for SBO this spectrum is wide but not flat. Especially large variation between the tuning curve and the RSV spectrum is observed at RSV values responsible for nonlinear diffraction in the longer wavelength part of the tuning curve.

Figure 22 presents spectrum of the fundamental radiation and that of the second harmonic, for convenience of comparison the latter being reduced to the wavelength scale of the fundamental. The fundamental wave spectrum o is measured at 840 mW average power in the beam that had passed the SBO crystal sample. Contrary to SBN, fundamental spectrum bears no signs of phase self-modulation and broadening caused by it. Spectrum of harmonic is clearly narrower than that of the fundamental, although this narrowing is very modest, of order of 10% - 20% for different wavelengths. When fundamental wave spectrum falls on a slope of the RSV spectrum, a small shift of the maximum of the harmonic spectrum is also observed. The narrowing of the harmonic spectrum is well understandable and is due to the non-flat RSV spectrum. Note that the extent of the narrowing is small, which evidences that real RSV spectrum, being not averaged over laser bandwidth, does not strongly differ from the averaged RSV spectrum in figure 21b. Relative smoothness of the harmonic spectrum as compared with the spectrum of the fundamental is caused by poorer resolution in the wave number scale of spectrometer in the shorter wavelength region.

During the measurements of the tuning curves, the duration of laser pulses at the entrance of the nonlinear medium could not be sustained absolutely constant and might have influenced the results. To verify the extent of this influence, we examined dependence of SHG coupling coefficient on the fundamental pulse duration. The latter was varied by changing the spectrum of the Tsunami laser radiation, as the design of this laser enables generation of transform-limited pulses, duration of them being monitored via calibration curves supplied by the producer. As can be seen from dependences in figure 23 taken at 790 and 850 nm fundamental wavelengths, with the account for variation of average power, this dependence is very weak in the range from 40 to 90 fs. As expected both from theory [29] and known experiments for collinear femtosecond harmonic generation, this dependence must follow approximately the inverse proportional law with the variation of duration pulse. Explanation of observed experimental dependence, in our opinion, can be found in non-collinear character of SHG, however, this possibility needs additional extensive theoretical studies. However, the results presented in figure 23 allow us to conclude that unintended variation of pulse duration during wavelength tuning does not seriously affect the tuning curves measurements plotted in figure 21. Duration of the second harmonic pulses in the conversion process is expected to increase, both due to the narrowing of the spectrum and, to a greater extent, due to the group velocity dispersion. A typical value of the latter in SBO equals to $2.8 \cdot 10^{11}$ cm/sec, leading to the harmonic pulse duration calculated for non-stationary regime [29] of order of 2 ps for the crystal length used in our experiment. However, this value can be decreased when shorter crystals are used, with a compromise between shorter pulse duration and lower efficiency. For a 1 mm crystal one may expect 350 fs pulses of the second harmonic.

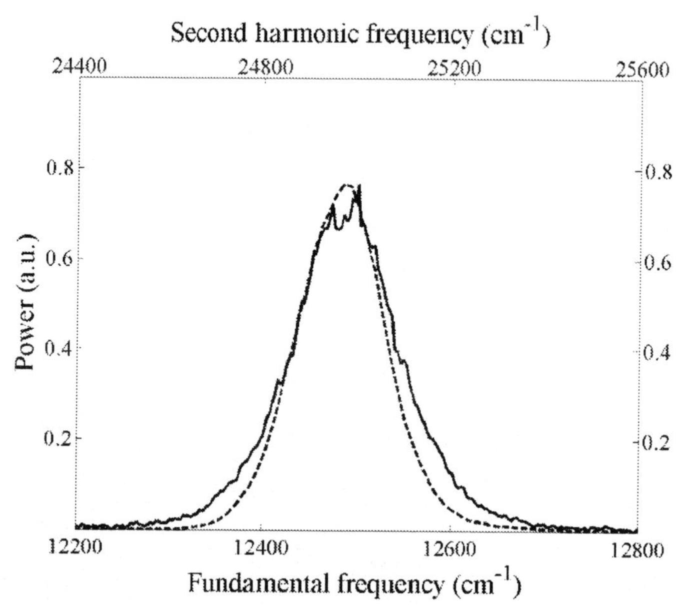

Figure 22.[ii] Spectra of fundamental (solid line) and second harmonic (dashed line). Ocean Optics HR4000, spectral resolution 0.75 nm.

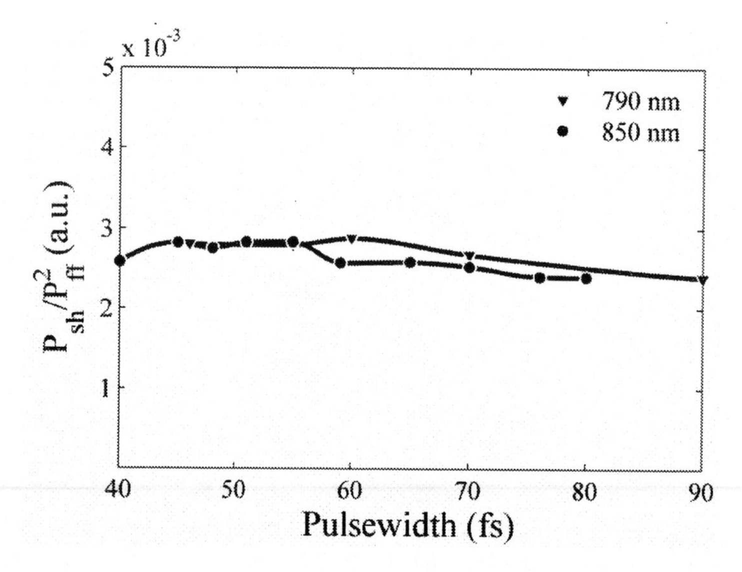

Figure 23[ii] Variation of second harmonic coupling coefficient with the duration of the fundamental pulses at the entrance of nonlinear medium.

Results obtained in this subsection can be transferred into shorter-wavelength region, leading to creation of compact tunable source of femtosecond radiation in the region of wavelengths 250 - 175 nm. Two conditions must be satisfied for this to be done, namely, 1) the second harmonic of a Ti:sapphire oscillator introduced into SBO crystal must be of order of 1 W, and 2) the characteristic domain sizes in a nonlinear photonic crystal structure must be scaled down by value of order of 2. The first condition seems to be realistic at the present state of the art of laser technology, while the second one needs further studies.

Conclusion

Strontium tetraborate single crystals containing alternate oppositely poled domain structure were grown by Czochralski method. Domains are visualized by etching and are shown to have the form of sheets with domain walls perpendicular to *a* axis. The structure is rather well ordered along *b* and *c* crystallographic axes of the crystal but highly randomized along *a* axis. Mechanism of domains formation is proposed that is based on instability of unit cell formation in the presence of longitudinal component of the static polarization field. The domain structures are shown to be reproducible during the repeated growths.

Nonlinear diffraction was observed in strontium tetraborate crystal at the frequencies of the second and third harmonics of Nd:YAG laser. This diffraction allowed admitting the presence of partially ordered alternate oppositely poled domains suitable for the use in nonlinear optics. The shape and orientation of domains found from nonlinear optical studies confirm the results of optical microscopy. They are the same as in the ferroelectric KTP crystal belonging to the same symmetry class *mm2*. The thickness of domains is a random quantity; the spatial Fourier spectrum of the reciprocal superlattice wave vector is measured in the range from $\pi/7.6$ to $\pi/0.18\ \mu m^{-1}$.

Random quasi-phase-matching of the second-harmonic generation process is investigated experimentally and theoretically in a strongly randomized spontaneously grown nonlinear photonic crystal structure of strontium tetraborate. The thickness of domains in the structure is measured and employed in calculations of angular and spectral dependences of random quasi-phase-matching. Despite the great scale of randomness, the RQPM effect is preserved. The calculated enhancement factor due to RQPM equals to 500 and is in excellent agreement with experiment, and good agreement between calculations and experiment is obtained for the shape of the angular dependence of RQPM. The calculated spectral dependence of RQPM for a typical nonlinear photonic structure in SBO consists of a number of relatively narrow peaks, and maximum enhancement factors of order of several thousands are expected to be attainable for fundamental wavelengths lying in the near ultraviolet.

An efficient tunable SHG of femtosecond pulses from a Ti:sapphire oscillator in randomized nonlinear photonic crystal structure of strontium tetraborate has been demonstrated. Tuning range of second harmonic is from 355 to 460 nm, with maximum efficiency 0.57% per two diffracted beams. The RSV spectrum of nonlinear photonic structure is not completely flat but enables the second harmonic generation with minimal narrowing of the spectrum.

Further investigations of the possibility of controllable domain growth aiming to creation of regular domain structures and structures with a desired RSV spectra in SBO for further applications in nonlinear optics of UV and VUV spectral range are of great importance.

Acknowledgments

The work was supported by the Grant of the President of the Russian Federation for the support of leading scientific schools (No. SS-4645.2010.2), Grant No. RNP.2.1.1.3455, and the Projects 2.5.2 and 3.9.1 of PSB RAS, №5 and 27.1 of SB RAS. We are grateful to IOP Publishing for the permission to reproduce tables 1 and 2 and figures 7, 8 and 9 from J. Opt. A, 9, pp. S334-338. A.M.Vyunishev is grateful for the support from Krasnoyarsk Regional Science Fund.

References

[1] Berger, V. (1998). Nonlinear photonic crystals. *Phys. Rev. Lett.*, Vol.81, 4136-4139.

[2] Fejer, M. M., Magel, G. A., Jundt, D. H. & Byer, R. L. (1992). Quasi-Phase-Matched Second Harmonic Generation: Tuning and Tolerances. *IEEE J. of Quant. Electron*, Vol.28, 2631- 2654.

[3] Dmitriev, V. G., Gurzadyan, G. G. & Nikogosyan, D. N. (1999). *Handbook of Nonlinear Optical Crystals.* Third Edition, Berlin: Springer.

[4] Buchter, S. C., Fan, T. Y., Liberman, V., Zayhowski, J. J., Rothschild, M., Mason, E. J., Cassanho, A., Jenssen, H. P. & Burnett, J. H. (2001). Periodically poled BaMgF 4 for ultraviolet frequency generation, *Opt. Lett*, Vol.26, 1693-1695.

[5] Zaitsev, A. I., Aleksandrovsky, A. S., Vasiliev, A. D. & Zamkov, A. V. (2008). Domain structure in strontium tetraborate single crystal. *Journal of Crystal Growth*, vol. 310, 1-4.

[6] Block, S., Perloff, A. & Weir, C. E. (1964). The Crystallography of Some M^{2+} Borates, *Acta Crystallogr*, vol. 17, 314-315.

[7] *International Tables for Crystallography*. Vol *A*. Dordrecht/London/Boston:Kluwer, 2002.

[8] Bohaty, L., Libertz, J. & Stähr, S. (1985). Strontiumtetraborat SrB_4O_7: Einkristallzűchtung, electrooptishe und electrostriktive Eigenschaften, Zeitschrift fűr Kristallographie, vol. 172, 135-138.

[9] Oseledchik, Y. S., Prosvirnin, A. I., Starshenko, V. V., Osadchuk, V., Pisarevsky, A. I., Belokrys, S. P., Korol, A. S., Svitanko, N. V., Krikunov, S. A. & Selevich, A. F. (1995). New nonlinear optical crystals: strontium and lead tetraborates. *Opt. Mater*, Vol.4, 669-674.

[10] Pan, F., Shen, G. Q., Wang, R., Wang, X. Q. & Shen, D. (2002). Growth, characterization and nonlinear optical properties of SrB_4O_7 crystals, *J. Cryst. Growth*, Vol.241, 108-114.

[11] Knyrim, J. S., Becker, P., Johrendt, D. & Huppertz, H. (2006). A new non-centrosymmetric modification of BiB_3O_6. Angew. *Chem. Int. Ed.*, Vol. 45, 8239- 8241.

[12] Aleksandrovsky, A. S., Vasiliev, A. D., Zaitsev, AI. & Zamkov, A. V. (2008). Growth, optical and electromechanical properties of single-crystalline orthorhombic bismuth triborate. *J. Cryst. Growth*, Vol. 310, 4027-4030.

[13] Blistanov, A. A; Bondarenko, V. S; Perelomova, N. V; Strizhevskaia, F. N; Chkalova, V. V; Shaskolskaia, M. P. (1982). Acoustic crystals. Moscow: *Nauka*, (In Russian).

[14] Machida, K., Hata, H., Okuno, K., Adachi G. & Shiokawa., J. (1979). Synthesis and characterization of divalent-europium (Eu^{2+}) compounds, EuB_4O_7, EuB_2O_4 and $Eu_2B_2O_5$. *J. Inorg. Nucl. Chem*, Vol. 41, 1425-1430.

[15] Aleksandrovsky, A. S., Malakhovskii, A. V., Zabluda, V. N., Zaitsev, A. I. & Zamkov, A. V. (2006). Optical and magneto-optical spectra of europium-doped strontium tetraborate single crystals. *J. Phys. Chem. Solids*, Vol. 67, 1908-1912.

[16] Petrov, V., Noack, F., Dezhong Shen, Feng Pan, Guangqui Shen, Xiaoqing Wang, Komatsu, R. & Alex, V. (2004). Application of the nonlinear crystal SrB_4O_7 for ultrafast diagnostics converting to wavelengths as short as 125 nm, *Optics Letters*, Vol.29, 373-375.

[17] Zaitsev, A. I., Aleksandrovsky, A. S., Zamkov, A. V. & Sysoev, A. M. (2006). Nonlinear optical, piezoelectric and acoustic properties of SrB_4O_7. *Inorganic Materials*, Vol.42, 1360-1362.

[18] Pack, M. V., Armstrong, D. J. & Smith, A. V. (2004). Measurement of the $\chi^{(2)}$ tensors of $KTiOPO_4$, $KTiOAsO_4$, $RbTiOPO_4$, and $RbTiOAsO_4$ crystals, *Applied Optics*, Vol.43, 3319-3323.

[19] Evlanova, N. F., Naumova, I. I., Chaplina, T. O., Lavrishchev, S . V. & Blokhin, S. A. (2000). Periodic domain structure in Czochralski-grown LiNbO3: Y crystals, *Physics of the Solid State,* Vol. 42, 1727-1730.

[20] Nai-ben Ming, Jing-fen Hong, Duan Feng, (1982). The growth striations and ferroelectric domain structures in Czochralski-grown $LiNbO_3$ single crystals, *J. Materials Sci.* Vol, 17, 1663-1670.

[21] Dekker, P. & Dawes, J. M. (2004). Characterization of nonlinear conversion and crystal quality in Nd- and Yb-doped YAB, *Opt. Express,* Vol.12, 5922-5930.

[22] Freund, I. (1968). Nonlinear diffraction. *Phys. Rev. Lett*, Vol. 21, 1404-1406.

[23] Morozov, EYu., Kaminskii, A. A., Chirkin, A. S. & Yusupov, D. B. (2001). *Second Optical Harmonic Generation in Nonlinear Crystals with a Disordered Domain Structure JETP Letters*, Vol.73, 647-650.

[24] Baudrier-Raybaut, M., Haïdar, R., Kupecek, Ph., Lemasson, Ph. & Rosencher, E. (2004). Random quasi-phase-matching in bulk polycrystalline isotropic nonlinear materials, *Nature*, Vol.432, 374-376.

[25] Vidal, X. & Martorell, J. (2006). Generation of Light in Media with a Random Distribution of Nonlinear Domains, *Phys. Rev. Lett.*, Vol.97, 013902.

[26] Aleksandrovsky, A. S., Vyunishev, A. M., Shakhura, I. E., Zaitsev, A. I. & Zamkov, A. V. (2008). Random quasi-phase-matching in nonlinear photonic crystal structure of strontium tetraborate, *Phys. Rev. A*, Vol.78, 031802.

[27] Bechthold, P. S. & Haussuhl, S. (1977). Nonlinear optical properties of orthorhombic barium formate and magnesium barium fluoride, *Appl. Phys.*, Vol.14, 403-410.

[28] Fischer, R., Saltiel, S. M., Neshev, D. N., Krolikowski, W. & Kivshar, Yu. S. (2006). Broadband femtosecond frequency doubling in random media, *Appl. Phys. Lett*, Vol.89, 191105.

[29] Akhmanov, A. S., Vysloukh, V. A. & Chirkin, A. S. (1992). Optics of femtosecond laser pulses, *AIP*, Sec. 3.2.

[30] Aleksandrovsky, A. S., Vyunishev, A. M., Slabko, V. V., Zaitsev, A. I. & Zamkov, A. V. (2009). Tunable femtosecond frequency doubling in random domain structure of strontium tetraborate, *Opt. Comm.*, Vol.282, 2263-2266.

In: Photonic Crystals
Editor: Venla E. Laine, pp. 277-292

ISBN: 978-1-61668-953-7

Chapter 12

OPTICAL PROPERTIES OF SELF-ASSEMBLED COLLOIDAL PHOTONIC CRYSTALS DOPED WITH CITRATE-CAPPED GOLD NANOPARTICLES

Marco Cucini, Marina Alloisio, Anna Demartini and Davide Comoretto*

Dipartimento di Chimica e Chimica Industriale, Università degli Studi di Genova, via Dodecaneso 31, 16146 – Genova,Italy.

Abstract

Photonic crystals are composite materials possessing a periodical modulation of their dielectric constant on the scale length of the visible wavelength. This dielectric periodicity affects light propagation creating allowed and forbidden photon energy regions. Among different types of photonic crystals, we addressed our attention to artificial opals i.e. three dimensional self-assembled arrays of spheres packed in a face centred cube crystal lattice. In order to tune the optical properties of such photonic crystals, we inserted gold nanoparticles in the interstices between the spheres. We call this process doping. In this work we report on the optical properties of artificial opals doped at various levels with gold nanoparticles (NpAu) stabilized with citrate molecules. By increasing the NpAu doping level, the opal photonic stop band was bathochromically shifted as well as its full width half maximum and intensity were reduced. No effects of nanoparticle coalescence was observed as previously detected for NpAu prepared by laser ablation. In order to improve the mechanical stability of NpAu doped opals, we started a preliminary thermal annealing study with in situ monitoring of optical properties. We found that thermal degradation of doped samples was reduced with respect to bare opals. In addition, the annealing temperature of NpAu doped opals was much higher than that observed for bare ones.

Keywords: Photonic crystals, artificial opals, gold nanoparticles, optical properties.

* E-mail address: comorett@chimica.unige.it. (Corresponding author)

1. Introduction

Nanostructured materials are attracting much attention due to their application in photonics, electronics, biology and catalysis. Among these variegated systems, materials possessing a periodical modulation of the dielectric constant on the wavelength scale of the visible light (photonic crystals, PC) are currently extensively investigated since they behave as light semiconductors and then find application in photonics for low threshold laser action, high bending angle waveguides, superprism effect, sensors and optical switches [1-4]. Photonic crystals are grown both with top-down and bottom-up approaches. The first method is very powerful and allows detailed and variegated structures to be prepared [4]. However, it is very expensive since requires an extensive application of UV or electron beam lithographic techniques. On the other hand, bottom-up approaches are very interesting, cheap and easy methods since exploit the self-assembling properties of molecules, supramolecules and meso-objects like colloids which move spontaneously towards quasi-equilibrium steps, building an ordered structure [5]. Bottom-up methods are extensively adopted in nature to create nanostructures possessing unusual optical properties suitable for biomimetic purposes [6,7] or mating [8,9]. For instance, it has been noticed that the presence of ripples on arthropoda ommatidia or on large wings of diurnal moths provides a sort of anti-reflection layer, which increases transparency thus improving vision or develops mimetism against predators attacks [10]. Very well known is the PC nature of nanostructures inside butterflies wings [6,8-14] or peacock feathers [15], which are responsible for their wonderful colours. In sea mouse, the iridescent needles observed at scanning electron microscope (SEM) display a photonic fiber-like structure [16,17] also observed in edelweiss bracts [18]. The chitin exocuticol of Chrysochroa vittata is an example of one dimensional (1D) PC [19], but very surprising is the case of an Australian beetle, which artificially produces opals over its exoskeleton in order to improve its mimetism [20]. As reported in the Sanders' pioneer work, the structure of opals is made of silica nanospheres packed in a face centred cubic (FCC) structure. Diffraction of light from crystallographic planes provides the amazing colour of these gemstones [21,22]. Different preparation methods of opals starting from monodisperse silica or polymeric microspheres ad hoc synthesized are currently extensively pursued in the scientific community [23,24]. As a matter of fact, opals and inverse opals are very interesting since provide a useful playground for photonic band gap engineering [23,25-28]. This can be done both by introducing controlled structural defects [29,30] into the opal PC structure or by infiltration of sphere interstices with photoactive materials [23,31-40].

Recently, the use of metallic nanoparticles as a medium for opal infiltration or inverse opal preparation has attracted much attention [29,41-47]. As a matter of fact, nanoparticle optical properties due to surface plasmon resonance are strongly dependent on the environment, thus being used for sensing and sensitizing applications [48]. Moreover, NpAu also show a high and fast nonlinear optical (NLO) response [49]. In our previous works [28,50], NpAu used for opal infiltration were fabricated by laser ablation [31,51], a technique which requires the use of a surfactant dissolved in the liquid where NpAu are prepared in order to prevent their coalescence. Those opals showed interesting nanosecond optical switching properties when resonantly photoexcited [28]. In this work, we would like to extend our experience to NpAu obtained by chemical reduction of a gold salt and stabilized with a citrate outer shell in order to improve the anti-aggregation effects even in the solid state.

The paper is organized with an experimental section where NpAu synthesis and opal preparation are described together with the experimental techniques used for their morphological and optical characterization. In the results and discussion section, spectroscopic properties of bare and NpAu doped opals are discussed in the framework of the dielectric response. Suggestions to improve the mechanical stability of these systems by thermal annealing are also reported.

2. Experimental

NpAu were prepared following a modified version of the method developed by Turkevich and co-workers [52]. Briefly, a weighed amount of $HAuCl_4$ was dissolved in MilliQ H_2O, heated at boiling temperature under vigorous stirring and added dropwise of a fixed amount of citrate reducing/stabilizing agent. The final reagent concentrations were 0.28 and 2.8 mM for the metal and citrate ions, respectively. The aqueous mixture was let to boil until the solution colour turned from yellow to red (about 1 hour) and then maintained under stirring at room temperature for other 3 hours. At reaction completed, a red-purple solution composed by gold nanoparticles stabilized with citrate (citrate-NpAu, Figure 1a) was collected and stored in the dark at room temperature. Insignificant modifications were observed for over 6 months, indicating that the as-prepared colloids had good stability in aqueous solutions. All chemical reagents and solvents were spectroscopic grade commercial products used as received. $AuCl_3$, HCl and trisodium citrate were purchased from Lancaster.

Electronic absorption spectra were recorded at room temperature on a Perkin-Elmer Lambda 9 spectrophotometer in the UV-Vis range with spectral resolution ±0.2 nm).

Bright-field transmission electron microscopy (TEM) images were obtained with a Jeol electron microscope, model IEM-2010, operating at 200 kV. Specimens were prepared by evaporating a drop on a 300 mesh carbon-coated copper grid (Lacey). Qwin software V3 for digital image processing and analysis was used for measurement of particle size employing multiple pictures from different areas.

Opals were grown from commercial polystyrene (PS) monodisperse microspheres (260, 300 and 340 nm diameters, standard deviation < 5%) water suspension 10% by weight (Duke Scientific; refractive index, n_{PS}=1.59). For each microsphere diameter, we prepared six different suspensions having different concentration of both PS microspheres and NpAu. We used two different concentrations of microspheres and three different concentrations of NpAu. We mixed 0.06 (0.3) ml microsphere suspension with 0.5 ml (Low Load, LL), 1.0 ml (Medium Load, ML) and 2.0 ml (High Load, HL) NpAu suspension and then we further diluted to 5.0 ml with distilled water. The amount of PS microspheres determined the average thickness of samples, namely thin (thick) samples. Samples were named according to sphere diameter, thickness and NpAu load (ex. 260 thin LL). With these six suspensions we grew opal films having different degrees of infiltration (NpAu doping) and different thickness by using the meniscus technique [43].

Transmittance (T) and reflectance (R) spectra were recorded with an Avantes Avaspec-2048 compact spectrometer (300-1100 nm spectral range, 1.4 nm spectral resolution). Light from a tungsten-halogen lamp was guided by an optical fiber to a proper focusing optics (spot diameter 100 μm or variable in the range 0.5-5 mm) on the sample, eventually mounted on a goniometer; transmitted light was then collected and driven by optical fibers to the spectrometer. For the 100 μm diameter spot, the sample was mounted on a XY micrometric translator and probed area was selected and observed by a Sony XCD-X710CR video-camera. Normal incidence reflectance was measured on a 2 mm diameter spot by an Y reflection probe bundle fiber. Sample heating was provided by a hot plate; the temperature was monitored with a Testo 110 thermometer equipped with a flat head probe Testo AG 630.

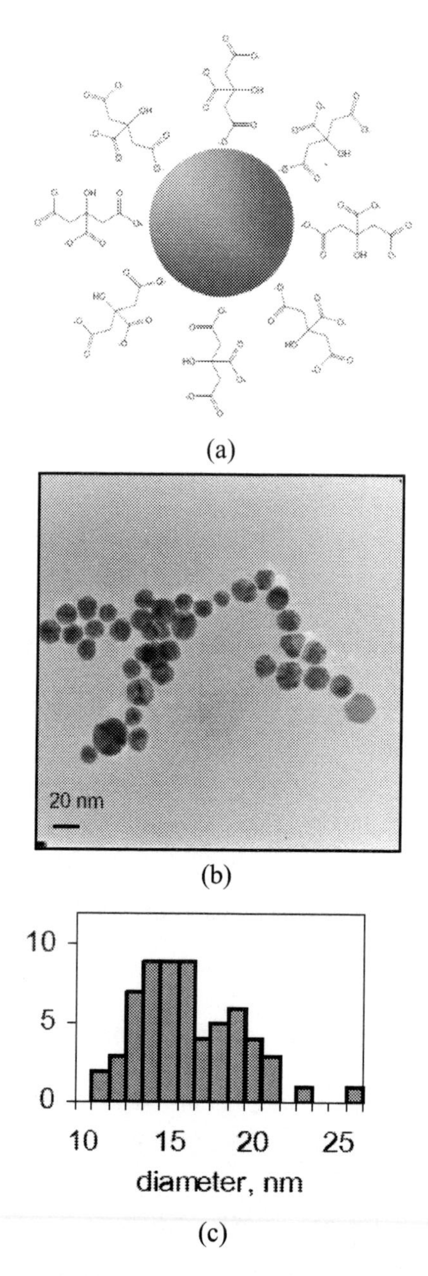

Figure 1. Schematic structure of gold nanoparticles (NpAu) stabilized with citrate (a); bright field TEM image (b) and its corresponding size distribution histogram (c) for citrate-capped NpAu.

3. Results and Discussion

The citrate-NpAu morphological and optical properties were investigated by TEM and UV-Vis techniques. Figure 1b presents a TEM micrograph of a drop-cast film of the colloids with the corresponding histogram (Figure 1c) of the particle size distribution alongside. The nanoparticles average diameter is 16.6±3.0 nm and the axial ratio is 1.14 as measured from the TEM image. It is evident that quite homogeneous, well-separated, nearly spherical gold nanoparticles are obtained. By assuming all particles as spherical, a final colloid concentration 1.94 nM was evaluated from the mean diameter value. As far as concerned the particles arrangement, the nanohybrids tend to hold together, despite the negatively charged citrate chains, in linear or two-dimensional structures quite ordered in shape.

The state of citrate-coated NpAu was confirmed by means of electronic absorption spectroscopy. In Figure 2a, the electronic absorption spectrum of the aqueous suspension is reported. A well-resolved, sharp surface plasmon band (SPB) at 524.0 nm, typical of gold nanosized spherical clusters and responsible for the red-purple colour of the colloidal solution, is observed. The band is attributable to the SPB transverse mode, perpendicular to the particle long axis [53]. The absence of distinct absorptions in the 650-700 nm range, corresponding to the longitudinal plasmon resonance band, confirms that in these conditions the colloids do not aggregate, in agreement with the TEM data.

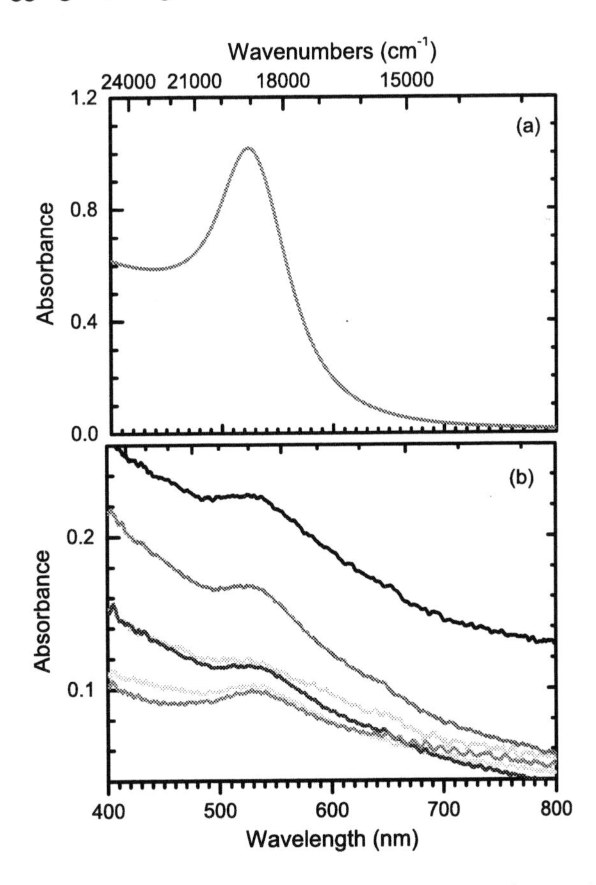

Figure 2. Electronic absorption spectra of citrate-NpAu aqueous suspension used in this work (a) and of a 300 nm dissolved doped opal film in different positions (b).

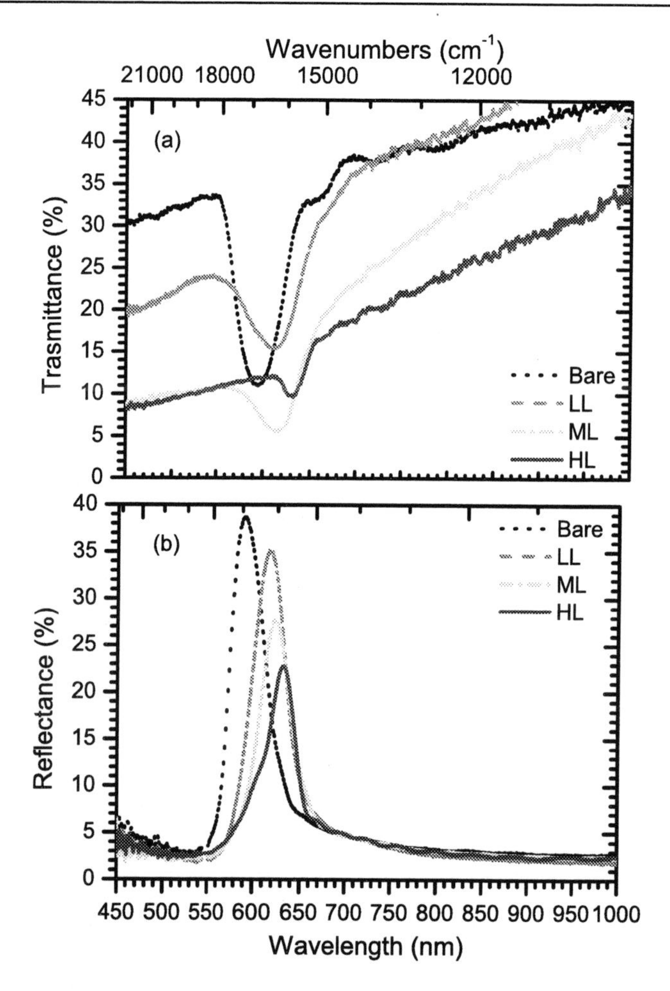

Figure 3. Transmittance (a) and reflectance (b) spectra of thin 260 nm opals bare (black dot line) and NpAu doped at LL (red dash line), ML (green dot dash line) and HL (blue solid line) levels.

We turn now to the discussion of the optical properties of opals infiltrated with citrate-capped NpAu. Figure 3 shows the T (a) and R (b) spectra of opals infiltrated with NpAu at three different doping levels compared with those of a bare one.

The of transmission/reflectance spectrum of bare opals presents a minimum/maximum at 595/590 nm. Its origin is due to the photonic stop band of the photonic crystal which does not allow the light to propagate inside the opal. As a matter of fact, light of such wavelength is backward diffracted by opals (111) crystallographic planes giving rise to a prominent transmission dip/reflectance peak. It is evident that the spectral position of these structures bathochromically shifts upon increasing the NpAu doping level. In particular, for LL, ML, and HL samples, the transmission spectra stop band shifts to 610, 615 and 630 nm respectively (Figure 3a), while in reflectance peaks are observed at 615, 625 and 630 nm. Notice that in these spectra, no evidence of NpAu plasmon absorption is detected. The minor wavelength difference between the stop bands recorded in T spectra with respect to the R ones is probably due to light scattering which particularly affects the background of transmittance. However, we cannot exclude that the sample area probed by the two set-ups could be slightly different.

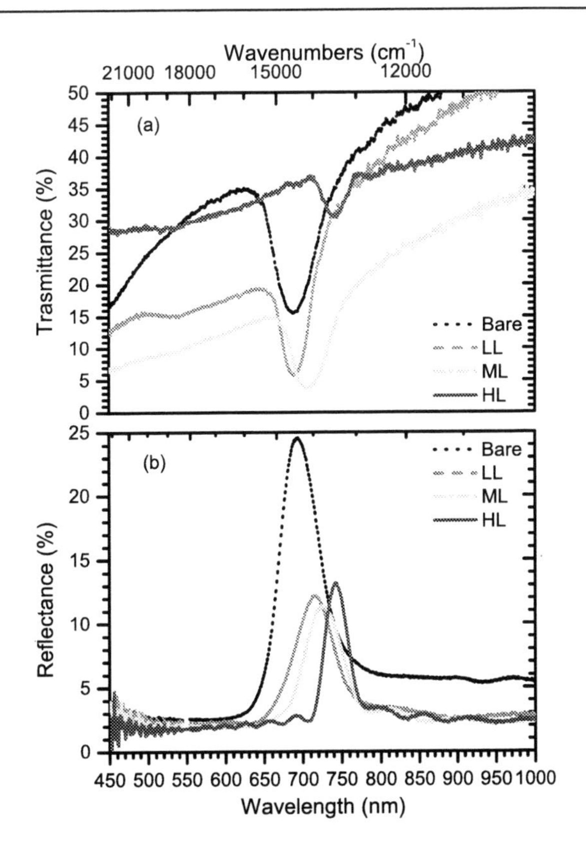

Figure 4. Transmittance (a) and reflectance (b) spectra of thin 300 nm opals bare (black dot line) and NpAu doped at LL (red dash line), ML (green dot dash line) and HL (blue solid line) levels.

Reflectance spectra, in particular for thin samples, show a wavy background due to interference fringes pattern caused by multiple reflection inside the sample. The presence of interference fringes in opal spectra is a signature for their good optical quality.

Additional effects of NpAu doping level are related to the peak width and intensity. From spectra reported in Figure 3 we observe a reduction of both the full width half maximum (FWHM) and signal intensity upon increasing the doping level. This is particularly evident in the R spectra since they show a flat background. In T spectra, even if the effect is still detectable, the background is less regular since strongly dependent on the surface roughness and scattering processes. Deviation from these trends are due to sample inhomogeneities.

Figure 4 shows the transmittance (a) and reflectance (b) spectra for thin 300 nm opals doped with NpAu at the same levels previously used for the 260 ones. Bare opals show the transmittance (reflectance) stop band at 690 nm (695 nm), while LL, ML and HL NpAu infiltrated opals at 690 (720), 710 (730) and 740 (745) nm, respectively.

If we further increase the sphere diameter at 340 nm (Figure 5) a more relevant bathochromic shift is observed in T/R spectra for bare (780/785 nm), LL (820/820 nm), ML (830/835 nm), and HL (850/850 nm) opals. Moreover, interference fringes in the transparent region are found for all samples. Notice also that for 300 and 340 nm opals a progressive intensity and FWHM reduction of the stop band upon the increasing doping level is observed, as previously detected for the 260 nm ones. Table I summarized the main optical and morphological parameters evaluated for all prepared samples.

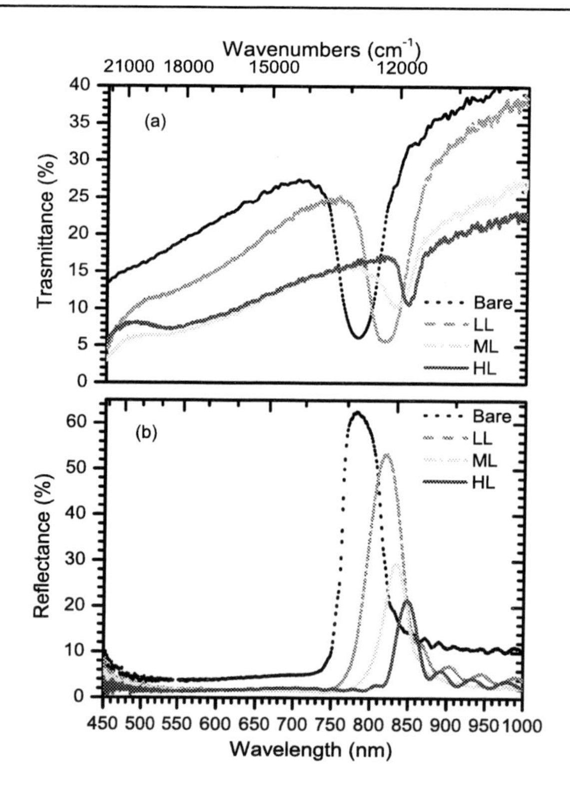

Figure 5. Transmittance (a) and reflectance (b) spectra of thin 340 nm opals bare (black dot line) and NpAu doped at LL (red dash line), ML (green dot dash line) and HL (blue solid line) levels.

Table I. λ_{Bragg}, FWHM, n_{eff} (both from shift and dispersion when available), n_i, ε_i, T, R, thickness (d), and number of sphere layers (p)deduced from the optical spectra.

Sample	λ_{Bragg} (nm)	FWHM (cm^{-1})	n_{eff} shift	n_{eff} disp	n_i	ε_i	T (%)	R (%)	d (μm)	p
260 bare	594	1537	1.40		1.0	0.92	24	36	-	
260 thin LL	613	1713	1.45	1.45	1.1	1.21	12	33	-	
260 thin ML	616	1084	1.45		1.1	1.21	7	26	5.5	11
260 thin HL	630	509	1.49		1.2	1.49	3	20	5.5	11
260 thick LL	588	1284	1.39		0.9	0.86	17	33	-	
260 thick ML	590	1265	1.39		0.9	0.86	16	31	-	
260 thick HL	637.5	411	1.50		1.3	1.56	2	21	7.5	15
300 bare	688	1312	1.41		1.0	0.96	23	22	-	
300 thin LL	689	779	1.41		1.0	0.96	16	10	-	
300 thin ML	706	1015	1.44		1.1	1.14	13	10	-	
300 thin HL	741	575	1.52		1.3	1.74	7	11	-	
300 thick LL	705	1010	1.44		1.1	1.14	5	27	-	
300 thick ML	726	1060	1.48		1.2	1.42	3	25	8.5	15
300 thick HL	740	288	1.51		1.3	1.66	3	17	-	
340 bare	780	976	1.39		0.9	0.86	24	58	-	-
340 thin LL	818	839	1.46		1.1	1.28	23	52	7	11
340 thin ML	832	675	1.48		1.2	1.42	8	28	-	
340 thin HL	848	338	1.51		1.3	1.66	8	20	7	11

The bathochromic shifts observed in spectra reported in Figs. 3-5 can be generated by two factors, one of which is the increase of the microsphere diameter and the other is the infiltration process. We now discuss these important features in details.

The shift of the stop band upon changing the sphere diameter can be interpreted according to the scaling laws of photonic crystals [54]. The master equation governing the properties of photonic crystals is

$$\nabla \times \left(\frac{1}{\varepsilon(\mathbf{r})} \nabla \times \mathbf{H}(\mathbf{r}) \right) = \left(\frac{\omega}{c} \right)^2 \mathbf{H}(\mathbf{r}) \tag{1}$$

where $\varepsilon(\mathbf{r})$ is the periodical real part of the dielectric constant, H(r) is the macroscopic magnetic field, ω the frequency, and c the speed of light. When the spatial distribution of the dielectric constant of the system is scaled through a generic factor s, the dielectric constant assumes the form

$$\varepsilon'(\mathrm{r}) = \varepsilon(\mathrm{r}/\mathrm{s}) \qquad (2).$$

By simply changing variables ($\mathbf{r}'=s\mathbf{r}$) the new master equation can be solved with mode profile H'(r')=H(r'/s) and frequency becomes

$$\omega' = \omega/s \qquad (3).$$

Since in our case s>1, we expect a reduction (increase) of energy (wavelength) of the photonic band structure upon increasing the PS microsphere diameter as experimentally observed.

Let's now discuss the effects of infiltration. The spectral position of the stop band or Bragg peak (λ_{Bragg}) as well as its dependence on the incidence angle off from the normal (θ) and effective refractive index ($n^2_{eff} = \varepsilon_{eff}$, effective dielectric constant) of the system depends on the details of the photonic band structure, i.e. on the eigenvalues of Eq. 1 [25]. If we are interested only in the low energy properties of this band structure (close to the stop band around the Γ-L direction) we can use the much simpler Bragg-Snell formula [55] which can be derived by the Bragg law joined to the Snell one

$$m\lambda_{\mathrm{Bragg}} = 2D\sqrt{n_{eff}^2 - \sin^2 \vartheta} \tag{4}$$

where m is the diffraction order, D is the interplanar spacing in the [111] growth direction which is perpendicular to the opal surface. We can notice that Eq. 4 predicts a bathochromic shift of λ_{Bragg} upon increasing n_{eff}. This explains the spectral position of the Bragg peak upon increasing the NpAu doping level of opals. As a matter of fact, upon adding NpAu into the interstices, their dielectric constant is increased respect to the value for void (ε=1) and then λ_{Bragg} increases. However, Bragg-Snell formula (Eq. 4) cannot be used to understand the nature of infiltrated material. This is a very important item since NpAu might give rise to aggregates or to bulk gold, if they loose their nanostructure. Since properties of variously (nano)structured gold are significantly different each other, it is fundamental to address this point. In order to investigate on the real state of the infiltrated NpAu, we dissolved our doped opals with few drops of toluene. The absorption spectra for a 300 nm doped dissolved opal film were recorded in different sample positions and are reported in Figure 2b. We notice that the stop band at 690 nm disappears since the PC has been destroyed, while a new not previously seen absorption peak is now evident at 530 nm. The peak position corresponds to that of surface plasmon for spherical NpAu in solution (Figure 2a). First of all, this result demonstrates that infiltration of opals was actually achieved with spherical nanoparticles. Moreover, since no evidence of absorption band at 650-700 nm is detected, no NpAu aggregates or bulk gold are formed during opal preparation because the citrate-capping shell preserves the NpAu individuality. We also remark that when similar studies were carried out for opals doped with NpAu obtained by laser ablation, where the capping layer is provided by a surfactant dissolved in the solution, strong aggregation effects have been observed instead [28].

We now turn back our attention to n_{eff}. The value of n_{eff} determined form the optical spectra is important to obtain a quantitative evaluation of the dielectric constant (ε_i) or refractive index (n_i) of interstices. As a matter of fact, it has been recently shown [28] that n_{eff} is connected to n_i by the Lorentz-Lorenz formula for the effective medium [44]

$$\frac{\varepsilon_{eff}-1}{\varepsilon_{eff}+2}=f_{PS}\frac{\varepsilon_{PS}-1}{\varepsilon_{PS}+2}+(1-f_{PS})\frac{\varepsilon_{i}-1}{\varepsilon_{i}+2} \quad (5)$$

where $\varepsilon_{eff}=(n_{eff})^2$, $\varepsilon_{PS}=2.53$ is the dielectric constant of polystyrene sphere, ε_i is the dielectric constant of the interstices ($\varepsilon_i=1$ for bare opals), and f_{PS} is the volume fraction of the unitary cell filled with nanospheres (0.74 for a close packed FCC crystal). By inverting Eq. 5, we can obtain ε_i (n_i) values. As an input for Eq. 5 we used n_{eff}, which can be obtained in two ways from Eq. 4. We can first fit n_{eff} in order to match the observed position of the stop band(λ_{Bragg}). On the other hand, we can modify the incidence angle, record the spectra and then fit the dispersion data with Eq. 4. This last procedure is well established [28], but in this work was used only for 260 thin LL sample. The calculated data for n_{eff}, n_i and ε_i are reported in Table I. We notice that both procedures provide results in good agreement thus confirming its internal consistency as well as the quality of our samples.

It is also interesting to give a look to the absolute values of n_{eff} (ε_{eff}) or n_i (ε_i). For bare opals n_{eff} ranging from 1.39 to 1.41 is obtained in good agreement with the value of 1.41 deduced from the effective medium theory. Upon increasing the doping level ε_i increases and its average values, calculated among the available samples, become 1.09, 1.21 and 1.62 for LL, ML, and HL samples, respectively. This means that our doping procedure provides a very strong variation of ε_i thus deeply modifying the dielectric properties of the system.

As far as n_{eff} is known, we can obtain additional information from their spectroscopic characterization. First of all, the interference fringes pattern observed in the R spectra allows the determination of the sample thickness d from the formula [56]

$$d=\frac{N_f}{\frac{1}{\lambda_k}-\frac{1}{\lambda_{k+N_f}}} \quad (6)$$

where N_f is the number of fringes observed in the spectral range λ_k-λ_{k+Nf} , k labelling the generic fringe. On this basis, starting from the spectra in Figure 3 to 5 and from Eq. 6, we derived the thickness of our samples that is also reported in Table I. Thickness d can be related to the number of sphere planes (p) in the samples. As a matter of fact, for FCC packing the interplanar spacing D along the [111] direction is given by $D=a(2/3)^{0.5}$ where a is the PS microsphere radius. As a consequence p=d/D corresponds to 11 or 15 microsphere planes for thin and thick samples, respectively.

Additional interesting properties of the NpAu doping process are the clear reduction of the Bragg peak FWHM and intensity. They can be interpreted as the effect of the reduction of the dielectric contrast between nanospheres and interstices, where the presence of few NpAu strongly modifies ε_i (9-62 %) with respect to the case of bare opals (ε_i =1). The high sensitivity of the photonic band structure upon partial infiltration of interstices with highly polarizable materials envisage their use not only for photonics but also for sensing [49]. The dependence of the FWHM on the NpAu doping level can be understood recalling the expression for the stop band width (W) for a one dimensional (1D) photonic crystal [57]

$$W = \frac{4hc}{\pi}\frac{1}{\lambda_{Bragg}}\frac{|n_i - n_{PS}|}{n_i + n_{PS}} \tag{7}$$

Even though Eq. 6 cannot be directly used to determine the FWHM of infiltrated opals, since they posses a much more complicated photonic structure with respect to a one-dimensional PC, it is interesting to analyze the behaviour of reduced FWHM (W/W_0), i.e. the FWHM for infiltrated opals divided by that for a bare one (W_0). The reduced FWHM as a function of n_{eff} or n_i for all our samples is reported in Figure 6. Both n_{eff} and n_i are representative of the NpAu doping level, through Eq. 5. It is evident that a very good agreement between experimental data for all samples and theory is achieved.

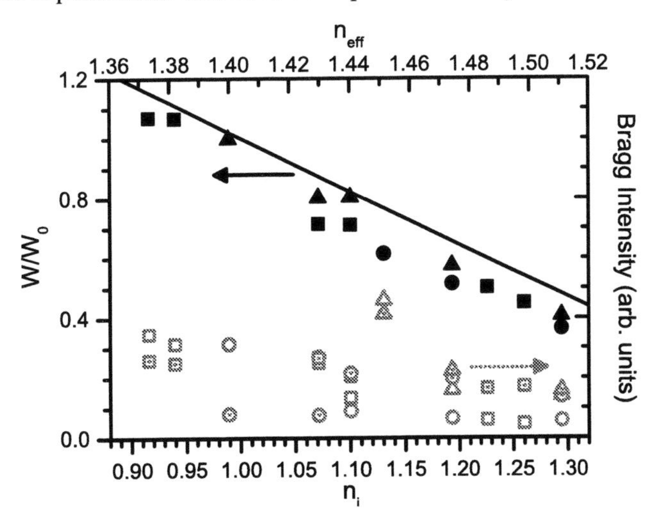

Figure 6. Reduced FWHM (left axis, full symbols) and intensity of the Bragg peak (right axis) as measured both in transmittance (open symbols) and reflectance (marked open symbols) spectra for bare and infiltrated opals of different diameters (squares, 260 nm; circles, 300 nm; diamonds, 340 nm) as a function of n_i (bottom axis) or n_{eff} (top axis). Full line shows theoretical values for reduced FWHM derived from Eq. 7.

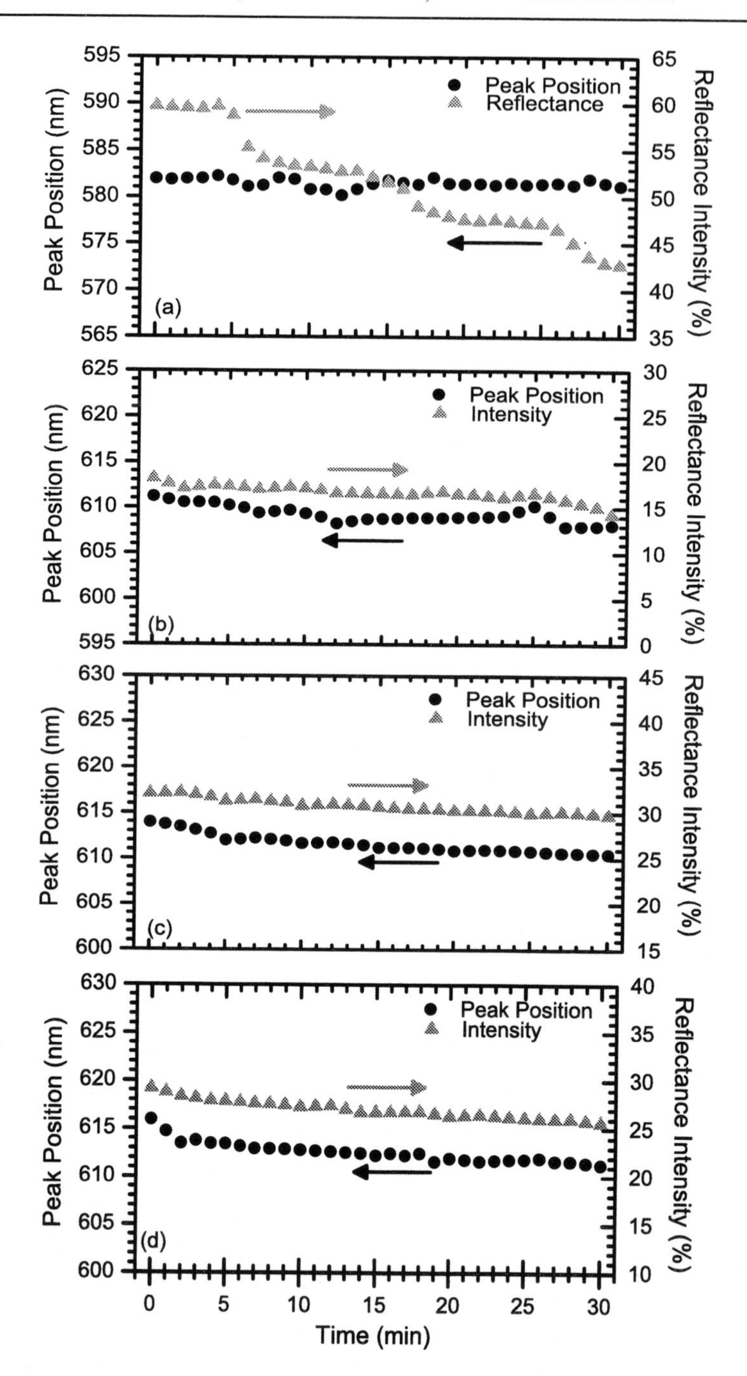

Figure 7. Time evolution of the reflectance Bragg peak wavelength (circles) and intensity (triangles) during thermal annealing at nominal temperature 75 °C for 260 nm bare (a), LL (b), ML (c), and HL (d) opals.

An additional spectral effect observed in Figs. 3-5 and again related to the infiltration process concerns the intensity of the reflectance peak. The increase of n_i upon increasing the doping level reduces the dielectric contrast with respect to polystyrene spheres. This implies a reduction of light diffraction efficiency by the sphere planes along the [111] direction thus decreasing the intensity of the Bragg peak. Figure 6 also shows the reflectance and transmittance intensities measured at the Bragg peak for our samples as a function of n_i. Again a clear trend of intensity reduction upon increasing doping is observed. We cannot however exclude that an additional effect might contribute to the experimental intensity reduction. In fact, doping with absorbing media induces a strong light scattering due to NpAu, which increases upon increasing their loading. Photons scattered in all directions reduces the signal detected in specular reflectance and transmittance configurations. However, at better of our knowledge however, no analytical theoretical models accounting for these effects have been reported so far.

As previously observed for bare opals, these photonic crystals are brittle (in particular when infiltrated from solutions) and their mechanical robustness sometimes limited [40]. In order to improve the mechanical properties of these opals, a specific annealing procedure has been developed. Briefly, a differential scanning calorimetric characterization of bare opals showed a thermal behaviour similar to that of standard polystyrene i.e. a glass transition temperature (T_g) at about 80 °C and a melting temperature (T_m) of 110 °C, independently on the sphere diameter. On the basis of these results, annealing was performed in three temperature ranges, namely $T<T_g$, $T \leq T_g$ and $T_g<T<T_m$. Bare opals annealed at $T_g<T \leq T_m$ suddenly loose their photonic properties even though they preserve a pale reminiscence of their original nanostructure [40]. Opals annealed at $T<T_g$ (about 50 °C) increases their robustness and sometimes an improvement in the optical properties (evaluated both as a reduction of the Bragg peak FWHM and of the scattering background) is observed [40]. In Figure 7a we report the effect of annealing at $T \leq T_g$ (nominal temperature 75°C) for bare opals on the Bragg peak spectral position and intensity. Notice that these measurements are in situ performed during annealing on a hot plate. It is well known that hot plates does not provide a stable temperature; as a matter of fact we measured a periodical (about 10 minutes) variation of temperature in the 65-80 °C range. Data in Figure 7a show a dramatic change of the bare opals optical properties following the periodical heating of the hot plate. In particular, even though λ_{Bragg} is almost insensitive to temperature, within the spectral resolution of our set-up, the reflectance intensity during the heating phase of the hot plate suddenly decreases thus demonstrating a progressive and irreversible damaging of the opal. Damaging was also confirmed by scanning electron microscopy analysis showing an anisotropical deformation of PS microspheres that destroys the periodical nanostructure of the photonic crystal. Notice that annealing effects are similar for any opals at a given temperature and do not depend on the polystyrene microsphere diameter.

Since even NpAu doped opals show the poor mechanical stability observed in the bare ones, we repeated a thermal annealing study on these samples in order to address the problem. Again for $T_g<T \leq T_m$ the photonic crystal get lost even though not as suddenly as observed for bare ones. For $T< T_g$ no change in the optical properties and no significant variation of mechanical robustness are observed. In Figure 7b, c, and d the effect of annealing at $T \leq T_g$ (T=75°C) on 300 nm LL, ML, and HL opals, respectively, is reported. While in bare opals the heating cycle generated irreversible degradation of the photonic crystal, no such an effect is observed for NpAu doped opals independently on the doping level even after several cycles. Moreover, the mechanical stability of these opals seems to be slightly increased. From these results, we deduce that the in doped opals heating flux does not affect polystyrene microspheres but it is mainly directed through NpAu which become an heating sink preventing polystyrene melting. As a consequence, a proper annealing procedure for metallic nanoparticles doped polymeric opals different with respect to that used for bare ones have to be developed in order to optimize their mechanical properties.

4. Conclusion

We demonstrated that artificial opals can be successfully infiltrated in a controlled manner (doped) by citrate-capped NpAu. The citrate shell around NpAu prevents nanoparticles coalescence previously observed in opals doped with NpAu obtained by laser ablation. A detailed optical characterization of opals doped with NpAu shows a dramatic change of their dielectric properties which induces a bathochromic shift of the stop band as well as a reduction of its FWHM and intensity with respect to bare ones. These effects have been accounted for by a simple dielectric model. NpAu also affect thermal behaviour of doped opals. Preliminary results on the temperature dependence of the optical properties of NpAu doped opals indicates that their annealing temperature is significantly higher than that of bare ones.

Acknowledgments

This work is supported by the Italian Ministry of Instruction, University and Research through projects PRIN 2006031511 and by Progetti di Ricerca di Ateneo 2006 (Università di Genova).

References

[1] Painter, O; Lee, RK; Scherer, A; Yariv, A; O'Brien, JD; Dapkus, PD; Kim, I. *Science*, 1999, 284, 1819.

[2] Mekis, A; Chen, JC; Kurland, I; Fan, S; Villeneuve, PR; Joannopulos, JD. *Phys. Rev. Lett*, 1996, 77, 3787.

[3] Serbin, J; Gu, M. *Adv. Mater*, 2006, 18, 221.

[4] *Photonic Crystals: Advances in Design, Fabrication, and Characterization*; K; Busch, S; Lölkes, RB; Wehrspohn, H. Föll, Eds; Wiley: Weinheim, 2004.

[5] Jin, F; Li, CF; Dong, XZ; Chen, WQ; Duan, XM. *Appl. Phys. Lett*, 2006, 89, 241101.

[6] Huang, J; Wang, X; Wang, ZL. *Nan Lett*, 2006, 6, 2325.

[7] Vukusic, P; Sambles, R. *Nature*, 2003, 424, 852.

[8] Sweeney, A; Jiggins, C; Johnsen, S. *Nature*, 2003, 423, 31.

[9] Biro´, LP; Ba´lint, Z; Kerte´sz, K; Ve´rtesy, Z; Ma´rk, GI; Horva´th, ZE; Bala´zs, J; Me´hn, D; Kiricsi, I; Lousse, V; Vigneron, JP. *Phys. Rev. E*, 2003, 67, 021907.

[10] Vukusic, P; Sambles, JR; Lawrence, CR; Wootton, RJ. *Nature*, 2001, 410, 36.

[11] Vukusic, P; Hooper, I. *Science*, 2005, 310, 1151.

[12] Ghirardella, H. *Appl. Opt.*, 1991, 30, 3492.

[13] Kertész, K; Bálint, Z; Vértesy, Z; Márk, GI; Lousse, V; Vigneron, JP; Rassart, M; Biró, LP. *Phys. Rev. E*, 2003, 74, 021922.

[14] Vukusic, P; Sambles, JR; Lawrence, CR. *Nature*, 2000, 404, 457.
[15] Zi, J; Yu, X; Li, Y; Hu, X; Xu, C; Wang, X; Liu, X; Fu, R. *Proceedings of the National Academy of Science*, 2003, 100, 12576.
[16] Tayeb, G; Gralak, B; Enoch, S. *Opt. Phot. News*, 2003, February, 38.
[17] McPhedran, RC; Nicorovici, NA; McKenzie, DR; Rouse, GW; Botten, LC; Welch, V; Parker, AR; Wohlgennant, M; Vardeny, V. *Physica B*, 2003, 338.
[18] Vigneron, JP; Rassart, M; Vértesy, Z; Kertész, K; Sarrazin, M; Biró, LP; Ertz, D; Lousse, V. *Phys. Rev. E*, 2005, 71, 011906.
[19] Vigneron, JP; Rassart, M; Vandenbem, C; Lousse, V; Deparis, O; Biró, LP; Dedouaire, D; Cornet, A; Defrance, P. *Phys. Rev. E*, 2006, 73, 041905.
[20] Parker, AR; L.Welch, V; Driver, D; Martini, N. *Nature*, 2003, 426, 786.
[21] Sanders, JV. *Nature*, 1064, 204, 1151.
[22] Sanders, JV; Darragh, P. *J. Min. Rec.*, 1971, 2, 261.
[23] López, C. *Adv. Mater*, 2003, 15, 1679.
[24] Jiang, P; Bertone, JF; Hwang, KS; Colvin, VL. *Chem. Mat*, 1999, 11, 2132.
[25] Pavarini, E; Andreani, LC; Soci, C; Galli, M; Marabelli, F; Comoretto, D. *Phys. Rev. B*, 2005, 72, 045102.
[26] Palacios-Lidon, E; Galisteo-Lopez, JF; Juarez, BH; Lopez, C. *Adv. Mater*, 2004, 16, 341.
[27] Tetreault, N; Mihi, A; Miguez, H; Rodriguez, I; Ozin, GA; Meseguer, F; Kitaev, V. *Adv. Mater*, 2005, 17, 1912.
[28] Morandi, V; Marabelli, F; Amendola, V; Meneghetti, M; Comoretto, D. *Adv. Funct. Mater*, 2007, 17, 2770.
Pasquazi, A.; Stivala, S.; Assanto, G.; Amendola, V.; Meneghetti, M.; Cucini, M.;Comoretto, D. In-situ All-Optical Tuning of a Photonic Band-Gap with Laser Pulses *Appl. Phys. Lett.* 93, 091111 (2008).
Morandi, V.; Marabelli, F.; Amendola, V.; Meneghetti, M.; Comoretto, D. Light Localization Effect on the Optical Properties of Opals Doped with Gold Nanoparticles *J. Phys. Chem. C*, 112, 6293-6298 (2008)
[29] Tan, Y; Qian, W; Ding, S; Wang, Y. *Chem. Mater*, 2006, 18, 3385.
[30] Antoine, R; Brevet, PF; Girualt, HH; Bethell, D; Schiffrin, D. *Chem. Commun*, 1997, 1901.
[31] Amendola, V; Polizzi, S; Meneghetti, M. *J. Phys. Chem. B*, 2006, 110, 7232.
[32] Markowicz, PP; Tiryaki, H; Pudavar, H; Prasad, PN; Lepeshkin, NN; Boyd, RW. *Phys. Rev. Lett*, 2004, 92, 083903.
[33] Markowicz, P; Friend, C; Shen, Y; Swiatkiewicz, J; Prasad, PN; Toader, O; John, S; Boyd, RW. *Opt. Lett*, 2002, 27, 351.
[34] Polson, RC; Chipouline, A; Vardeny, ZV. *Adv. Mater*, 2001, 13, 760.
[35] Shkunov, MN; Vardeny, ZV; DeLong, MC; Polson, RC; Zakhidov, AA; Baughman, RH. *Adv. Funct. Mater*, 2002, 12, 21.
[36] Eradat, N; Sivachenko, AY; Raikh, ME; Vardeny, ZV; Zakhidov, AA; Baughman, R. *Appl. Phys. Lett*, 2002, 80, 3491.
[37] Sumioka, K; Nagahama, H; Tsutsui, T. *Appl. Phys. Lett*, 2001, 78, 1328.
[38] Marabelli, F; Comoretto, D; Bajoni, D; Galli, M; Fornasari, L. *Mat. Res. Soc. Symp. Proc.*, 2002, 708, BB10.19.1.

[39] Comoretto, D; Marabelli, F; Soci, C; Galli, M; Pavarini, E; Patrini, M; Andreani, LC. *Synth. Met*, 2003, 139, 633.

[40] Cucini, M; Narizzano, R; Comoretto, D; Morandi, V; Marabelli, F. *Paper presented at 7th International Conference on Optical Probes of p-conjugated polymers and functional self-assemblies, Turku (FI)* 11-15 *June* 2007. *Manuscript in preparation.*

[41] Kulinowski, KM; Jiang, P; Vaswani, H; Colvin, VL. *Adv. Mater*, 2000, 12, 833.

[42] Rodriguez-Gonzalez, B; Salgueiriño-Maceira, V; Garcia-Santamaria, F; Liz-Marzán, LM. *Nano Lett*, 2002, 2, 471.

[43] Miclea, PT; Susha, AS; Liang, Z; Caruso, F; Sotomayor-Torres, CM; Romanov, SG. *Appl. Phys. Lett*, 2004, 84, 3960.

[44] Romanov, SG; Susha, AS; Sotomayor-Torres, CM; Liang, Z; Caruso, F. *J. Appl. Phys.*, 2005, 97, 086103.

[45] Wang, D; Salgueirño-Maceira, V; Liz-Marzán, LM; Caruso, F. *Adv. Mater*, 2002, 14, 908.

[46] Wang, D; Li, J; Chan, CT; Salgueirño-Maceira, V; Liz-Marzán, ĹM; Romanov, S; Caruso, F. *Small*, 2005, 1, 122.

[47] Tessier, PM; Velev, OD; Kalambur, AT; Lenhoff, AM; Rabolt, JF; Kaler, EW. *Adv. Mater*, 2001, 13, 396.

[48] Garcia-Santamaria, F; Galisteo-Lopez, JF; Braun, PV; Lopez, C. *Physical Review B (Condensed Matter and Materials Physics)* 2005, 71, 195112.

[49] Xia, Y; Gates, B; Li, ZY. *Adv. Mater*, 2001, 13, 409.

[50] Comoretto, D; Morandi, V; Marabelli, F; Amendola, V; Meneghetti, M. SPIE: 2006, Vol. 6182, 6182D.

[51] Amendola, V; Rizzi, GA; Polizzi, S; Meneghetti, M. *J. Phys. Chem. B*, 2005, 109, 23125.

[52] Turkevich, J; Stevenson, PL; Hillier, J. *Discuss. Faraday Soc.*, 1951, 11, 55.

[53] Yu, YY; Chang, SS; Lee, CL; Wang, CRC. *J. Phys. Chem. B* 1997, 101, 6661.

[54] Joannopulos, JD; Meade, RD; Win, JN. *Photonic Crystals: Molding the Flow of the Light*; Princeton University Press: Princeton, 1995.

[55] Lee, W; Pruzinsky, S.A; Braun, P.L. *Adv. Mater.* 2002, 14, 271.

[56] Comoretto, D; Dellepiane, G; Marabelli, F; Cornil, J; dos Santos, DA; Bredas, JL; Moses, D. *Phys. Rev. B* 2000, 62, 10173.

[57] Haroche, S. In *Fundamental Systems in Quantum Optics*; JDJMRJ., Zinn-Justin, Ed; Elsevier: Amsterdam, 1992.

In: Photonic Crystals
Editor: Venla E. Laine, pp. 293-295

ISBN: 978-1-61668-953-7

Short Communication

MATHEMATICAL REPRESENTATION OF THE GROUP INDEX IN A PHOTONIC CRYSTAL

*M.A. Grado-Caffaro** *and M. Grado-Caffaro*
Madrid, Spain

Abstract

In this communication, we analyze the mathematical structure of the refractive group index (briefly, group index) of a photonic crystal starting from the fact that, in a photonic crystal, the index of refraction as a function of wavelength is a periodic function whose period is the atomic lattice spacing of the crystal so that the expansion in a Fourier series of the refractive-index function is considered. In particular, the case in which the wavelength is sufficiently small (geometrical optics) is examined.

Keywords: photonic crystal; group index; periodic functions; geometrical optics.

1. Introduction

Photonic crystals are periodic dielectric or metallo-dielectric nanostructures that affect the motion of photons. Indeed, the propagation of light waves is affected by the nanostructures in question so that there are allowed modes of propagation forming bands and disallowed modes forming band gaps. Both the physics and technology of photonic crystals present very attractive features which, to date, have not been fully understood. In particular, the fundamental physics underlying photonic crystals presents very relevant aspects [1-3] which should be elucidated sufficiently in order to understand, for example, issues concerning band structure [1-3]. At this point, issues relative to photonic band gaps have a notorious relevance (see, for instance, refs.[1,2]). A photonic crystal exhibits a wavelength-dependent refractive index which is a periodic function whose period equals the atomic lattice spacing of the crystal. Therefore, this function is expandable in a Fourier series so the corresponding

* E-mail address: ma.grado-caffaro@sapienzastudies.com. C/ Julio Palacios 11, 9-B, 28029-Madrid (Spain) www.sapienzastudies.com (Corresponding author)

refractive group index or simply group index can be expressed in terms of Fourier series. As a matter of fact, the aim of the present communication lies on studying the group index of a photonic crystal starting from the Fourier-series expansion of the refractive index.

2. Theory

The group index obeys the following relationship:

$$n_g(\lambda) = n(\lambda) - \lambda \frac{dn(\lambda)}{d\lambda} \tag{1}$$

where λ denotes wavelength, $n_g(\lambda)$ is the group index, and $n(\lambda)$ is the refractive index.

On the other hand, the refractive index of a photonic crystal is a periodic function whose period coincides with the atomic lattice spacing of the crystal so that $n(\lambda + a) = n(\lambda)$ where a is the atomic lattice spacing of the crystal. Therefore, the above function admits the following Fourier-series expansion (see ref.[4]):

$$n(\lambda) = 2\sum_{k=0}^{\infty} n_k \cos\left(\frac{2\pi k\lambda}{a}\right) + \tilde{n}_k \sin\left(\frac{2\pi k\lambda}{a}\right) \tag{2}$$

where the coefficients of the series read:

$$n_0 = \frac{1}{2a}\int_0^a n(\lambda) d\lambda \tag{3}$$

$$n_k = \frac{1}{a}\int_0^a n(\lambda)\cos\left(\frac{2\pi k\lambda}{a}\right) d\lambda \tag{4}$$

$$\tilde{n}_k = \frac{1}{a}\int_0^a n(\lambda)\sin\left(\frac{2\pi k\lambda}{a}\right) d\lambda \tag{5}$$

where now $k = 1,2,...$ Note that the above Fourier coefficients expressed by formulae (3) to (5) refer to the interval $0 \le \lambda \le a$ so that the lower limit of the involved integrals is zero wavelength which implies an extrapolation to $\lambda = 0$ (see ref.[4]). At this point, we remark that the limit $\lambda \to 0$ means geometrical optics.

By replacing (2) into (1), one gets:

$$n_g(\lambda) = 2\sum_{k=0}^{\infty}\left(n_k + \frac{2\pi k\lambda}{a}\tilde{n}_k\right)\cos\left(\frac{2\pi k\lambda}{a}\right) + \left(\tilde{n}_k - \frac{2\pi k\lambda}{a}n_k\right)\sin\left(\frac{2\pi k\lambda}{a}\right) \tag{6}$$

At the limit $\lambda \to 0$ (geometrical optics), eq.(6) reduces to:

$$n_g(\lambda)_{\lambda \approx 0} - \langle n \rangle \approx 2\sum_{k=1}^{\infty} n_k + \frac{2\pi k \tilde{n}_k \lambda}{a} \qquad (7)$$

where $\langle n \rangle$ is the average value of the refractive index when $0 \le \lambda \le a$ so that $\langle n \rangle = 2n_0$ (see formula (3)).

By eq.(1), one has that $n_g(\lambda) \approx n(\lambda)$ if $\lambda \approx 0$. Then this fact in conjunction with relationship (7) indicate that the approximate deviation of the refractive index with respect to its average value over the range $0 \le \lambda \le a$ is an affine function of wavelength when it approaches zero.

3. Conclusion

In summary, we have investigated the main mathematical-physics aspects of the group index of a photonic crystal as a function of wavelength starting from the fact that the corresponding index of refraction is a periodic function (and also, in practice, an analytic function [4]) whose period is the atomic lattice spacing so this function is expandable in a Fourier series whose uniform convergence is assured (see, for instance, refs.[4,5]). By the way, from that $n(\lambda + a) = n(\lambda)$ we obtain recursively that $n(\lambda + ka) = n(\lambda)$ for k = 1,2,... and consequently we find out, by extrapolation to zero wavelength, that $n(ka) = n(0)$. In addition, looking at relation (1), it is clear that the group index is a quasi-periodic function for $\lambda \approx 0$. From the Fourier-series expansion of $n_g(\lambda)$ expressed in relationship (6), we have obtained, as a final result, formula (7) within the context of geometrical optics.

References

[1] Sakada, K. *Optical properties of photonic crystals* (Second Edition, Springer-Verlag, 2004).

[2] Vahala, KJ. Optical microcavities, *Nature*, 2003, 424, 839.

[3] Böttger, G; Schmidt, M; Eich, M; Boucher, R; Hubner, U. Photonic crystal all-polymer slab resonators, *J. Appl. Phys.*, 2005, 98, 103101.

[4] Grado-Caffaro, MA; Grado-Caffaro, M. The coefficient of material dispersion of a photonic crystal as a Fourier series, *Optik*, 2009, 120, 449-450.

[5] Churchill, RV. *Fourier Series and Boundary Value Problems* (Mc Graw-Hill, New York, N.Y., 1970).

INDEX

D

E

F

G

H

I

J

K

L

M

N

O

P

Q

R

S

T

U

V

W

X